D1515314

PHARMACOCHEMISTRY LIBRARY – VOLUME 1

THE HYDROPHOBIC FRAGMENTAL CONSTANT

Its Derivation and Application
A Means of Characterizing Membrane Systems

PHARMACOCHEMISTRY LIBRARY, edited by W.Th. Nauta and R.F. Rekker

Volume 1   The Hydrophobic Fragmental Constant. Its Derivation and Application.
A Means of Characterizing Membrane Systems
by R.F. Rekker

PHARMACOCHEMISTRY LIBRARY

Editors: W.Th. Nauta and R.F. Rekker

Volume 1

THE HYDROPHOBIC FRAGMENTAL CONSTANT

Its Derivation and Application
A Means of Characterizing Membrane Systems

ROELOF F. REKKER

Gist-Brocades R & D, Haarlem, and Vrije Universiteit, Amsterdam

ELSEVIER SCIENTIFIC PUBLISHING COMPANY

Amsterdam — Oxford — New York   1977

ELSEVIER SCIENTIFIC PUBLISHING COMPANY
335 Jan van Galenstraat
P.O. Box 211, Amsterdam, The Netherlands

Distributors for the United States and Canada:

ELSEVIER/NORTH-HOLLAND INC.
52, Vanderbilt Avenue
New York, N.Y. 10017

Cover design by Dr. C. van der Stelt

ISBN: 0-444-41548-3

Printed in Great Britain

PHARMACOCHEMISTRY LIBRARY

A Series of Monographs covering various aspects of
Biological Activity in terms of structural features

During the compilation of the monograph on "The Hydrophobic Fragmental Constant", it occurred to us that a demand might exist for a series of special publications presenting, with no more than one theme
per volume, those topics from the field of pharmacochemistry which are
related to the various aspects of biological activity in terms of structural features.

The Editors consider the following topics and services to be suitable for inclusion:

(1) Improvements in the various parameters used in structure-activity
studies. The present monograph may be considered as a contribution
in this respect.

(2) Improvements to the "counter side" of the equations modelling structure-activity relations, i.e., attempts to achieve a better approach
to the wide range of biological activities presented to the medicinal chemist; in particular, the search for parameters that are
meaningful for the understanding of biological action.

(3) Presentation of structurally related series of compounds investigated, with positive or negative results for structure-activity
relationships.

(4) Permanent service for publishing the Proceedings of (Q)SAR symposia
and seminars; especially of interest are those meetings where the
various aspects of SAR are considered from more than one discipline.

(5) Implications of (Q)SAR for patentability of inventions in the
chemical-pharmaceutical field.

(6) Publication of basic text-books for the teaching of (Q)SAR.

W.Th. NAUTA
R.F. REKKER

Department of Medicinal Chemistry, Vrije Universiteit, De Boelelaan 1083, Amsterdam-Buitenveldert
The Netherlands

A man would do nothing if he waited
until he could do it so well that no
one would find fault with what he
had done

                    Cardinal NEWMAN

                In Commemoration of MEYER and OVERTON
                                    who started,

                Dedicated to CORWIN HANSCH
                                    who continued,

                To Encourage all those
                        who will try to finish.

CONTENTS

PART II

MUTUAL DIFFERENCES BETWEEN PARTITIONING SOLVENT SYSTEMS AND
THE CHARACTERIZATION OF MEMBRANE SYSTEMS

Chapter VI.   SYSTEMS OTHER THAN OCTANOL — WATER

Chapter VII. DERIVATION OF APPROPRIATE SETS OF HYDROPHOBIC
             FRAGMENTAL CONSTANTS FOR SEVERAL SOLVENT SYSTEMS

FOREWORD

Pharmaceutical Chemistry is now a subject to be treated in its own right, and no longer should be considered as a branch of applied organic chemistry. The discovery and development of drugs relies on the application of certain principles that have been formulated from studies of the mode of action of biologically active compounds.

These studies have been necessary at three levels of organisation; extracellular, cellular and molecular. At the extracellular level, studies have been made of the absorptions, distribution, metabolism, and excretion of drugs. At the cellular level, variations in affinity for cell membranes and for such organelles as ribosomes or nuclei are relevant. At the molecular level, we are interested in interaction with biopolymers such as the proteins which make up ribosomal sub-units, the enzymes and their coenzymes and cofactors which control the rate of cellular activity and the nucleic acids and associated regulatory molecules which differentiate one cell from another, or one species from another.

PHARMACOCHEMISTRY LIBRARY is not just another series of books reflecting the role of chemical structure in biological activity. It is a series of Monographs which lay emphasis on studies at the molecular level of organisation, and as such the series will be of particular value to those whose interest lies in a physicochemical approach to understanding the mode of action of drugs, and in a rational approach to drug design.

The first book in the series is by Professor R.F. REKKER, and details both his derivation and some applications of THE HYDROPHOBIC FRAGMENTAL CONSTANT. It is an appropriate first volume, for two reasons.

First, the studies of HANSCH, beginning in the early 1960's rapidly led medicinal chemists to appreciate the value of establishing correlations between biological response and various physicochemical parameters describing molecular structure. Such Quantitative Structure – Activity Relations (QSAR) have both a diagnostic and predictive value to the drug designer. The physicochemical parameter most often relevant to biological QSAR is one describing hydrophobicity, such as the HANSCH log P or π, or the new f constant of REKKER. Secondly, the volume is appropriate because it is not merely a reference book on partition data, but is

a springboard to further research. This scholarly treatment not only
does much to clarify the nature and relevance of hydrophobic bonding,
but poses further questions and suggests avenues of research, particularly into the nature of drug—membrane interactions. The book is in
commemoration of MEYER and OVERTON who started (QSAR), dedicated to
CORWIN HANSCH who continued, and will most certainly encourage those
who wish to finish.

The drug designer, wishing to study the effect of variation in hydrophobicity of his molecules, would be well advised to make use of the
extensive compilation of Fragmental Hydrophobicity Constants, collected
together as an appendix to this volume. He may intend to use values
from the octanol — water system, but be stimulated to try parameters
from other systems as being perhaps more relevant to the biological
membrane(s) involved. No medicinal chemist can fail to be stimulated by
the concepts developed in this Monograph. Of particular interest is the
suggestion that partition values may be quantised. This is based on the
finding that integral "key numbers", multiplied by a "magic constant",
derived by sound statistical procedures, may be used to calculate hydrophobicity values with uncanny accuracy. That the magic constant is related to a fundamental property of structured water, and that the key
number relates to an integral number of displaced water molecules is a
speculation inherent in the text.

How does one account for the effects of such complex functional
groups as ureides and amides, where substitution of an acidic hydrogen,
or steric modification of resonance, can make the proper choice of hydrophobic parameter most awkward? Another perplexing problem relates to
the anomalous behaviour of cations and quaternary amines, where substitution of hydrogen by a methyl group causes a decrease in lipophilicity.
These and other problems relating to the hydrophobicity parameter of
QSAR are dealt with in a lucid style.

It is intended that later volumes of the series will include similar
treatment of other parameters, and the discussion of themes such as the
quantitative measurement of biological response, or the role of quantum
chemistry in drug design. Surely, the PHARMACOCHEMISTRY LIBRARY is des-
tined to become the standard work both for the practice, and for teaching the principles, of drug design by the QSAR method.

December, 1976

M.S. TUTE
PFIZER Ltd, Sandwich, Kent

PREFACE

The purpose of this monograph is to acquaint medicinal and pharmaceutical chemists, biochemists and those who work in allied disciplines,
with the application of lipophilicity values in the study of biologically active molecules.

A brief historical review and some practical details cannot be dispensed with, but Chapter I, in which these are presented, has been kept
short, giving only some essential information on the problems that are
frequently encountered in practice.

Chapter II treats part of the work of HANSCH and his school as far
as the derivation of partition coefficients in the octanol – water system on the basis of the π-system is concerned.

Chapter III gives a detailed review of the development of a novel
set of hydrophobic fragmental constants (f values). These were chiefly
derived in order to obtain decisive information on the occurrence of
"folding" in certain organic molecules, as proposed by HANSCH and co-
-workers.

Once available, the fragmental constants proved suitable for building up the total lipophilicity of a structure with surprising accuracy
and, by comparing calculated partition values of structures where folding was thought to be possible with experimentally determined values,
it could be demonstrated that folding, if any, does not result in a
significant change in partitional behaviour.

In the remaining sections of this Chapter III, it is demonstrated
how the concept of fragmentation, the additivity of fragmental constants, after having started with aliphatic structures, could be gradually developed into a system comprising a varied number of structural
types including the most important aromatic (carbocyclic and heterocyclic) compounds.

As the f system was nearing completion, evidence was accumulated
which suggested that not only f values but also the partition values themselves would be associated with a quantifiable constant. Chapter IV presents a variety of reasons why we were forced to pursue a course leading to concepts such as "alignment of partition values", "magic constant" and "key number". Admittedly, a certain speculative element remains, but there appeared to be no other way to account for a number of
curious facts that occur in lipophilic behaviour.

Chapter V begins with the treatment of the influence which the de-coupling of resonance exerts on the lipophilicity in an aromatic structure. The quantifiability dealt with in Chapter IV also seems to be involved here. Next, structures with an N^+ centre are discussed. All medicinal chemists will frequently come across these structures, and it is hoped that the treatment proposed in Chapter V will encourage other investigators to attempt to unravel many of the problems that exist.

Thus far, the decrease in the lipophilicity of N^+ by changing N^+-H into N^+-CH_3 (while an increase of 0.50 on the logarithmic scale is to be expected) has remained unexplained. The present study, however, provides evidence that the hydrophilic potency of an N^+ centre cannot be fully developed unless its positive charge is partially moved from the N^+ centre via substitution of its hydrogen atoms by alkyl groups.

The author thanks most cordially Dr. T. BULTSMA, Miss G. BIJLOO, Mrs. L. ELGERSMA and Mr. H.H. HARMS (Department of Medicinal Chemistry, Vrije Universiteit, Amsterdam) and Mr. H.M. de KORT, Mr. G.G. NYS and Mr. W.F.C. VLAARDINGERBROEK (Gist-Brocades, Research & Development, Haarlem) for experimental and computational assistence in assembling the material necessary for preparing Chapter V. Without the devoted help of these teams, the realization of this chapter would have been impossible.

A few practical applications in the field of SAR are dealt with at the end of Chapter V.

Part II treats a variety of solvent systems along the lines developed in Part I, and indicates a way to characterize membrane systems. In practical SAR work, octanol − water is usually the system of choice, as it is believed that in partitioning a structure between its two liquid phases this system will indicate correctly the consequences of the introduction of that structure into the lipoid part of a membrane via an aqueous biophase. By a careful comparison of different solvent systems with the standard octanol − water system, a general formulation for the solvent regression equation could be proposed. Although of the COLLANDER type, the equation accounts for differences in the hydration/solvation of hydrophilic functional groups. These differences were incorporated in terms of the magic constant and key numbers.

In principle, a membrane lipoid, which in combination with an aqueous biophase is actually no more than a common solvent system, may be characterized by following the above reasoning. Only the data obtained by SEEMAN for human erythrocyte membrane lipoid were sufficiently exact for the treatment proposed. The human erythrocyte membrane lipid appears to differ significantly from the generally accepted pattern. If only lipophilic groups are taken into consideration it can be regarded as being almost _iso_discriminative relative to the system octanol – water, i.e., in a series of structures offered for partition it will give the same spread of partition values as does the octanol – water system. With respect to functional groups with hydrophilic character the membrane lipid is _hyper_discriminative relative to the octanol – water system, i.e., in a series of structures offered for partition it will give a larger spread of partition values than does the system octanol – water. Some possible implications of these findings for the currently used procedure for multiple regression analysis of biological data via the HANSCH approach are discussed in Chapter X.

The monograph includes detailed tables of all of the partition values used to build up the system of hydrophobic fragmental constants and an extensive index makes these partition values readily accessible.

Thanks are due particularly to Dr. Eike DRUCKREY for patiently listening to the first very tentative ideas about what was to become the hydrophobic fragmental constant, and for fruitful discussions later.

I am also indebted to Gilbert NYS for his cordial cooperation in the first stages of the lipophilicity project and, to express my feelings in SAR-like terms, I regret the day that he decided to leave the right-hand side of the scenery for its left-hand side.

Thanks are also due to Bart de KORT for his indispensible aid in many experiments and endless computations which proved necessary in drawing up this monograph. I would also like to thank Dr. A.F. HARMS for reading and commenting on the manuscript and Mr. J.C. van GERVEN for his valuable assistance in the translation.

In conclusion, thanks are extended to a great number of colleagues for encouraging and providing much valuable feedback in the past few years.

Amsterdam, Fall 1976   Roelof F. Rekker

1

PART I

THE HYDROPHOBIC FRAGMENTAL CONSTANT AND ITS APPLICATION IN THE CALCULATION OF PARTITION COEFFICIENTS OF ORGANIC STRUCTURES IN THE OCTANOL — WATER SYSTEM

CHAPTER I

THE PARTITION COEFFICIENT

INTRODUCTION; THE IMPORTANCE OF PARTITION COEFFICIENTS IN STRUCTURE — ACTIVITY RELATIONSHIPS

The question of whether there would be some relationship between chemical structure and biological activity arose as early as the last century, and once it appeared possible to extend the range of drugs gradually by synthetic routes, many efforts were made to find an answer.

Fundamental work was carried out by RICHETT [1], who found that the toxic effects of ethers, alcohols, aldehydes and ketones are inversely proportional to their solubility in water; by TRAUBE [2], who established a linear relationship between surface tension and activity for a series of narcotics; and by FUEHNER [3,4,5], who noted the possibility of there being a quantitative relationship between a biological property, for example the narcotic activity, and the number of carbon atoms.

All of this research may be viewed in conjunction with the postulation of CRUM — BROWN and FRASER [6,7] that the physiological action, Φ, of a molecule is a function of its chemical constitution (C):

$$\Phi \;=\; f(C) \tag{I-1}$$

While recognizing the work of other pioneers, it may be said that the research of MEYER [8] and OVERTON [9,10,11] at the turn of the century marked a milestone in studies on structure — activity relationships (SAR), especially when considered in the light of the develop-

Refs. pg 22

ments that have been made in the last few decades.

MEYER and OVERTON discovered that most organic compounds foreign to
the organism penetrate tissue cells as though the membranes were lipid
in nature. They also found that the passage across these barrier systems
and the subsequent narcotic action parallel the oil – water partition
coefficients of the structures investigated.

While it has become apparent that biological activity can rarely be
coupled to a single parameter but should rather be envisaged as the result of the interplay of various parameters, the partition coefficient
did remain the major parameter in SAR studies.

The period between the research of MEYER and OVERTON and the present situation is about half a century. For a review of the most important events during this period, reference is made to a monograph by
PURCELL, BASS and CLAYTON [12].

A brief account of the SAR model, which has been developed by
HANSCH and has attracted much interest over a number of years as the
HANSCH Approach [13,14,15], will suffice.

The HANSCH equation can be regarded as an expanded HAMMETT equation
[16,17]:

$$\log K_s = \rho\sigma + \log K_o \qquad (I-2)$$

$$\log (1/C) = k_1 \log P + \rho\sigma + k_2 \qquad (I-3)$$

In eqn. I-2, K_s and K_o represent the equilibrium constants for the
reactions of substituted and unsubstituted compounds, respectively, σ is
an electronic constant that depends on the nature and the position of
the substituent and ρ is a constant associated with the type of reaction and the conditions under which the reaction takes place.

In eqn. I-3, C represents the molar concentration of a congener necessary to give a defined biological response in any set of structures
investigated; P is the partition coefficient of that congener in the
octanol – water system, σ is the HAMMETT substituent constant and k_1, ρ
and k_2 are constants for the set investigated; the last three constants
are generated by regression analysis.

The great usefulness of eqn. I-3 has been adequately demonstrated
in practice [18]. Two examples are given here: the 50% inhibition of
Arbacia egg cell division by barbiturates and of Avena cell elongation
by phenoxyacetic acids [19,20]. It appears that inhibition can be described well simply with P as an independent parameter (eqns I-4 and

I-5):

$$\log (1/C) \ = \ 0.801 \log P + 1.076 \qquad\qquad (I-4)$$
$$n \ = \ 19; \quad r \ = \ 0.960$$

$$\log (1/C) \ = \ 0.778 \log P + 1.971 \qquad\qquad (I-5)$$
$$n \ = \ 22; \quad r \ = \ 0.928$$

The equal slopes in these equations indicate the same dependence of inhibition on the hydrophobic character of the drug, whereas the different intercepts indicate the greater sensitivity of the Avena test [21].

It soon appeared necessary to extend eqn. I-3 by including additional parameters. An interesting extension is that involving a quadratic log P term (eqn. I-6):

$$\log (1/C) \ = \ - k_1 (\log P)^2 + k_2 \log P + \rho\sigma + k_3 \qquad\qquad (I-6)$$

Initially, this $(\log P)^2$ term was introduced [13,14] with the reasoning that somewhere between $P = 0$ and $P = \infty$ an ideal partition value (P_o) will exist for a given set of congeners in a given biological system such that those members which have this value P_o will find the sites of action via a random walk process in the minimum time [22].

A subsequent investigation by PENNISTON et al. [23] has lent support to the validity of this assumption, while McFARLAND [24] has shown, on the basis of probability concepts, that the relationship is not entirely but sufficiently parabolic to justify fully the use of a quadratic log P term in eqn. I-6.

An example of a regression equation with a quadratic log P term is given below; the minimum anaesthetic dose for 50% of a population of mice by ethers (MARSH and LEAKE [25]) can be expressed by eqn. I-7 (HANSCH and CLAYTON [26]):

$$\log (1/C) \ = \ - 0.22 (\log P)^2 + 1.01 \log P + 2.19 \qquad\qquad (I-7)$$
$$n \ = \ 27; \quad r \ = \ 0.960; \quad s \ = \ 0.107$$

from which a value of 2.34 can be derived for log P_o.

There are two disputable points in eqns. I-3 and I-6:
(a) It is assumed that the octanol – water system is the one of choice in giving a correct picture of what happens to a structure when it is introduced into a biological system. Is this supposition correct?
(b) In eqn. I-8:

$$\log (1/C) \ = \ - k_1 (\log P)^2 + k_2' \log P + k_2'' \log P'' + \rho\sigma + k_3 \qquad (I-8)$$

which is a modification of eqn. I-6, the original log P is split into two terms, k'_2 log P and k''_2 log P". Combined with the quadratic term k_1 (log P)^2, the former describes the transport possibilities of the drug and the latter reflects the hydrophobic interaction on a receptor. In practice, eqn. I-6 is invariably preferred to eqn. I-8; in other words, it is assumed that the partition values of the octanol – water system provide an adequate description of the two indicated phenomena – transport and hydrophobic interaction – and that P = P". This means that the lipophilic merits of all membranes that the drug will meet while being transported are basically not different, and are also equal to the lipophilic features of the receptor system. Again, is this assumption correct?

Various aspects of the octanol – water system as the preferential solvent pair will be discussed in detail in Part II.

THE NERNST LAW

When a solute partitions at constant temperature between two solvents, which are immiscible or partially miscible, there exists a constant ratio between the solute concentration in the two phases as soon as equilibrium has been attained. A prerequisite for equilibrium is that as many molecules pass through the interphase from phase 1 to phase 2 as from phase 2 to 1. In thermodynamic terms, the equilibrium is characterized by equality of the chemical potentials, μ_1 and μ_2, of the solute in the two phases (eqns. I-9 and I-10):

$$\mu_1 = \mu_1^o + kT \ln C_1 \qquad (I-9)$$

$$\mu_2 = \mu_2^o + kT \ln C_2 \qquad (I-10)$$

If $\mu_1 = \mu_2$, then

$$C_2/C_1 = e^{-(\mu_2^o - \mu_1^o) / kT} = e^{-\Delta\mu^o / kT} = P \qquad (I-11)$$

In these equations, C_1 and C_2 represent the equilibrium concentrations of the solute, μ_1 and μ_2 the chemical potentials at these concentrations, μ_1^o and μ_2^o the chemical potentials at $C_1 = C_2 = 0$, k the gas constant per molecule (BOLTZMANN constant), T the temperature (^{o}K) and P the partition coefficient.

This simple relationship is usually attributed to NERNST [27], but was first applied by BERTHELOT and JUNGFLEISCH [28] as early as 1872.

One of the solvents used is always water, and there is a wide selection of organic compounds to choose as the other solvent.

The partition coefficient can now be written as

$$P = C_s / C_w \qquad\qquad (I-12)$$

where C_s and C_w are the equilibrium concentrations of the solute in the organic and aqueous phase, respectively.

A solute with high P is regarded as lipophilic and a solute with low P as hydrophilic. As the P scale covers a range of more than 10^{10}, logarithmic P values are usually preferred. This transforms eqn. I-12 into eqn. I-13:

$$\log P = \log C_s - \log C_w \qquad\qquad (I-13)$$

MEASUREMENTS OF PARTITION COEFFICIENTS

Without pretentions to completeness, a number of facts of practical importance in measuring partition coefficients are listed below.

Partitioning Procedure

The solute is weighed accurately and dissolved in the most appropriate phase of the solvent pair. Next, the second phase is added in an amount adapted to both the partition coefficient and to the analytical facilities available. The solute concentrations as such are of no importance in determining the partition coefficient [29]. See also KAKEMI et al., who measured the influence of concentration on the partition coefficient for a series of barbiturates [30]. Mechanical shaking is usually recommended but some workers prefer a manual shaking procedure [12]. The Griffin flask shaker (Griffin & George Ltd.) is very suitable for this purpose, as four shaking bottles can be treated at a time. Different shaking times have been reported in the literature; about half an hour appears to be most appropriate. Shaking should, of necessity, be followed by centrifugation at 2000 rpm for 1 - 2 h. This centrifugation is the more important, the higher are the partition values. It is not strictly necessary to establish the total amount of solute partitioned, but it does facilitate a check of the analytical results (mass balance). It is advisable to carry out at least four partition experiments for any one solute.

Temperature

The temperature need not be kept within very narrow limits, except in the case of the sec. butanol - water system, which has a fairly sub-

Refs. pg 22

stantial mutual miscibility. It is, therefore, recommended that when this system is used, the fluctuations in temperature should not exceed about 0.5 $^{\circ}$C.

Purity of Solutes

Solute purity is of great importance, and needs checking first. Chromatography, if possible reversed-phase thin-layer chromatography (TLC), is the method of choice here. This technique provides useful information on the approximate lipophilicity of any one substance, especially if one or more other compounds of known lipophilicity are eluted on the plate at the same time. In this way, additional information of value is very easily obtained on the lipophilicity of any contaminating materials relative to that of the main component.

The procedure is particularly useful when a substance is available in only a very small amount. If there is a contaminant with a greater log P value than that of the main component, the organic phase can be discarded after an initial shaking and be replaced by fresh organic phase for the final shaking procedure. A check of the analytical results, based on the total amount of weighed material, is no longer possible but the reliability of the log P value is enhanced considerably in many cases.

Dissolution

With poorly soluble compounds, it is useful to pulverize the material thoroughly before weighing. It may also be useful to employ a small amount of an auxiliary solvent such as methanol or dimethylformamide so as to ensure optimum solubility. Next, the solvent pair of choice is added. A few per cent of methanol or dimethylformamide will have no influence on the partition values, at least if no extremely high partition values have to be determined.

Volatile Solutes

Measurements with volatile materials present another type of difficulty. The procedure necessary to compensate for losses due to vaporization and the correction needed with very volatile material for the amount partitioned into the area above the liquid phases have been described by FUJITA et al. [31].

Purity of Solvents

As spectrophotometry is a frequently used technique for analyzing

the two phases, spectroscopic grade solvents are most suitable for use
in partition work. For octanol, which is the most widely applied solvent in partition studies, such a quality is not yet available. This
also applies to pentanol, sec. butanol, etc., and hence one has to re-
sort to some commercial-grade solvent. Even superior grades of commer-
cial octanol still contain impurities that make it difficult to use
in analysis at lower wavelength absorption bands. Many grades of octanol
are contaminated with small amounts of material that absorb at about
240 nm. These impurities can be removed by purification, but it seems
that in many instances they re-appear in the last phase of purification,
the vacuum distillation.

Table I,1 shows some cut-off values for 50% transmission of some
spectroscopic grade solvents in a 1-cm cell [32].

TABLE I,1

CUT-OFF VALUES FOR 50% TRANSMISSION OF
SOME SOLVENTS IN A 1-cm CELL

Solvent	λ (nm)
methanol	217
ethanol	214
iso propanol	216
diethyl ether	230
chloroform	250
carbon tetrachloride	270
hexane	206
heptane	207
cyclo hexane	220
benzene	284

From the values presented for some alcohols, it can be concluded,
especially if the three saturated hydrocarbons are taken into account,
that the cut-off for a comparatively pure grade of octanol will not differ much from 216 nm. In practice, this may have several consequences
for the analytical possibilities. Thus, an excellent quality of octanol
permits the analysis of benzoic acid on its C band (228 nm with ϵ =
10,000), while insufficiently pure octanol necessitates the use of the
B band (272 nm with ϵ = 850), which means that a factor of 12 is lost
in the detection.

Solvents such as octanol and pentanol are best purified by washing

successively with 10% sodium hydroxide solution, dilute sulphuric acid
and hydrogen carbonate solution, followd by drying over magnesium sul-
phate and vacuum distillation in which at least 10% of both the forerun
and tail should be discarded.

   All water used in partition experiments should be distilled. Mutually saturated solvent phases are usually preferred, especially in the
determination of high or low partition values.

Analysis

Spectrophotometric Analysis

   Spectrophotometry should be applied when possible, as it is easy to
use and accurate. At least a benzene ring or one of its equivalents is
required as a chromophoric group [*], and if isolated from groups such as
NH_2, COOH or NO_2, UV absorption can be expected at ca. 255 nm (B band)
with ϵ ranging from 200 to 600 for each benzene ring present. This
means that partition systems with chloroform, carbon tetrachloride and
benzene as the organic phase are not suitable here for spectrophotometric analysis (see Table I,1). Diethyl ether, however, can be used.
Structures with an isolated pyridine ring lend themselves more to spectrophotometric analysis than the corresponding benzene derivatives, but
the presence of a thiophene or furan ring leads to a sharply diminished
absorption pattern.

   A very favourable effect is produced by the conjugative coupling of
an aromatic ring to one of the groups referred to above. The C band,
which for isolated benzene systems lies at ca. 200 nm, will thus be
shifted to a favourable wavelength range with reasonable extinction values ($\epsilon \sim$ 10,000). Although such C bands are situated in the 230 – 270
nm range, most of the conjugated structures in question do not lend
themselves to measurements in carbon tetrachloride or benzene-like partition phases. In most instances, chloroform – water will be applicable
and there are no difficulties with octanol – water.

   It is recommended that a calibration graph (one with five points is
usually sufficient) should be prepared for both phases. It is also

_____

[*] "Chromophoric group" refers to the localized part of the molecule
which gives rise to an absorption band in the visable or ultraviolet
part of the spectrum.
Structures such as $(C = C)_n - (C = X) - R$ may also offer suitable
wavelengths for accurate detection.
In cases where no suitable chromophore is present and one insists on
spectrophotometric analysis special pre-treatment is necessary.

strongly advisable not to restrict the UV measurements to the determination of an absorbance at the absorption maximum but to run the complete UV curve also. A comparison of the two curves (one in the organic
phase and one in water) often provides meaningful information on con-
taminents and aids the decision to follow the procedure mentioned under
"Purity of Solutes".

Gas-Liquid Chromatographic Analysis

Gas-liquid chromatography (GLC) can often be applied successfully
when the spectrophotometric method is unsuitable, although it is approximately two times less accurate. For details on its practical application, reference is made to McNAIR and BONELLI [33].

Other Methods of Analysis

Compounds from which an NH_3 molecule can be split off, such as amides and carbamates, can be analyzed by using NESSLER's reagent [34].
They are subjected to hydrolysis and the ammonia so released reacts
with NESSLER's reagent (K_2HgI_4) to give $(NH_2)Hg_2I_4$, which can be determined spectrophotometrically at 410 nm. The method is extremely sensitive, and only applicable in ammonia-free surroundings. It will be obvious that in the presence of a phenyl group or another reasonably ab-
sorbing chromophore in the molecule, spectrophotometry is the method of
choice. For details on the NESSLER method, see also PURCELL et al. [12].

With penicillin and its derivatives, the method described by BUND-
GAARD and ILVER [35] can be used. It consists of conversion of the penicillin derivative into penicillic acid mercurimercaptide by allowing
it to react with imidazole and mercury(II)chloride. In the first step
the lactam ring is opened and in the second the 5-ring is also ruptured:

N-penicilloylimidazole

Refs. pg 22

$$ClHg - S \diagdown_{C}\diagup^{CH_3}_{CH_3}$$

$$R - C \diagdown^{N} \diagup C = CH - NH - CH$$

penicillic acid mercurimercaptide

The reaction is highly specific, and the mercaptide formed (with a
light absorption ranging from 325 to 345 nm, depending on the type of
penicillin) can be determined down to a concentration of 0.5 μg/ml expressed as penicillin.

Quaternary ammonium compounds, which cannot be detected spectrophotometrically owing to the lack of a suitable chromophore, can be subjected to pyrolysis and subsequent GLC detection. This method, reported
by VIDIC et al. [36], appears to be less accurate than the demethylation
described by JENDEN et al. [37], which removes a methyl group from the
quaternary structure, leaving a tertiary amine. A mixture of sodium
phenylthiolate and thiophenol in anhydrous butanone is used to achieve
demethylation.

Micellar Properties of some Compounds

The formation of micelles results primarily from intimate interactions between molecules in which hydrophobic binding plays an important
role. Aggregation may have far-reaching consequences both on the level
of biological activity and in measuring lipophilicity. Recent papers on
these aspects are those by ATTWOOD et al.[38] and MUKERJEE [39].
It is advisable always to operate below the critical micellar concentration and to evaluate a dilution profile for the partition value.

INFLUENCE OF DIMERIZATION AND DISSOCIATION ON THE PARTITION VALUE *

One may encounter deviations from eqn. I-11 in certain solvent systems, such as when the partition coefficient of an organic acid is measured in the benzene − water system. It will be noted in this instance
that the partition coefficient increases with increasing concentration,

---

*This section partly follows the treatment given on this subject by
MOELWYN − HUGHES [80].

owing to the fact that organic acids in a solvent such as benzene dim-
erize:

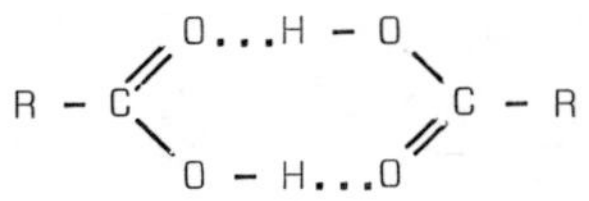

If, for simplicity, the acid is represented by HA, the partition e-
quilibrium can be written as

$$HA_w \rightleftharpoons HA_{benzene} \qquad (I-14)$$

and the partition coefficient is given by

$$P' = [HA_{benzene}] / [HA_w] \qquad (I-15)$$

The equation for the equilibrium of dimerization is

$$(HA)_{2 \ benzene} \rightleftharpoons 2 \ HA_{benzene} \qquad (I-16)$$

and the relevant dimerization constant can be written as

$$K_D = [HA_{benzene}]^2 / [(HA)_{2 \ benzene}] \qquad (I-17)$$

The equilibrium concentrations for the two phases are

$$C_w = [HA_w] \qquad (I-18)$$

and

$$C_{benzene} = [HA_{benzene}] + 2 [(HA)_{2 \ benzene}] \qquad (I-19)$$

It follows from eqns. I-15, I-17, I-18 and I-19 that

$$C_{benzene} / C_w = P = P'^2 + (2 P'^2 / K_D)C_w \qquad (I-20)$$

The figures shown in Table I,2 were reported by HENDRIXSON [40] and
refer to the partition of acetic acid in the benzene − water system.

TABLE I,2

PARTITION OF ACETIC ACID IN BENZENE − WATER

$C_{benzene}$	C_w	$C_{benzene} / C_w$		residual
		obs.	est.	
0.00303	0.2291	0.01323	0.01326	−0.00003
0.00776	0.4328	0.01793	0.01789	0.00004
0.01551	0.6661	0.02328	0.02319	0.00009
0.02480	0.8855	0.02801	0.02817	−0.00016
0.03586	1.0900	0.03290	0.03282	0.00008

Regression analysis of these data affords the following equation:

$$P = 0.00805 + 0.02272 \, C_w \qquad (I-21)$$

$$n = 5; \quad r = 0.999; \quad s = 0.000121; \quad F = 16,590; \quad \underline{t} = 129$$

Eqns. I-20 and I-21 can be combined to give $P' = 0.00805$ and $2 \, P'^2 / K_D = 0.02272$, from which it follows that $K_D = 5.704 \times 10^{-3}$.

The K_D values reported in the literature often differ as shown in Table I,3, which lists different K_D values for benzoic acid and gives an impression of the effect of the nature of the organic phase.

TABLE I,3

DIMERIZATION OF BENZOIC ACID

Solvent	K*	Ref.
benzene	3.39×10^{-3}	[41]
benzene	3.36×10^{-3}	[42]
benzene	9.26×10^{-3}	[43]
toluene	1.27×10^{-2}	[44]
xylene	7.00×10^{-4}	[45]
chloroform	8.33×10^{-3}	[46]
chloroform	3.03×10^{-2}	[44]
diethyl ether	∞	[45]

* K defined according to eqn. I-17.

The above calculation of the partition coefficient neglects the effect of ionic dissociation in the aqueous phase. This effect can be allowed for by correcting C_w in eqn. I-20 for dissociation:

$$C_{benzene} / C_w(1 - \alpha) = P = P'^2 + (2 \, P'^2 / K_D) C_w(1 - \alpha) \qquad (I-22)$$

where α represents the degree of dissociation of the acid.

With the aid of the expression

$$\alpha = \sqrt{K_a / C_w} \qquad (I-23)$$

where K_a represents the dissociation constant of the acid, the regression analysis affords

$$P = 0.00819 + 0.02280 \, C_w \qquad (I-24)$$

$$n = 5; \quad r = 0.999; \quad s = 0.000123; \quad F = 16,030; \quad \underline{t} = 127$$

and the combination of eqns. I-22 and I-23 results in $P' = 0.00819$ and $2 \, P'^2 / K_D = 0.02280$, from which it follows that $K_D = 5.884 \times 10^{-3}$.

The results show that it is valid to neglect dissociation in this instance. When compounds with definitely stronger acidic reactions are investigated by using eqn. I-22, α should be taken as

$$\alpha = (-K_a + \sqrt{K_a^2 + 4 K_a C_w}) \, / \, 2 C_w \qquad (I\text{-}25)$$

Applied to picric acid with $K_a = 0.222$ [48], eqn. I-22 affords P' = 47.89 and indicates the non-significance of the dimerization of picric acid.

If one is sure of the absence of dimerization in the organic phase, the problem can also be approached as follows:

$$P' = [HA_s] \, / \, [HA_w] = C_s \, / \, C_w (1 - \alpha) \qquad (I\text{-}26)$$

from which it follows that

$$\alpha = 1 - C_s/P'C_w \qquad (I\text{-}27)$$

The combination of eqn. I-27 with the dissociation constant of the acid provides

$$K_a = ([H_w^+][A_w^-]) \, / \, [HA_w] = C_w^2 \alpha^2/(C_s/P') \qquad (I\text{-}28)$$

from which it follows that

$$\alpha = \sqrt{K_a/P'} \, . \, \sqrt{C_s}/C_w \qquad (I\text{-}29)$$

By combining eqns. I-27 and I-28, the following equation is obtained:

$$C_s/C_w = P' - \sqrt{K_a P'} \, . \, \sqrt{C_s}/C_w \qquad (I\text{-}30)$$

Table I,4 shows the results of an investigation by ROTHMUND and DRUCKER [49] on the partition of picric acid in the benzene – water system.

Incorporation of the data in Table I,4 into eqn. I-30 leads to

$$C_s/C_w = 32.716 - 2.199 \sqrt{C_s}/C_w \qquad (I\text{-}31)$$
$$n = 10; \quad r = 0.994; \quad s = 0.177; \quad F = 708; \quad \underline{t} = -26.6$$

so that P' = 32.716 and $\sqrt{K_a P'}$ = 2.199, from which it follows that K_a = 0.148. This dissociation constant is slightly different from that obtained by DIPPY et al. (K_a = 0.222) from conductivity measurements [48], while P' also differs from the value in eqn. I-22. The latter equation, in which a priori it is assumed that dimerization is absent, appears to be more accurate than eqn. I-29.

When partition measurements are performed in alcohol – water systems, dimerization of a carboxylic acid in the alcohol phase can probably be neglected; the equilibrium

$$(RCOOH)_2 \ + \ 2n \ ROH \ \rightleftharpoons \ 2 \ (RCOOH) \cdot (ROH)_n$$

is shifted almost completely to the right (attributable to the very
large excess of ROH).

TABLE I,4

PARTITION OF PICRIC ACID IN BENZENE — WATER

C_s	C_w	$\sqrt{C_s}/C_w$	C_s/C_w obs.	C_s/C_w est.	residual
0.1772	0.03342	12.596	5.3022	5.0123	0.2899
0.1235	0.02759	12.737	4.4763	4.7009	−0.2246
0.06996	0.01994	13.265	3.5085	3.5409	−0.0324
0.05225	0.01701	13.438	3.0717	3.1597	−0.0880
0.03453	0.01357	13.694	2.5446	2.5969	−0.0523
0.01993	0.01011	13.964	1.9713	2.0035	−0.0322
0.01647	0.00913	14.057	1.8039	1.7996	0.0043
0.01011	0.00701	14.343	1.4422	1.1682	0.2740
0.002248	0.003273	14.486	0.6868	0.8547	−0.1679
0.000932	0.002079	14.684	0.4483	0.4188	0.0295

Partition coefficients are frequently measured in a buffer solution
for which a pH similar to that under physiological conditions, namely
7.2 − 7.3, is chosen. With structures in which a dissociable acidic or
basic group is lacking, this procedure seems useless as no effect of the
buffer on the partitional behaviour of a neutral compound occurs. Measurement of a partition coefficient of a dissociable structure in some
buffer system may be of advantage, not because such a log P value would
have any particular meaning but because a carefully chosen pH makes it
possible to measure partition coefficients that normally lie at or just
outside the limit of analytical detectability.

The mathematical relationship between P, pK_a and pH can be simply
derived [50]. For a base

$$P \ = \ [B_s]/[B_w] \tag{I-32}$$

and

$$K_a \ = \ [B_w][H^+]/ \ [BH^+] \tag{I-33}$$

The apparent partition coefficient, which is the value established
by making measurements at the chosen pH, can be represented by

$$P' = [B_s] / ([B_w] + [BH^+]) \qquad (I-34)$$

Eqn. I-34 assumes that $[BH_s^+]$ can be neglected in comparison wih $[B_s]$. Table I,5 lists some log P values for the two species of a few types of acid and base, indicating how far such a neglection is justified.

TABLE I,5

EFFECT OF SALT FORMATION ON LOG P VALUES OF AMINES AND ACIDS

Compound	measured as	log P	ref.	Δ log P
3-phenylpropylamine	free base	1.83	[51]	2.92
	HCl-salt	-1.09	[52]	
diphenhydramine	free base	3.30	[53]	3.42
	HCl-salt	-0.12	[52]	
aniline	free base	0.98	[54]	2.10
	HCl-salt	-1.12	[52]	
hexanoic acid	free acid	1.88	[55]	4.05
	Na-salt	-2.17	[56]	
p.biphenylcarboxylic acid	free acid	3.91[a]		4.18
	Na-salt	-0.27	[57]	

[a]   calculated log P value

As the degree of dissociation of the base is constant at constant hydrogen-ion concentration and independent of the base concentration, it follows from eqns. I-32, I-33 and I-34 that

$$P/P' = ([B_w] + [BH^+])/ [B_w]$$

$$= 1 + [BH^+] / [B_w] = 1 + [H^+]/K_a$$

or

$$P/P' = 1 + \text{antilog} (pK_a - pH) \qquad (I-35)$$

In an analogous manner, the following equation is found for an acid structure:

$$P/P' = 1 + \text{antilog} (pH - pK_a) \qquad (I-36)$$

When, for example, the lipophilicity of a base with an expected log P value of about 4.0 is to be established, and not more then one unconjugated C_6H_5 group is present in the structure, UV measurement of the aqueous phase will be of no use. However, eqn. I-35 makes it possible to reduce significantly the partition value (log P') to be measured, thus

Refs. pg 22

TABLE I,6

CALCULATIONS OF PERCENTAGES IONIZED FROM GIVEN pK_a AND pH VALUES

pK_a − pH	if Anion	if Cation
−6.0	99.99990	0.0000999
−5.0	99.99900	0.0009999
−4.0	99.9900	0.0099990
−3.5	99.968	0.0316
−3.4	99.960	0.0398
−3.3	99.950	0.0501
−3.2	99.937	0.0630
−3.1	99.921	0.0794
−3.0	99.90	0.09991
−2.9	99.87	0.1257
−2.8	99.84	0.1582
−2.7	99.80	0.1991
−2.6	99.75	0.2505
−2.5	99.68	0.3152
−2.4	99.60	0.3966
−2.3	99.50	0.4987
−2.2	99.37	0.6270
−2.1	99.21	0.7879
−2.0	99.01	0.990
−1.9	98.76	1.243
−1.8	98.44	1.560
−1.7	98.04	1.956
−1.6	97.55	2.450
−1.5	96.93	3.07
−1.4	96.17	3.83
−1.3	95.23	4.77
−1.2	94.07	5.93
−1.1	92.64	7.36
−1.0	90.91	9.09
−0.9	88.81	11.19
−0.8	86.30	13.70
−0.7	83.37	16.63

TABLE I,6 (continued)

pH − pK$_a$	if Cation	if Anion
−0.6	79.93	20.07
−0.5	75.97	24.03
−0.4	71.53	28.47
−0.3	66.61	33.39
−0.2	61.32	38.68
−0.1	55.73	44.27
0	50.00	50.00

From ALBERT and SERJEANT [58]. Reproduced with permission of the authors.

extending the possibility of UV detection. When it is desirable to lower log P, for example, from $\sim$ 4.0 to $\sim$ 3.3, log P − log P' = 0.7 and P/P' = 5.0. Eqn. I-35 shows that pK_a − pH = 0.6, which means that log P should be measured in the presence of a buffer that reduces the pH to about 0.6 below the pK_a of the base. As can be seen in Table I,6, the base species are in the ratio [B] / [BH$^+$] = 20.1 : 79.9 under these conditions. It should be noted that the difference log P − log P' should not be exaggerated too much, care being taken that the amount of free base (or undissociated acid when eqn. I-36 is used) is at least a few per cent. It should also be borne in mind that there is no need for strong buffering, as any excessively high buffer salt concentration will cut off the short UV wavelength side to an undesirable extent.

It is recommended that with dissociable structures, log P should be measured for both species. This entails a measurement in 0.1 N acid (preferably hydrochloric acid) and one in 0.1 N hydroxide (preferably sodium hydroxide). Together with the pK_a value − which, at any desired pH, determines the species ratio − these two log P values will give all relevant information for solving various SAR problems.

SPECIAL METHODS FOR MEASURING PARTITION VALUES

The AKUFVE System

In 1970, REINHARDT and RYDBERG described a rapid and continuous system for measuring the distribution ratios in solvent extraction [59]. The principle of their system, called the AKUFVE system, consists of a

Refs. pg 22

rapid sequence of continuous unit operations: mixing, separation and
on-line analysis. In a thermostatted chamber, mixing of the solute and
a two-phase system is achieved. The two-phase mixture then flows down
into a continuous flow centrifuge of special design, which permits a
complete and rapid separation. Finally, the pure phases flow through
measuring cells where the concentration of the solute is determined
continuously by means of suitable detectors.

DAVIS and ELSON [60] applied the AKUFVE method to find partition coefficients. The partition of propranolol (A) between water and octanol
resulted in a log P value of 3.56 ± 0.06, which was the mean value of

$$O - CH_2 - CH - CH_2 - NH - CH(CH_3)_2$$
$$OH$$

(A)

seven measurements carried out in the pH range 4.65 − 7.12 after transformation of the apparent log P values into those of the free base. The
shaking flask procedure gave values of 3.33 and 3.65.

The merits of the AKUFVE system seem to be rapidity (equilibrium is
attained within a few minutes) and perfection of separation of the two
phases so that pH profile studies require only a few hours. A disadvantage is the unduly large amount of organic phase consumed per determination (≈ 500 ml).

Simultaneous Determination of the Partition Coefficient and Acidity Constant of a Substance

Developed by SEILER [61] as an extension of the work of BRANDSTROM
[62], this method is based on the potentiometric titration of a solute
(either an acid or a base) in an aqueous medium accompanied by the addition of a small amount of an organic phase to the solution so as to at-
tain partition of the solute between the two phases.

The procedure can be described mathematically by eqn. I-46, which
can be derived as follows:

$$BH^+ \rightleftharpoons B + H^+ \tag{I-37}$$

$$K_a = [B][H^+] / [BH^+] \tag{I-38}$$

$$P = [B_s] / [B_w] \tag{I-39}$$

$$M = [B_s].V_s + ([B] + [BH^+]).(V_o + V_T) \tag{I-40}$$

$$[BH^+] + [K^+] + [H^+] = [OH^-] + [Cl^-] \tag{I-41}$$

$$K_w = [H^+][OH^-] = 10^{-14} \qquad (I\text{-}42)$$

$$P' = [BH_s^+]/[BH_w^+] \ll P \qquad (I\text{-}43)$$

In these equations, concentrations (given in square brackets) are expressed in moles/litre; volumes (V) are in litres; V_o = volume of the aqueous phase before titration; V_T = volume of titrant added; $V_s = V_o + V_T$; M = weighed amount of solute BH^+ or B (in moles); $[Cl^-]$ = hydrochloric acid added to dissolve base B; $[K^+]$ = titrant (KOH). Eqns. I-38 and I-39, denoting the acidity constant and partition coefficient, respectively, are introduced into eqn. I-40 (law of mass conservation) and then the equation obtained is solved for $[BH^+]$:

$$[BH^+] = M/(K_a/[H^+].(V + V_s.P) + V) \qquad (I\text{-}44)$$

It follows from eqn. 40 that

$$[BH^+] = [OH^-] + [Cl^-] - [K^+] - [H^+]$$

From eqns. I-42, I-44 and I-45, by substituting $[Cl^-].V = S$ (i.e., moles of HCl added plus x.M moles of HCl when a salt $B.(HCl)_x$ is used and by substituting $[K^+].V = n_T.V_T$ (n_T = normality of the titrant), one obtains

$$([H^+] - 10^{-14}/[H^+]).V + n_T.V_T$$

$$= - \frac{M}{\dfrac{K_a}{[H^+]}\left[\dfrac{P.V_s}{V} + 1\right] + 1} + S \qquad (I\text{-}46)$$

By using a non-linear least-squares procedure, both the partition coefficient and acidity constant can be calculated from a set of titrimetric data.

For chlorpromazine, the above procedure gave the following results: pK_a = 9.29 ± 0.13, literature values: 9.3 [63], 9.22 [64] and 9.30 [65]; log P (octanol) = 5.55 ± 0.14, literature values: 5.15 [66], 5.23 [67], 5.49 [68], 5.35 [55] and 5.29 [56].

Determination of Lipophilicities using Reversed-Phase Thin-Layer Chromatography

A simple and rapid method for the determination of the relative partitional behaviour of a series of compounds has been described by BOYCE and MILBORROW [70], and has frequently been used since, for example, by BIAGI et al. [71,72].

The structures under investigation are applied to glass plates cov-

Refs. pg 22

ered with silica gel impregnated with a liquid paraffin. Instead of silica gel, polyamides can be employed [73], while the paraffin can be replaced with a silicone oil.

The oil in the reversed—phase procedure is the stationary phase. A
mixture of water and acetone or another organic phase that is miscible
with water, such as methanol or ethanol (sometimes some ethyl acetate
may advantageously be added), is used as the polar mobile phase and is
first saturated with the stationary phase. The mixing ratio of water to
organic solvent depends on the nature of the compounds under investigation: the higher their lipophilicity, the lower the percentage of water
to be used in the mobile phase.

The plates are developed in a chromatographic chamber using an ascending technique, until the solvent front has advanced to about 15 cm
above the starting line. After drying, the plates are examined by an
appropriate detection method, and the R_F values determined.

MARTIN [74] gives the following relationship between partition coefficient and R_F value:

$$P = K(1/R_F - 1) \qquad (I-47)$$

where K is a constant for the system.

Eqns. I-48 and I-49, introduced by BATE — SMITH and WESTALL [75],
can be used in place of eqn. I-47:

$$R_M = \log (1/R_F - 1) \qquad (I-48)$$

$$\log P = \log K + R_M \qquad (I-49)$$

Like π, R_M for a substituent (ΔR_M) is a free energy-based constant. As a corollary of this, R_M and π usually have a linear correlation.

The results obtained with a series of seven
17-estersteroids (B, with R = acetate, butyrate, valerate, cyclopropylcarboxylate, cyclo-
hexylcarboxylate, adamantylcarboxylate or
myristate), including hydrocortisone (R = H),
can be used as an example:

$$R_M = 0.356 \, \pi - 0.136 \qquad (I-50)$$

$$n=8; \; r = 0.995; \; s = 0.082; \; F = 602; \; \underline{t} = 24.5$$

Acetone — water (4:3) was used as the mobile phase, and the silica
gel was impregnated with paraffin oil [76].

Very little material, often no more than a few micrograms, is required but a disadvantage of the method is that the spread in lipophilicity of any series studied is rather small, not exceeding a few log P units.

Determination of Partition Coefficients by High-Pressure Liquid Chromatography

The introduction of high-pressure pumps and the development of sensitive detectors have revived interest in column chromatography, which was becoming less important as a result of the significant advances being made in thin-layer chromatography. The new field is known as high-pressure liquid chromatography (HPLC), and, in principle, it implies nothing more than an improvement in the flow-rates, which used to be so low in "classical" column chromatography that they hindered routine application. By applying high pressures, it became feasible to obtain flow-rates of about 1 ml/min coupled with retention times that could be less than 1 min.

Further information on the instrumental aspects of HPLC can be found in an article by CHANDLER and McNAIR [77].

An interesting feature of these new developments is that HPLC is ideally suited to the determination of partition coefficients. For more details, reference is made to the publications of HAGGERTY and MURRILL [78], McCALL [79] and MIRRLEES et al. [81].

Corasil C-18 [*] is employed as the stationary phase in such determinations. It is a pellicular silica gel to which octadecyl chains are chemically bound. The material has good hydrolytic stability and it combines hydrophilic and hydrophobic properties (because of the silyl ether terminations and the alkyl chains, respectively), thus behaving as a reversed-phase system on application of an aqueous mobile phase.

An improvement that McCALL introduced for Corasil-18 consists in treatment with hexamethyldisilazane and trimethylsilyl chloride in hot pyridine with a view to blocking any residual active silanol sites that might interfere in liquid – liquid partition experiments [79].

In HPLC, the retention of a compound is routinely expressed in terms of k', which is defined as:

$$k' = (t_R - t_0)/t_0 \qquad (I-50)$$

where t_R and t_0 represent the elution times of a retained and a non-re-

_____

[*] Trademark, Waters Ass.

Refs. pg 22

tained peak, respectively. Because of the analogy between the terms k'
and $1/R_F - 1$ (from paper and thin-layer chromatography), eqn. I-49 can
be written as

$$\log P \;=\; \log K + \log k' \qquad\qquad (I-51)$$

which implies that the relative retention times, k', are linearly related
to measured liquid — liquid partition coefficients expressed as log P
values.

REFERENCES

1 M.C. Richet, C.R. Soc. Biol., 45(1893)775.

2 J. Traube, Arch. Ges. Physiol. (Pflügers), 105(1904)541.

3 H. Fühner, Arch. Exptl. Pathol. Pharmakol., 51(1904)1.

4 H. Fühner, Arch. Exptl. Pathol. Pharmakol., 52(1904)69.

5 H. Fühner and E. Weubauer, Arch. Exptl. Pathol. Pharmakol.,
   56(1907)333.

6 A. Crum — Brown and T.R. Fraser, Trans. Roy. Soc. Edinburgh,
   25(1868–1869)151.

7. A. Crum — Brown and T.R. Fraser, Trans. Roy. Soc. Edinburgh,
   25(1868–1869)693.

8 H.H. Meyer, Arch. Exptl. Pathol. Pharmakol., 42(1899)109.

9 E. Overton, Z. Physikal. Chem., 22(1897)189.

10 E. Overton, Vierteljahresschr. Naturforsch. Ges. Zürich,
    44(1899)88.

11 E. Overton, Studien über die Narkose, zugleich ein Beitrag zur
    allgemeine Pharmakologie, Fischer — Jena, 1901.

12 W.P. Purcell, G.E. Bass and J.M. Clayton, Strategy of Drug Design,
    A Guide to Biological Activity, Wiley Interscience — New York, 1973.

13 C. Hansch and T. Fujita, J. Amer. Chem. Soc., 86(1964)1616.

14 C. Hansch, R.M. Muir, T. Fujita, P.P. Maloney, F. Geiger and
    M. Streich, J. Amer. Chem. Soc., 85(1963)2817.

15 C. Hansch, Acc. Chem. Res., 2(1969)232.

16 L.P. Hammett, Chem. Rev., 17(1935)125.

17 L.P. Hammett, Physical Organic Chemistry, McGraw — Hill Book Comp.
    New York, 1940.

18 C. Hansch and W.J. Dunn III, J. Pharm. Sc., 61(1972)1.

19 C. Hansch and S.M. Anderson, J. Med. Chem., 10(1967)745.

20 R.M. Muir, T. Fujita and C. Hansch, Plant Physiol., 42(1967)1519.

21 C. Hansch, Ann. Rep. Med. Chem., (1967)348.

22 C. Hansch, Ann. Rep. Med. Chem., (1966)347.

23 J.T. Penniston, L. Becket, D.L. Bentley and C. Hansch,
    Mol. Pharmacol., 5(1969)333.

24 J.W. McFarland, J. Med. Chem., 13(1970)1192.

25 D.F. Marsh and C.D. Leake, Anesthesiology, 11(1950)455.

26 C. Hansch and J.M. Clayton, J. Pharm. Sc., 62(1973)1.

27 W. Nernst, Z. Physikal. Chem., 8(1891)110.

28 M. Berthelot and E. Jungfleisch, Ann. Chim. Phys., 4(1872)26.

29 C. Hansch and S.M. Anderson, J. Org. Chem., 32(1967)2583.

30 K. Kakemi, T. Arita, R. Hori and R. Konishi,Chem. Pharm. Bull.,
    15(1967)1705.

31 T. Fujita, J. Iwasa and C. Hansch, J. Amer. Chem. Soc.,
    86(1964)5175.

32 B. Hampel, in DMS — UV Atlas of Organic Compounds, Butterworths —
    — Londen, Verlag Chemie — Weinheim, 1971.

33 H.M. McNair and E.J. Bonelli, Basic Gas Chromatography,
    Consolidated Printers — Berkeley, Cal., 1968.

34 D.F. Boltz, Chemical Analysis, Vol. VIII: Colorimetric Determi-
    nations of Nonmetals, Interscience Publishers — New York, 1958.

35 H. Bundgaard and K. Ilver, J. Pharm. Pharmac., 24(1972)790.

36 H.-J. Vidic, H. Dross and H. Kewitz, Z. klin. Chem. u. klin.
    Biochem., 10(1972)156.

37 D.J. Jenden, I. Hanin and S.I. Lamb, Anal. Chem. 40(1968)125.

38 D. Attwood, A.T. Florence and J.M.N. Gillan, J. Pharm. Sc.,
    63(1974)988.

39 P. Mukerjee, J. Pharm. Sc., 63(1974)972.

40 W. Hendrixson, Z. Anorg. Chem., 13(1897)73.

41 H.W. Smith, J. Phys. Chem., 26(1922)256.

42 R. van Duyne, S.A. Taylor, S.D. Christian and H.E. Affsprung,
    J. Phys. Chem., 71(1967)3427.

43 N. Schilow and L. Lepin, Z. Phys. Chem., 101(1922)353.

44 H.W. Smith and T.A. White, J. Phys. Chem., 33(1929)1953.

45 H.W. Smith, J. Phys. Chem., 25(1921)204,605.

46 M. Davies and D.M.L. Griffiths, J. Chem. Soc., (1955)132.

47 M. Davies, P. Jones, D. Patnaik and E.A. Moelwyn — Hughes,
    J. Chem. Soc., (1951)1249.

48 J.F. Dippy, S.R. Hughes and J.W. Laxton, J. Chem. Soc., (1956)2995.

49 V. Rothmund and K. Drucker, Z. Physikal. Chem., 46(1903)827.

50 J. Cymerman — Craig and A.A. Diamantis, J. Chem. Soc., (1953)1619.

51 J. Iwasa, T. Fujita and C. Hansch, J. Med. Chem., 8(1965)150.

52 R.F. Rekker and H.M. de Kort, unpublished results.

53 M. Tute, private communication to C. Hansch (see ref.[69]).

54 M. Tichy and K. Bocĕk, private communication to C. Hansch (ref.[69]).

55 S. Anderson, unpublished results (see ref. [69]).

56 C. Church, unpublished results (see ref. [69]).

57 J. Schaeffer and C. Hansch, unpublished results (see ref. [69]).

58 A. Albert and E.P. Serjeant, Ionization Constants of Acids and
    Bases, London: Methuen & Co. Ltd., New York: John Wiley & Sons
    Inc., 1962.

59 H. Reinhardt and J. Rydberg, Chem. and Ind., (1970)488.

60 S.S. Davis and G. Elson, J. Pharm. Pharmac., 26S(1974)90P.

61 P. Seiler, Eur. J. Med, Chem., 9(1974)663.

62 A. Brandstrom, Acta Chem. Scand., 17(1963)1218.

63 A.L. Green, J. Pharm. Pharmacol., 19(1967)10.

64 L.G. Chatten and L.E. Harris, Anal. Chem., 34(1962)1495.

65 P.B. Marshall, Brit. J. Pharmacol., 10(1955)270.

66 H. Glasser and J. Krieglstein, N.-Schm.b. Arch. Pharmak., 265(1970)
                                                                    321.
67 M. Frisk — Holmberg and E. van der Kleyn, Eur. J. Pharmacol.,
    18(1972)139.

68 K.S. Murthy and G. Zografi, J. Pharm. Sc., 59(1970)1281.

69 A. Leo, C. Hansch and D. Elkins, Chem. Rev., 71(1971)525.
    See also the more extended file of lipophilicity data issued
    by Pomona College Medicinal Chemistry Project.

70 C.B.C. Boyce and B.V. Milborrow, Nature, 208(1965)537.

71 G.L. Biagi, A.M. Barbaro and M.C. Guerra, J. Chromatogr.
    41(1969)371.

72 G.L. Biagi, M.C. Guerra, A.M. Barbaro and M.F. Gamba,
    J. Med. Chem.,13(1970)511.

73 J.C. Dearden, A.M. Patel and J.H. Tubby, J. Pharm. Pharmacol.,
    26S(1974)74P.

74 A.J.P. Martin, Biochem. Soc. Symp. Camb., 3(1949)4.

75 E.C. Bate — Smith and R.G. Westall, Biochem. Biophys. Acta,
    4(1950)427.

76 D.J.C. Engel, A.F. Marx and R.F. Rekker, to be published.

77 C.D. Chandler and H.M. McNair, J. Chromatogr. Sc., 11(1973)468.

78 W.J. Haggerty and E.A. Murrill, Res. Dev., 25(1974)30.

79 J.M. McCall, J. Med. Chem., 18(1975)549.

80 E.A. Moelwyn — Hughes, Physical Chemistry, Pergamon Press —
    — Oxford, London, New York, Paris, 1961.

81 M.S. Mirrlees, S.J. Moulton, C.T. Murphy and P.J. Taylor,
    J. Med. Chem., 19(1976)615.

25

CHAPTER II

CALCULATION OF PARTITION COEFFICIENTS IN THE OCTANOL — WATER
SYSTEM BY USING THE π-METHOD

THE HANSCH EQUATION; THE HYDROPHOBIC SUBSTITUENT CONSTANT, π

The HAMMETT equation $(II,1)$ makes it possible to calculate a rate
or an equilibrium constant of a meta- or para-substituted derivative of
C_6H_5-R with R as the reacting centre [1,2,3]:

$$\log k_s/k_o = \rho\,\sigma \qquad (II-1)$$

In the left-hand term, k_o represents the rate constant of an unsubstituted structure and k_s that of a substituted derivative; this
left-hand term can be replaced by $\log K_s/K_o$, where K represents an e-
quilibrium constant. On the right-hand side, σ denotes a constant typical of the substituent, reflecting its ability to attract or repel e-
lectrons, and ρ is a constant characteristic of the type of reaction.

HAMMETT standardized eqn. II-1 to the dissociation of benzoic acid
in water at 25^oC, ρ being taken to be 1.000. Eqn. II-1 can thus be
written as

$$\log K_s/K_o = \sigma \qquad (II-2)$$

where K_s and K_o indicate the dissociation constants of a substituted
benzoic acid and benzoic acid itself in water at 25^oC.

By analogy with the HAMMETT equation, in 1962 HANSCH et al. proposed to describe lipophilicity as follows [4,5,6]:

$$\log P(SX) / P(SH) = \rho\,\pi(X) \qquad (II-3)$$

where $P(SH)$ and $P(SX)$ represent the partition coefficients of SH and
SX, respectively; $\pi(X)$ represents the hydrophobic substituent constant, i.e., the contribution of substituent X to the lipophilicity of
structure SH when X replaces an H atom in SH; and the constant ρ reflects the characteristics of the solvent pair used in determining the
partition coefficient.

Standardization here means choosing one of the several solvent systems as the reference, so that $\rho = 1.000$ and eqn. II-3 can be transformed into

Refs. pg 36

$$\log P(SX) \,/\, P(SH) \;=\; \pi(X) \tag{II-4}$$

Eqn. II-4 can, where appropriate, be transformed into eqn. II-5:

$$\log P(S'X_1X_2\ldots.X_n) \;=\; \log P(S'H_n) \;+\; \sum_1^n \pi(X_n) \tag{II-5}$$

with the practical implication that for a structure for which a partition value has not been or perhaps cannot be determined (because of unduly high or low log P values, whether or not combined with poor analytical detectability), any given moiety with known lipophilicity can serve
as a basic fragment from which the total lipophilicity can be construed
by simple addition of the proper π values.

Like the HAMMETT constant σ, π is a free-energy-related constant; it
is a measure of the difference in the free energy changes involved in
moving unsubstituted molecules from one solvent phase to another.

Octanol - water has been proposed by HANSCH and DUNN [7] as the solvent system of choice (ρ = 1.000). Its relatively poor capacity to
dissolve water (water-saturation concentration = 1.7 M), combined with
the presence of a hydroxyl group, which can act as a hydrogen-bond donor as well as an acceptor, is considered to constitute a reasonable

TABLE II,1

HYDROPHOBIC SUBSTITUENT CONSTANTS

function	π-value		correction for	$\Delta\pi$
	aliph.	arom.		
F	-0.17	0.14	Branching	
Cl	0.39	0.71	1. in C chain	-0.20
Br	0.60	0.86	2. of functional group	-0.20
I	1.00	1.12	3. ring closure	-0.09
OH	-1.16	-0.67	Double bond	-0.30
OCH_3	-0.47	-0.02	Folding	-0.60
SCH_3	0.45	0.61	Intramolecular H-bonding	0.65
CN	-0.84	-0.57	Ring joining	-0.20
COOH	-0.67	-0.28		
$COOCH_3$	-0.27	-0.01		
$COCH_3$	-0.71	-0.55		
NH_2	-1.19	-1.23		
$N(CH_3)_2$	-0.85	-0.28		
NO_2	-0.85	-0.28		
CH_3	0.50	0.56		

model of the average macromolecular pattern in which drug interaction takes place.

While eqn. II-3 was designed primarily by analogy with the HAMMETT equation and is thus applicable to aromatic structures, it may also serve in describing the lipophilicity of aliphatic structures, provided, of course, that a separate set of π values is needed.

Table II,1 shows the π values of the most important functional groups and the correction terms required by the π system in calculating lipophilicity. Most of the aliphatic π values tabulated were obtained from log P measurements for only one structure [8,9], so that some caution should be exercised in their practical application.

The π value given for OH holds for primary, secondary and tertiary alcohols, provided that the prescribed branching corrections are correctly made. Some examples are given to illustrate this point.

Log P(2-butanol)   =   4 π (CH$_3$)   + π (OH)   +   1 br.c.
                   =   2.00   +   (-1.16) +   (-0.20)   =   0.64
                   Observed [7]: 0.61

Log P(tert. butanol)   =   4 π (CH$_3$)   + π (OH)   +   2 br.c.
                        =   2.00   +   (-1.16) +   (-0.40)   =   0.44
                        Observed [7]: 0.37

Two branching corrections (br.c.) are needed in the last example, one for the C chain and one for the functional group. The same applies in the next example:

log P(tert. amyl alcohol)   =   5 π (CH$_3$)   + π (OH)   +   2 br.c.
                            =   2.50   +   (-1.16) +   (-0.40)   =   0.94
                            Observed [7]: 0.89

The branching correction for a functional group can also be accommodated in the π value of this group, as has been reported by TUTE [10], who proposed π values of -1.16, -1.39 and -1.43 for primary, secondary and tertiary hydroxyl groups, respectively. With these π values, the respective log P values obtained for the above three alcohols are 0.61, 0.41 and 0.87.

By far the most important correction in the π system is that for folding. A study on the lipophilicity of a number of structures of the type C$_6$H$_5$-CH$_2$-CH$_2$-CH$_2$-X, where X denotes OH, F, Cl, Br, I, COOH, COOCH$_3$, COCH$_3$, NH$_2$, CN, OCH$_3$ or CONH$_2$, has revealed that log P in this series is about 0.60 $\pm$ 0.10 lower than calculations would indicate [8,11].

Refs. pg 36

Example:

$$\text{Log P}(C_6H_5CH_2CH_2CH_2OH) = 2.13 + 1.50 + (-1.16) = 2.47$$

The experimental value is 1.88, indicating a drop in lipophilicity of
0.59. HANSCH et al. claim that folding is the cause of this decrease,
assuming that the dipole induced at the end of the side-chain $(C^{\delta+}-X^{\delta-})$
interacts with the π-electrons of the phenyl ring with the X atom or
group projecting away from the ring / side-chain complex. The net result will be a more compact structure for $C_6H_5CH_2CH_2CH_2X$ and in this
"balled up" situation a greater than expected water solubility might be
observed.

The correction for ring joining, -0.20, is of recent date: the role
of the phenyl group is not fully additive when attached to another aromatic ring [12].

Examples:

$$\text{Log P}(\text{biphenyl}) = 2 \times 2.13 + (-0.20) = 4.06$$
$$\text{Observed} [13]: 4.09$$
$$\text{Log P}(\text{4-phenylpyridine}) = 2.13 + 0.65 + (-0.20) = 2.58$$
$$\text{Observed} [14]: 2.45$$
$$\text{Log P}(\text{2-phenylthiophene}) = 2.13 + 1.81 + (-0.20) = 3.74$$
$$\text{Observed} [15]: 3.74$$
$$\text{Log P}(\text{2-phenylquinoline}) = 2.13 + 2.03 + (-0.20) = 3.96$$
$$\text{Observed} [16]: 3.90$$

This correction for ring joining can also be incorporated into the
value for $\pi(C_6H_5)$, which then becomes 1.93.

In order to obtain information on the π pattern of functional
groups attached to a phenyl ring, FUJITA et al. [9] determined the partition coefficients of 203 mono- and di-substituted benzene derivatives
in the octanol - water system. A collection of π values for 67 functional groups could be derived from these partition values. The compounds
were considered to belong to eight different systems: phenylacetic acid,
phenoxyacetic acid, benzoic acid, benzyl alcohol, phenol, aniline, nitrobenzene and benzene.

Table II,2 shows the π values, divided over these eight systems, for
six selected functional groups.

As can be seen in Table II,2, large differences may occur with a
few functional groups, and it will be clear that for practical purposes
the ultimate choice of the system to be used for the calculation of unknown lipophilicities depends on the mutual arrangements of the func-

TABLE II,2

EXAMPLES OF AROMATIC π VALUES APPLICABLE IN EIGHT DIFFERENT SYSTEMS

Function		$C_6H_5CH_2COOH$	$C_6H_5OCH_2COOH$	C_6H_5COOH	$C_6H_5CH_2OH$	C_6H_5OH	$C_6H_5NH_2$	$C_6H_5NO_2$	C_6H_6
CH_3	o		0.84						
	m	0.54	0.51	0.52	0.50	0.56	0.50	0.57	0.56
	p	0.45	0.60	0.42	0.48	0.48	0.49	0.52	
OH	o	-0.54							
	m	-0.52	-0.49	-0.38	-0.61	-0.66	-0.73	0.15	-0.67
	p		-0.61	-0.30	-0.85	-0.87	-1.07	0.11	
COOH	o								
	m	-0.27	-0.15	-0.19		0.04		-0.02	-0.28
	p					0.12		0.03	
NH_2	o								
	m				-1.15	-1.29		-0.48	-1.23
	p					-1.63		-0.46	
NO_2	o		-0.04						
	m	-0.01	0.11	-0.05	0.11	0.54	0.47	-0.36	-0.28
	p	-0.04	0.24	0.02	0.16	0.50	0.49	-0.39	
F	o	0.04	0.01			0.25			
	m	0.19	0.13	0.28		0.47	0.40		0.14
	p	0.14	0.15	0.19		0.31	0.25		
Cl	o		0.59			0.69			
	m	0.68	0.76	0.83	0.84	1.04	0.98	0.61	0.71
	p	0.70	0.70	0.87	0.86	0.93	0.80	0.54	
Br	o		0.75			0.89			
	m	0.91	0.94	0.99		1.17	1.13	0.79	0.86
	p	0.90	1.02	0.98		1.13	1.12		
I	o		0.92			1.19			
	m	1.22	1.15	1.28		1.47			1.12
	p	1.23	1.26	1.14		1.45	1.39		
$COCH_3$	o		0.01						
	m		-0.28			-0.07		-0.43	-0.55
	p		-0.37			-0.11	-0.11	-0.36	
CH_2OH	m					-1.02	-0.95	-0.65	-1.03
	p					-1.26		-0.60	

Refs. pg 36

tional groups present on the benzene ring.

The phenoxyacetic acid system can be reconciled best with most practical conditions, provided that there are no groups which affect each other strongly. When such groups do occur, one has to consider the character of their mutual interactions and chose one of the other systems.

The system of π constants has been extended to a total of 130 by the latest publication by the HANSCH group [12].

HANSCH's study of group contributions to partition coefficients is the latest in a series of three investigations. The first extensive study in this field was made by COLLANDER [17,18,19,20]. In Part II, we shall discuss in detail one of the important findings of COLLANDER [19], namely the logarithmic relationship between the partition coefficients of a homologous series of compounds partitioned between water and two different organic solvents. The second interesting study of group contributions is that of McGOWAN [21,22,23]. Based on LANGMUIR's concept [24] that the major factor in partition is the energy necessary to make a hole in the solvent into which a solute is to be accommodated, McGOWAN derived eqn. II-6 connecting the partition coefficient to the parachor [Par.] of the partitioning solute:

$$\log P = k \; [Par.] \qquad\qquad (II-6)$$

where k is a constant.

The energy required to make a hole in a solvent can be considered to be proportional to the volume of the solute and it is therefore understandable that parachor values, being a measure of the solute volume, can function as they are doing in eqn. II-6.

A prerequisite for the validity of eqn. II-6 is that the volume of the solute should be the same in both solvents. If this condition is not fulfilled, deviations will occur. McGOWAN was able to extend his original equation to correct for deviations that arise from hydrogen bonding between the solute and solvent by the addition of an interaction constant (E_A):

$$\log P = k \; [Par.] + E_A \qquad\qquad (II-7)$$

A number of these E_A values have been calculated by McGOWAN [23], McGOWAN et al. [25] and DENO and BERKHEIMER [26].

Table II,3 lists a number of group contributions calculated by McGOWAN with eqns. II-6 and II-7.

TABLE II,3

McGOWAN's ADDITIVE FACTORS FOR THE CALCULATION OF SOLUBILITIES
AND PARTITION COEFFICIENTS

substitution	group contribution	
CH_3	0.48	(0.50) [a]
CH_2	0.48	(0.50)
C_6H_5	2.10	(2.13)
F	0.12	(−0.17)
Cl	0.48	(0.39)
Br	0.64	(0.60)
I	0.90	(1.00)
N (amines)	−1.34	(−1.19)
OH (alcoholic)	−1.03	(−1.16)
OH (phenolic)	−0.40	(−0.67)

[a] Values in brackets are π values.

From Table II,3 the parallelism between McGOWAN's factors and
HANSCH's π values is evident. This is extra clear in eqn. II-8 where
McGOWAN's factors are correlated with π:

$$\text{"McG"} = 0.955\,\pi + 0.059 \qquad (II\text{-}8)$$

$$n = 10; \quad r = 0.990; \quad s = 0.147; \quad F = 395; \quad \underline{t} = 19.9$$

OBJECTIONS TO THE π SYSTEM

A remarkable feature in developing the system of π constants is the
assumption that the CH_3 and CH_2 contributions to lipophilicity are i-
dentical. What makes this even more noticeable is that the branching
correction required by the π system in a C chain can also be directly
incorporated into the $\pi(CH)$ value, as was done in the preceding section
for secondary and tertiary hydroxyl groups. The value for $\pi(CH)$ would
then be changed to 0.30, but logic would require a corresponding difference of $\sim$ 0.20 for $\pi(CH_3)$ and $\pi(CH_2)$.

An analogous situation is encountered with log $P(C_6H_6)$ and $\pi(C_6H_5)$.
Again, there is no difference in lipophilicity in the π system. In both
instances a value of 2.13 is prescribed.

A frequently employed equation in calculating partition values:

$$\log P = \sum_{1}^{n} \pi_n \qquad (II\text{-}9)$$

Refs. pg 36

proposed by TUTE [27], implies that a partition value can be build up
by simple addition of the π contributions of the constituent groups.
Eqn. II-9 is, however, an incorrect transformation of the original
HANSCH equation (II-5).

DAVIS [28,29] concluded from thermodynamic considerations that the terminal CH_3 group in an alkyl chain is distinctly different from a mid-
-chain CH_2; and, as in all instances investigated thus far, physico-
chemical data are pronouncedly larger for a terminal CH_3 than for a
mid-chain CH_2 group, it cannot easily be seen why CH_3 and CH_2 contributions to lipophilicity should be identical. With respect to eqn. II-9,
DAVIS pointed out that it essentially requires the addition of an unknown constant and, in fact, should be transformed into

$$\log P = \sum_1^n \pi_n + k \qquad (II-10)$$

Some selected examples will show what can be considered to be a
fundamental error in the π system. The equation

$$\log P(C_6H_5-CH_3) = \log P(C_6H_6) + \pi(CH_3) \qquad (II-11)$$

is in complete accordance with the definition of π laid down in eqn. II-
4 representing the introduction of lipophilicity by a substitution that
replaces H. This procedure is, however, incorrectly applied in eqn. II-
12:

$$\log P(C_6H_5-CH_2-CH_3) = \log P(C_6H_6) + \pi(CH_2) + \pi(CH_3) \qquad (II-12)$$

and even more incorrect is the application of eqn. II-13:

$$\log P(C_6H_5-CH_2-CH_3) = \pi(C_6H_5) + \pi(CH_2) + \pi(CH_3) \qquad (II-13)$$

There can be no doubt that the only correct route for calculating
the lipophilicity of ethylbenzene is the following:

$$\log P(C_6H_5-CH_2-CH_3) = \log P(C_6H_5-CH_3) + \pi(CH_3)$$
$$= \log P(C_6H_6) + 2\ \pi(CH_3) \qquad (II-14)$$

Provided that the π system is operated appropriately, $\pi(CH_2)$ simply
does not exist. The same is true for $\pi(CH)$ and $\pi(C)$:

$$\log P(C_6H_5-CH(CH_3)_2) = \log P(C_6H_5-CH_3) + 2\ \pi(CH_3)$$
$$= \log P(C_6H_6) + 3\ \pi(CH_3) \qquad (II-15)$$

It will be clear that the correct application of the definition of
π also permits the use of eqn. II-16 as an equivalent of eqn. II-11:

$$\log P(C_6H_5-CH_3) = \log P(CH_4) + \pi(C_6H_5) \qquad (II-16)$$

and similarly eqn. II-17 can be regarded as the equivalent of eqn. II-14:

$$\log P(C_6H_5-CH_2-CH_3) = \log P(CH_4) + \pi(CH_3) + \pi(C_6H_5) \qquad (II-17)$$

It can be concluded from the above equations that the implicit incorporation of the following suppositions is responsible for the incorrectness of equations such as eqn. II-12 and II-13:

$$\log P(C_6H_6) = \pi(C_6H_5) \qquad (II-18)$$

$$\log P(CH_4) = \pi(CH_3) = \pi(CH_2) \qquad (II-19)$$

$$\pi(H) = 0 \qquad (II-20)$$

Eqn. II-20 itself is not incorrect, in that if in any given structure H is replaced by H, no change in lipophilicity occurs. However, it can by no means be concluded from eqn. II-20 that hydrogen, either atomic or molecular, has a partition value of zero!

The assumptions in eqns. II-18 and II-19 are inadmissable in that a log P value is equalized with a π value and the definition of this constant is ignored.

This definition requires that in any equation used for the calculation of a log P value the left-hand member of the equation must contain exactly the same number of log P terms as the right-hand member.

It would be equally wrong to derive an equilibrium constant exclusively from a set of σ constants via eqn. II-21:

$$\log K = \sum_1^n \sigma_n \qquad (II-21)$$

Other objections to the π system refer to the folding of structures of the $C_6H_5-CH_2-CH_2-CH_2-X$ type. The geometric relations in these structures are such that the complexation resulting from the interaction of the C – X dipole with the phenyl nucleus would not seem free from strain. Strainless folding is illustrated in Fig.II,1(A), which shows that the distance between the positive side of the dipole and the π-field of the ring is too large (no less than 3.24 Å to the centre of the phenyl ring). The structures investigated by HANSCH and ANDERSON [8] certainly vary in dipole moment and, as a consequence, the extent to which the complexes, if real, are stabilized would differ from case to case. Folding corrections derived thus far are fairly consistent, however: 0.60 ± 0.10.

Refs. pg 36

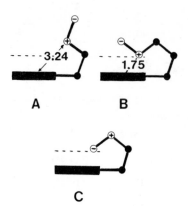

Fig.II,1: Optimal conditions for folding of structures
$C_6H_5-(CH_2)_3-X$ (A) and $C_6H_5-(CH_2)_4-X$ (B). C represents
an alternative way of folding for $C_6H_5-(CH_2)_3-X$.
The broken line indicates the distance to which the
π–orbital system of the phenyl ring extends.

When the $C_6H_5-CH_2-CH_2-CH_2-X$ structure is folded, a structure such
as $C_6H_5-CH_2-CH_2-CH_2-CH_2-X$ will certainly also be folded. Allowance being made for the correct steric relationships, the latter structure
will lend itself far more to folding: see Fig.II,1 (B), in which the
positive side of the dipole approaches the π–field. For $C_6H_5-CH_2-CH_2-$
$-CH_2-X$ this situation can be realized only by an interaction of the negative side of its dipole, as can be seen in Fig.II,1 (C).

In the concept of HANSCH and ANDERSON, the lipophilic behaviour of
structures such as $C_6H_5-CH_2-CH_2-X$ and $C_6H_5-CH_2-X$ will be completely
normal. Table II,4 shows the extent to which these considerations agree
with practical observations.

TABLE II,4

PARTITIONAL BEHAVIOUR OF $C_6H_5(CH_2)_nOH$

compound	log P			
	obs.	ref.	clc.	res.
$C_6H_5-CH_2-OH$	1.10	[9]	1.47	−0.37
	1.10	[30]		−0.37
$C_6H_5-CH_2-CH_2-OH$	1.36	[31]	1.97	−0.61
$C_6H_5-CH_2-CH_2-CH_2-OH$	1.88	[31]	2.47	−0.59
$C_6H_5-CH_2-CH_2-CH_2-CH_2-OH$	?		2.97	

No log P value is known for $C_6H_5-CH_2-CH_2-CH_2-CH_2-X$. LEO et al. [11], however, assume that the conditions for folding in a structure such as diphenhydramine (D) or chlorpromazine (E), both with four atoms con- necting N and phenyl, are analogous to those in $C_6H_5-CH_2-CH_2-CH_2-CH_2-X$, as appears from the following calculations.

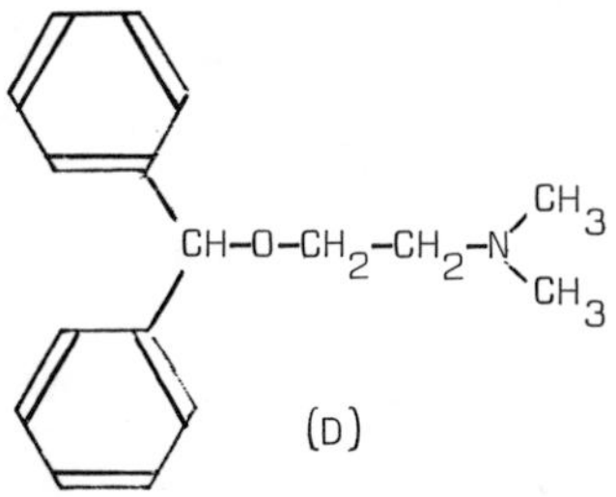

(D)

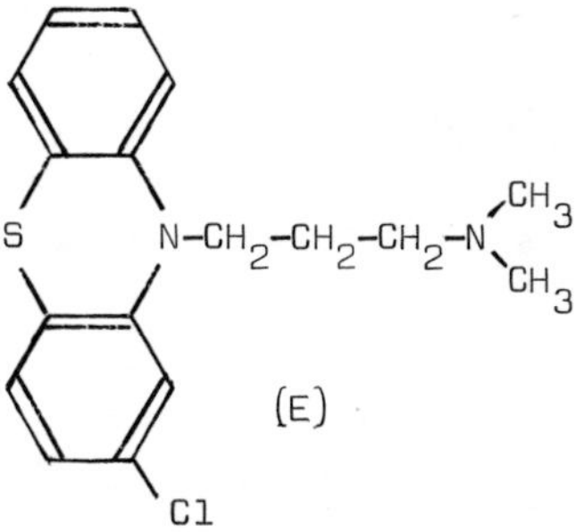

(E)

(1) Diphenhydramine

$$Log\ P = log\ P(C_6H_6) + \pi(C_6H_5) + \pi(CH_2)_b + [log\ P(C_2H_5-O-C_2H_5)$$
$$- 3\pi(CH_3)] + \pi(CH_2) + \pi[N(CH_3)_2]$$
$$= 4.26 + 0.30 + (-0.73) + 0.50 + (-0.90) = 3.43.$$

$\pi(CH_2)_b$ represents a π value of a CH_2 on which branching occurs. The value for $\pi[N(CH_3)_2]$ can be derived from the following equation:

$$\pi[N(CH_3)_2] = log\ P(C_6H_5-CH_2-CH_2-CH_2-N(CH_3)_2) - \pi(C_6H_5) - 3\pi(CH_2)$$
$$= 2.73 - 2.13 - 1.50 = -0.90.$$

As the folding in this model structure is well established (!), the folding correction is included in the calculation of $\pi[N(CH_3)_2]$ and if its value -0.90 is applied to the calculation of log P(diphenhydramine), one will obtain automatically the correct partition value without any further corrective procedure. The experimental log P value for diphen- hydramine is 3.27 [32], and is consistent with the above estimate.

(2) Chlorpromazine

$$Log\ P = log\ P(phenothiazine) + \pi(Cl_{ar}) + 3\pi(CH_2) + \pi[N(CH_3)_2]$$
$$= 4.15 + 0.70 + 1.50 + (-0.90) = 5.45.$$

The experimental value of log P is 5.35 [33].

These two examples, to which several others might be added, illus- trate the operation of the π system and the essential part played by the concept of folding. Evidently, acceptable log P values may be cal-

culated in this way, but this did not take away our suspicion that the
introduction of folding might have been erroneous as a consequence of
incorrect usage of the basic π eqn. II-4.

In this connection, mention should be made of a study by BIAGI et
al. [34] on the haemolytic activities of testosterone and testosterone
esters; in this series the phenyl propionate ester shows abnormal lipophilic behaviour on the thin-layer plate. This would suggest that the
considerations on benzyl alcohol (Table II,4) are also fully applicable
to phenylpropionic acid esters.

We shall revert to this problem in the following chapter.

REFERENCES

1   L.P. Hammett, Chem. Rev., 17(1935)125.

2   L.P. Hammett, Physical Organic Chemistry, McGraw - Hill Book Comp.
    New York, 1940.

3   H.H. Jaffé, Chem. Rev., 53(1953)191.

4   C. Hansch, P.P. Maloney, T. Fujita and R.M. Muir,
    Nature, 194(1962)180.

5   C. Hansch, R.M. Muir, T. Fujita, P.P. Maloney, F. Geiger and
    M. Streich, J. Amer. Chem. Soc., 85(1963)2817.

6   C. Hansch and T. Fujita, J. Amer. Chem. Soc., 86(1964)1616.

7   C. Hansch and W.J. Dunn III, J. Pharm. Sci. 61(1972)1.

8   C. Hansch and S.M. Anderson, J. Org. Chem. 32(1967)2583.

9   T. Fujita, J. Iwasa and C. Hansch, J. Amer. Chem. Soc.,
    86(1964)5175.

10 M.S. Tute in N. J. Harper and A.B. Simmonds (Editors), Advances
    in Drug Research, Vol. 6, Academic Press - London, New York, 1970.

11 A. Leo, C. Hansch and D. Elkins, Chem. Rev., 71(1971)525.

12 C. Hansch, A. Leo, S.H. Unger, K.H. Kim, D. Nikaitani and E.J. Lien,
    J. Med. Chem., 16(1973)1207.

13 C. Church, unpublished analysis (see ref.[11]).

14 D. Soderberg and C. Hansch, unpublished analysis (see ref.[11]).

15 K. Kim and C. Hansch, unpublished analysis (see ref.[11]).

16 D. Nikaitani and C. Hansch, unpublished analysis (see ref.[11]).

17 R. Collander, Acta Chem. Scand., 3(1949)717.

18 R. Collander, Acta Chem. Scand., 4(1950)1085.

19 R. Collander, Acta Chem. Scand., 5(1951)774.

20 R. Collander, Physiol. Plant., 7(1954)420.

21 J.C. McGowan, J. Appl. Chem., 2(1952)323.

22 J.C. McGowan, J. Appl. Chem., 2(1952)651.

23 J.C. McGowan, J. Appl. Chem., 4(1954)41.

24 I. Langmuir in H.N. Holmes (Editor), Colloid Symposium Monograph, Chemical Catalog Comp., 1925.

25 J.C. McGowan, P.N. Atkinson and L.H. Ruddle, J. Appl. Chem., 16(1966)99.

26 N.C. Deno and H.E. Berkheimer, J. Chem. Eng. Data, 5(1960)1.

27 M. S. Tute, Advances in Drug Research, 6(1971)1.

28 S.S. Davis, J. Pharm. Pharmac., 25(1973)1.

29 S.S. Davis, J. Pharm. Pharmac., 25(1973)293.

30 G.G. Nys and R.F. Rekker, unpublished analysis.

31 J. Iwasa, T. Fujita and C. Hansch, J. Med. Chem., 8(1965)150.

32 S. Anderson and C. Hansch, unpublished analysis (see ref.[11]).

33 S. Anderson, unpublished analysis (see ref.[11]).

34 G.L. Biagi, M.C. Guerra and A.M. Barbaro, J. Med. Chem., 15(1970)944.

CHAPTER III

THE HYDROPHOBIC FRAGMENTAL CONSTANT, f

INTRODUCTION

Our objections to the π system for the calculation of lipophilicity values, as discussed in the preceding chapter, mainly concerned the folding proposed by HANSCH and ANDERSON [1] and the connected incorrectness in handling the hydrophobic substituent constant. DAVIS [2,3] also criticized the system but with more emphasis on its thermodynamic aspects.

While dealing with these objections, we discovered that it was possible to abandon the basic π equation (eqn. II-4) and its extension (eqn. II-5) and to replace them successfully by eqn. III-1 and III-2, respectively:

$$\log P(SX) \;=\; f(S) \;+\; f(X) \qquad\qquad (III-1)$$

$$\log P(S'X_1X_2\ldots\ldots X_n) \;=\; f(S') \;+\; \sum_1^n f(X_n) \qquad\qquad (III-2)$$

Eqn. III-2 can also be written as

$$\log P \;=\; \sum_1^n \underline{a}_n f_n \qquad\qquad (III-3)$$

In these equations, f represents the <u>hydrophobic fragmental constant</u>, the lipophilicity contribution of a constituent part of a structure to the total lipophilicity, and $\underline{a}$ is a numerical factor indicating the incidence of a given fragment in the structure.

Two applications of eqn. III-3 are illustrated below:

1. 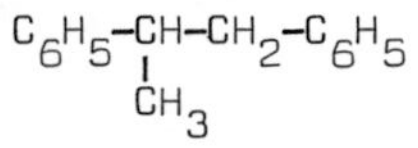 $C_6H_5\!-\!\overset{}{\underset{\underset{\textstyle CH_3}{|}}{CH}}\!-\!CH_2\!-\!C_6H_5$

2. $CH_3\!-\!\overset{\overset{\textstyle CH_3}{|}}{\underset{\underset{\textstyle CH_3}{|}}{C}}\!-\!CH_2\!-\!O\!-\!\overset{\overset{\textstyle O}{||}}{C}\!-\!CH_3$

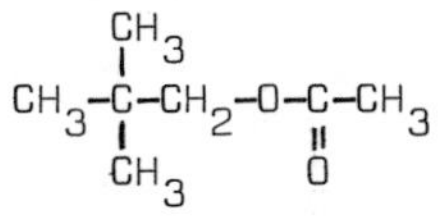

Log P (1) $=$ 2 $f(C_6H_5)$ + $f(CH_3)$ + $f(CH_2)$ + $f(CH)$

Log P (2) $=$ 4 $f(CH_3)$ + $f(CH_2)$ + $f(C)$ + $f(COO)$

It is to be noted that CH_3 attached to C is not differentiated from CH_3 attached to COO, and that the oxycarbonyl group (COO) is left intact and not split further into O and CO. Such rules were, of course,

Refs. pg 103

not yet established at the outset of our computations; they had to be discovered in a process of many trial and error calculations.

And in this way, continuously learning from faulty decisions, we were able to develop gradually a system of hydrophobic fragmental constants that was valid for a large and a varied number of structural types including a number of aromatic structures important from the medicinal chemical point of view.

The statistical framework of multiple regression analysis proved most suitable for the resolution of sets of equations of the type of eqn. III-3 to obtain the desired sets of hydrophobic fragmental constants.

The calculations were performed on an IBM 1130 Computing System with the Stepwise Multiple Regression Program EPL 1130-13.6.001. The numerical factors a were introduced as independent parameters and lipophilicity as a dependent parameter. The f values functioned as regression coefficients in the computer output. Computation was completed by statistical analysis [4].

The solution of a set of equations of the type of eqn. III-3 is obtained in practice in the form of eqn. III-4:

$$\log P = \sum_{1}^{n} a_n f_n + c \qquad (III-4)$$

where c = the constant term (intercept), which should have been zero but does, in fact, differ from zero. It is possible to reduce this constant term to a negligible value, i.e., to lead the curve belonging to eqn. III-4 through the origin of the coordinate system as closely as possible by application of forcing: the introduction of a few extra data points 0 = 0, 0 .... 0 into the computation through the required number of zero-cards.

One should be careful not to exaggerate this forcing; it is sufficient to adjust the constant term to an acceptable level, i.e., a value that does not exceed the average standard deviation of the regressor values.

ALIPHATIC STRUCTURES

Primary Set of f Values

For the calculations, a representative set of structures were selected, mainly from the literature. Only directly measured octanol — water partition values were included and log P values that could have

been made available by conversion from other solvent pairs were excluded. Further, there was no pre-selection; in other words, not even values that seemed suspicious at the outset were discarded, and not until
the statistical procedure had been completed were any outliers identi-
fied. The criterion applied here was whether the residual (= observed
minus estimate) surpassed double the value of the standard error of estimate. Any outliers were no longer removed from the data set,however,
as soon as the standard error of estimate decreased below 0.10. We consider that further "cleaning" of the regression equation would be useless because in many instances the reproducibility of log P measurements between two different laboratories is not good enough to warrant
any sharper selection.

The total number of log P values introduced was 128, referring to
87 different structures distributed over more than 40 structural types;
from this population, the following eleven f values could be computed:
CH_3, CH_2, CH, NH_2,NH, N, C_6H_5, OH, O, COOH and COO.

The reason for including the phenyl group in this selection
was that one of the major aims of the original study was to prove or
disprove any folding of a number of structures. Care was taken, however,
that no structures were included when C_6H_5 was connected directly with
a functional group because any resonance interaction between this functional group and the phenyl nucleus would certainly enhance lipophilicity (see Table II,1).

It is evident that a well balanced distribution of the fragments o-
ver the structures investigated would have been ideal. By definition,
however, aliphatic organic structures always contain CH_3 and other hydrocarbon fragments and this means that such fragments will always pre-
dominate and that, as a result, the standard deviations of the non-C
fragments will exceed those of the C fragments. Among the latter, $f(CH_2)$
is the most frequently occurring species and for that reason it emerges
with the smallest standard deviation.

A number of important functional groups are missing in the above
collection: carbonyl, carbonamide, halogen, etc. Structures with such
groups have been subjected to so few partitioning studies that their inclusion would have served no useful purpose but would only have increased
the risk of impairing the statistical data.

The most recent edition of the "Pomona Listing on Lipophilicity Data" [5] provides information on a collection of "small" hydrocarbons
and halogenated alkanes. These structures are not included in our com-

Refs. pg 103

TABLE III,1

STRUCTURES INTRODUCED INTO STATISTICAL TREATMENT

(I)   Values based on f <u>without</u> correction for proximity effects.
(II) Values based on π.
(III) Values based on f <u>with</u> correction for proximity effects.
The indication T. III,3 (no.) in the columns 4 and 5 (lower lines)
means that on the basis of the second regression analysis, where proximity effects were taken into account, the structure concerned had to
be qualified as an outlier. Such a structure will be found in Table III,
3 with an indication of its original Table III,1 number.
The indication T. III,3 (no.) in columns 4 and 5 (upper lines) indicates
that the structure concerned no longer functioned as an outlier in the
second regression analysis, and was transferred from Table III,3
to Table III,1 with an indication of its original Table III,3 number.

no	Compound	log P values				
		obs.	clc.I	res.I	clc.II	res.II
		ref.	clc.III	res.III		
1	CH_3-OH	-0.66	-0.74	0.08	-0.66	0.00
		[1]	-0.73	0.07		
		-0.82	-0.74	-0.08	-0.66	-0.16
		[6]	-0.73	-0.09		
2	CH_3-NH_2	-0.57	-0.69	0.12	-0.69	0.12
		[6]	-0.67	0.10		
3	CH_3-COOH	-0.17	-0.29	0.12	-0.17	0.00
		[7]	-0.29	0.12		
		-0.31	-0.29	-0.02	-0.17	-0.14
		[6]	-0.29	-0.02		
4	CH_3-CH_2-OH	-0.32	-0.22	-0.10	-0.16	-0.16
		[6]	-0.20	-0.12		
5	$CH_3-CH_2-NH_2$	-0.13	-0.16	0.03	-0.19	0.06
		[8]	-0.14	0.01		
6	$CH_3-COO-CH_3$	0.18	0.13	0.05	0.23	-0.05
		[6]	0.13	0.05		
7	CH_3-CH_2-COOH	0.25	0.23	0.02	0.33	-0.08
		[6]	0.24	0.01		
		0.33	0.23	0.10	0.33	0.00
		[7]	0.24	0.09		
8	$CH_3-(CH_2)_2-OH$	0.34	0.31	0.03	0.34	0.00
		[1]	0.33	0.01		
9	$CH_3-(CH_2)_2-NH_2$	0.48	0.37	0.11	0.31	0.17
		[8]	0.39	0.09		
10	$(CH_3)_2CH-NH_2$	0.26	0.27	-0.01	0.11	0.15
		[7]	0.27	-0.01		
11	$CH_3-CH_2-NH-CH_3$	0.15	0.07	0.08	0.31	-0.16
		[8]	0.08	0.07		

TABLE III,1 (continued)

12	$(CH_3)_3N$	0.27 [9]	0.27 T.III,3(20)	0.00	0.20	0.07
13	$CH_3-(CH_2)_2-COOH$	0.25 [6]	0.23 0.24	0.02 0.01	0.33	-0.08
		0.33 [7]	0.23 0.24	0.10 0.09	0.33	0.00
14	$CH_3-COO-CH_2-CH_3$	0.73 [1]	0.65 0.66	0.08 0.07	0.73	0.00
		0.63 [6]	0.65 0.66	-0.02 -0.03	0.73	-0.10
15	$CH_3-(CH_2)_3-OH$	0.88 [10]	0.84 0.86	0.04 0.02	0.84	0.04
16	$(CH_3)_2CH-CH_2-OH$	0.65 [10]	0.74 0.74	-0.09 -0.09	0.64	0.01
		0.83 [6]	0.74 0.74	0.09 0.09	0.64	0.19
17	$CH_3-CH_2-CH(CH_3)-OH$	0.61 [1]	0.74 0.74	-0.13 -0.13	0.64	-0.03
18	$CH_3-CH_2-O-CH_2-CH_3$	0.77 [7]	0.94 0.93	-0.17 -0.16	1.03	-0.26
		0.83 [6]	0.94 0.93	-0.11 -0.10	1.03	-0.20
		0.74 [11]	0.94 0.93	-0.20 -0.19	1.03	-0.29
19	$CH_3-CH(OH)-CH(OH)-CH_3$	-0.92 [6]	-0.97 T.III,3(21)	0.05	-0.72	-0.20
20	$CH_3-CH_2-O-(CH_2)_2-OH$	-0.54 [6]	-0.67 T.III,3(22)	0.13	-0.13	-0.41
21	$CH_3-(CH_2)_3-NH_2$	0.68 [6]	0.90 T.III,3(19)	-0.22	0.81	-0.13
		0.88 [9]	0.90 0.92	-0.02 -0.04	0.81	0.07
		0.97 [8]	0.90 0.92	0.07 0.05	0.81	0.16
		0.81 [7]	0.90 0.92	-0.09 -0.11	0.81	0.00
22	$(CH_3)_2CH-CH_2-NH_2$	0.88 [8]	0.79 0.80	0.09 0.08	0.61	0.27
		0.73 [8]	0.79 0.80	-0.06 -0.07	0.61	0.12
23	$CH_3-CH_2-CH(CH_3)-NH_2$	0.74 [8]	0.79 0.80	-0.05 -0.06	0.61	0.13
24	$(CH_3-CH_2)_2NH$	0.57 [9]	0.60 0.61	-0.03 -0.04	0.70	-0.13

TABLE III,1 (continued)

no	Compound	obs. ref.	clc.I clc.III	res.I res.III	clc.II	res.II
		0.58 [8]	0.60 0.61	-0.02 -0.03	0.70	-0.12
		0.43 [6]	0.60 0.61	-0.17 -0.18	0.70	-0.27
25	$CH_3-CH_2-COO-CH_2-CH_3$	1.21 [1]	1.18 1.19	0.03 0.02	1.23	-0.02
26	$CH_3-(CH_2)_4-OH$	1.40 [10]	1.37 1.38	0.03 0.02	1.34	0.06
27	$(CH_3)_2CH-(CH_2)_2-OH$	1.16 [10]	1.27 1.27	-0.11 -0.11	1.20	-0.04
28	$CH_2 \overset{CH_2-CH_2}{\underset{CH_2-CH_2}{\diagup \diagdown}} NH$ (piperidine)	0.85 [7]	0.79 0.78	0.06 0.07	1.11	-0.26
		0.81 [8]	0.79 0.78	0.02 0.03	1.11	-0.30
29	$CH_3-(CH_2)_4-NH_2$	1.49 [8]	1.43 1.44	0.06 0.05	1.31	0.18
30	$CH_3-(CH_2)_3-NH-CH_3$	1.33 [7]	1.12 1.13	0.21 0.20	1.20	0.13
31	$CH_3-CH_2-NH-CH(CH_3)_2$	0.93 [8]	1.02 1.02	-0.09 -0.09	1.00	-0.09
32	$CH_2 \overset{CH_2-CH_2}{\underset{CH_2-CH_2}{\diagup \diagdown}} CH-OH$ (cyclohexanol)	1.23 [1]	1.46 1.44	-0.23 -0.21	1.55	-0.32
33	$CH_3-(CH_2)_4-COOH$	1.88 [7]	1.82 1.82	0.06 0.06	1.83	0.05
		1.92 [7]	1.82 1.82	0.10 0.10	1.83	0.09
34	$CH_2 \overset{CH_2-CH_2}{\underset{CH_2-CH_2}{\diagup \diagdown}} CH-NH_2$ (cyclohexylamine)	1.49 [8]	1.52 1.51	-0.03 -0.02	1.52	-0.03
35	$CH_3-(CH_2)_5-OH$	2.03 [12]	1.90 1.91	0.13 0.12	1.84	0.19
36	$CH_3-CH_2-O-(CH_2)_3-CH_3$	2.03 [1]	1.99 1.99	0.04 0.04	2.03	0.00
37	$CH_3-(CH_2)_2-O-(CH_2)_2-CH_3$	2.03 [13]	1.99 1.99	0.04 0.04	2.03	0.00
38	$CH_3-(CH_2)_5-NH_2$	1.98 [8]	1.95 1.97	0.03 0.01	1.81	0.17
		2.06 [8]	1.95 1.97	0.11 0.09	1.81	0.25
39	$(CH_3-CH_2-CH_2)_2NH$	1.67 [8]	1.65 1.66	0.02 0.01	1.70	-0.03

TABLE III,1 (continued)

		1.73 [7]	1.65 1.66	0.08 0.07	1.70	0.03
40	$(CH_3)_2N-(CH_2)_3-CH_3$	1.70 [7]	1.86 1.57	-0.16 0.13	1.70	0.00
41	$C_6H_5-CH_3$	2.69 [14]	2.61 2.61	0.08 0.08	2.63	0.06
		2.73 [12]	2.61 2.61	0.12 0.12	2.63	0.10
		2.80 [15]	2.61 2.61	0.19 0.19	2.63	0.17
42	$C_6H_5-CH_2-OH$	1.10 [14]	1.00 0.99	0.10 0.11	1.47	-0.37
43	$C_6H_5-CH_2-NH_2$	1.09 [16]	1.04 1.06	0.05 0.03	1.44	-0.35
44	$CH_3-(CH_2)_6-NH_2$	2.57 [8]	2.48 2.50	0.09 0.07	2.31	0.26
45	$CH_3-(CH_2)_2-NH-(CH_2)_3-CH_3$	2.12 [8]	2.18 2.19	-0.06 -0.07	2.20	-0.08
46	$CH_3-CH_2-\underset{\underset{CH_3}{\mid}}{CH}-NH-(CH_2)_2-CH_3$	1.91 [8]	2.08 2.07	-0.17 -0.16	2.00	-0.09
47	$CH_3-(CH_2)_2-NH-CH_2-CH(CH_3)_2$	2.07 [8]	2.08 2.07	-0.01 0.00	2.00	0.07
48	$C_6H_5-CH_2-CH_3$	3.15 [16]	3.13 3.14	0.02 0.01	3.13	0.02
49	$C_6H_5-CH_2-COOH$	1.41 [14]	1.45 1.43	-0.04 -0.02	1.96	-0.55
50	$C_6H_5-(CH_2)_2-OH$	1.36 [16]	1.53 1.52	-0.17 -0.16	1.97	-0.61
51	$CH_3-(CH_2)_7-OH$	3.15 [6]	2.95 2.97	0.20 0.18	2.84	0.31
52	$C_6H_5-(CH_2)_2-NH_2$	1.41 [16]	1.59 1.58	-0.18 -0.17	1.94	-0.53
53	$CH_3-(CH_2)_4-CH(CH_2-CH_3)-NH_2$	2.82 [8]	2.91 2.91	-0.09 -0.09	2.11	0.71
54	$(CH_3-CH_2-CH_2-CH_2)_2NH$	2.68 [13]	2.71 2.72	-0.03 -0.04	2.70	-0.02
		2.83 [8]	2.71 2.72	0.12 0.11	2.70	0.13
55	$C_6H_5-CH(CH_3)_2$	3.66 [12]	3.56 3.55	0.10 0.11	3.63	0.03
		3.66 [17]	3.56 3.56	0.10 0.11	3.63	0.03
56	$C_6H_5-(CH_2)_2-COOH$	1.84 [16]	1.98 1.96	-0.14 -0.12	2.46	-0.62

Refs. pg 103

TABLE III,1 (continued)

no	Compound	obs. ref.	clc.I clc.III	res.I res.III	clc.II	res.II
			— log P values —			
57	$C_6H_5-(CH_2)_2-CH_3$	3.68 [16]	3.66 3.66	0.02 0.02	3.63	0.05
		3.57 [7]	3.66 3.66	−0.09 −0.09	3.63	−0.06
58	$C_6H_5-CH_2-COO-CH_3$	1.83 [16]	1.86 1.86	−0.03 −0.03	2.36	−0.53
59	$CH_3-COO-CH_2-C_6H_5$	1.96 [16]	1.86 1.86	0.10 0.10	2.36	−0.40
60	$C_6H_5-(CH_2)_3-OH$	1.88 [16]	2.06 2.05	−0.18 −0.17	2.47	−0.59
61	$C_6H_5-(CH_2)_3-O-CH_3$	2.70 [16]	2.68 2.66	0.02 0.04	3.16	−0.46
62	$C_6H_5-CH_2-N(CH_3)_2$	1.98 [8]	2.01 T.III,3(18)	−0.03	2.33	−0.35
		1.91 [8]	2.01 T.III,3(18)	−0.10	2.33	−0.42
63	$HOOC-(CH_2)_7-COOH$	1.57 [6]	1.72 1.70	−0.15 −0.13	2.16	−0.59
64	$C_6H_5-(CH_2)_3-COOH$	2.42 [16]	2.51 2.49	−0.09 −0.07	2.96	−0.54
65	$C_6H_5-(CH_2)_2-COO-CH_3$	2.32 [16]	2.40 2.38	−0.08 −0.06	2.86	−0.54
66	$CH_3-(CH_2)_8-COOH$	4.09 [7]	3.93 3.93	0.16 0.16	3.83	0.26
67	$C_6H_5-(CH_2)_3-COO-CH_3$	2.77 [16]	2.93 2.91	−0.16 −0.14	3.36	−0.59
68	$C_6H_5-C_6H_5$	4.04 [18]	3.82 T.III,3(2)	0.22	4.02	0.02
69	$C_6H_5-CH_2-C_6H_5$	4.14 [19]	4.35 4.33	−0.21 −0.19	4.52	−0.38
70	$C_6H_5-CH(OH)-C_6H_5$	2.67 [18]	2.63 2.60	0.04 0.07	3.16	−0.49
		2.67 [20]	2.63 2.60	0.04 0.07	3.16	−0.49
		2.64 [21]	2.63 2.60	0.01 0.04	3.16	−0.52
71	$C_6H_5-(CH_2)_2-C_6H_5$	4.79 [7]	4.86 4.86	−0.07 −0.07	5.02	−0.23
		4.82 [22]	4.86 4.86	−0.04 −0.04	5.02	−0.20

TABLE III,1 (continued)

No.	Structure	obs [ref]	clc.I	res.I	clc.II	res.II
72	$(C_6H_5)_2CH-O-(CH_2)_2-N(CH_3)_2$	3.27 [7]	3.21	0.06	4.05	-0.78
			3.29	-0.02		
		3.40 [7]	3.21	0.19	4.05	-0.65
			3.29	0.11		
73	C_6H_5-CH(-CH$_2$-OH)-COO-CH (bicyclic, N-CH$_3$)	1.83 [7]	1.81	0.02	2.92	-1.09
			1.80	0.03		
		1.79 [7]	1.81	-0.02	2.92	-1.13
			1.80	-0.01		
74	$HO-CH_2-COOH$	-1.11 [6]	T.III,3(8)		-1.33	0.22
			-1.10	-0.01		
75	$HO-CH_2-CH_2-NH_2$	-1.31 [6]	T.III,3(14)		-1.35	0.04
			-1.29	-0.02		
76	$CH_3-CH(OH)-COOH$	-0.62 [6]	T.III,3(9)		-0.83	0.21
			-0.69	0.07		
77	$HOOC-(CH_2)_2-COOH$	-0.59 [6]	T.III,3(12)		-0.34	0.25
			-0.48	-0.11		
78	$C_6H_5-CH(OH)-COOH$	0.56 [20]	T.III,3(10)		0.60	-0.04
			0.50	0.06		
79	$CH_3-CH_2-O-CH_2-O-CH_2-CH_3$	0.84 [7]	T.III,3(11)		0.56	0.28
			0.73	0.11		
80	$(CH_3-CH_2)_3N$	1.45 [8]	T.III,3(5)		1.57	-0.12
			1.57	-0.12		
		1.44 [7]	T.III,3(5)		1.57	-0.13
			1.57	-0.13		
81	1,4-dioxane (O$<$CH$_2$-CH$_2>$O ring)	-0.42 [6]	T.III,3(15)		0.06	-0.48
			-0.49	0.07		
82	$(HO-CH-CH_2)_2NH$ ($\vert$ CH$_3$)	-0.82 [6]	T.III,3(16)		-1.02	0.20
			-0.88	0.06		
83	$C_6H_5-CH-CH-NH-CH_3$ (OH CH$_3$)	0.93 [7]	T.III,3(13)		1.27	-0.34
			0.94			
84	$(C_6H_5)_2CH-COOH$	3.05 [20]	T.III,3(4)		3.65	-0.60
			3.04	0.01		

putations but will receive attention in the following chapter.

Table III,1 reviews all observations that remained in the MRA on removal of the outliers. Column "clc.I" records the estimates from our regression, column "clc.II" the log P values calculated on the basis of the HANSCH π calculation procedure (eqn. II-5) and columns "res.I" and "res.II" the residuals.

The *f* values, together with some important statistical data, are collected in Table III,2 in the columns "regression analysis **A**".

Refs. pg 103

TABLE III,2
PRIMARY SET OF f VALUES SOLVED BY MRA

(A before, and B after the Introduction of the Proximity Effect)

no	Fragment	regr. anal. A			regr. anal. B		
		f	st.dev.	t	f	st.dev.	t
1	CH_3	0.695	0.025	28.3	0.702	0.021	33.3
2	CH_2	0.528	0.007	71.6	0.527	0.006	82.0
3	CH	0.260	0.025	10.4	0.236	0.022	10.7
4	NH_2	-1.381	0.051	-26.9	-1.380	0.041	-33.5
5	NH	-1.849	0.050	-36.8	-1.864	0.037	-49.7
6	N	-1.809	0.053	-34.0	-2.133	0.052	-40.9
7	C_6H_5	1.912	0.033	58.8	1.896	0.027	69.5
8	OH	-1.438	0.046	-31.0	-1.440	0.036	-40.6
9	O	-1.508	0.047	-32.2	-1.536	0.031	-49.3
10	COOH	-0.987	0.048	-20.7	-1.003	0.035	-29.0
11	COO	-1.261	0.047	-26.8	-1.281	0.041	-31.4
12	p.e. 1[*]				0.80	0.08	
13	p.e. 2[**]				0.46	0.08	
	n	101			106		
	s	0.111			0.100		
	r	0.996			0.997		
	F	1,152			1,770		
constant term		-0.002			0.011		

[*]   p.e. 1: proximity effect for 1C separation of electronegative
           groups.
[**] p.e. 2: same, but 2C separation.

NOTE: The framed part of this Table is for practical purposes.

We wish to emphasize the following points:

(1) The quality of the MRA is extremely good: the standard error of
estimate is 0.111, the correlation coefficient is 0.996, which implies
that 99% of the variance in the data are accounted for and the F-test
value of 1,150 indicates significance on a very high level.

By narrowing the above indicated outlier criterion, it would have
been possible to obtain some further improvement in the statistical
data and to bring the standard error of estimate to an even lower level.
This procedure, however, would have removed an undesirable number of

structures from the computation. It would also suggest an unduly high accuracy of the partition measurement, especially when data collected from different institutes are pooled.

(2) The constant term is so small (-0.002) that it can be safely neglected in practical calculations of unknown partition coefficients. There was no need for forcing.

(3) The standard deviations of the regression coefficients (the f values) range from small (non C-fragments) to extremely small (CH_2 fragment).

(4) The high t values (Student test) invariably show that the values satisfy a severe criterion of significance.

(5) The objections formulated by DAVIS [2,3] on account of thermodynamic considerations are completely obviated by eqn. III-3. With this in mind, the constant term in the DAVIS equation (eqn. II-10) can be regarded simply as a corrective term transforming the total of π terms to a total of f terms.

(6) Although the t values of $f(NH_2)$, $f(NH)$ and $f(N)$ are all acceptable, the f values found for NH and N are not satisfactory; that for N should have been less than -2.0. It is important to realize, however, that the number of tertiary amines available for our regression analysis was small and that, in addition, several outliers were found among them.

Close inspection of Table III,1 reveals that, on the whole, the log P values calculated on the basis of f values differ less from the experimental values than those based on π values. The latter calculations showed as many as 35 unacceptable outliers.

Outliers removed from our computation are collected in Table III,3, which was compiled along the same lines as Table III,1.

Some partition values (see, e.g., 1, 2 (ref. 22) and 3 of Table III, 3) most probably involve errors made in the determination of the partition coefficient, as other workers have published values for these structures that do ensure a reasonable fit (see Table III,1).

A systematic anomaly, which we proposed to be attributable to a proximity effect, became apparent in Nos. 8 - 17: placed close together, two groups with distinctly electronegative properties (more or less hydrophilic in nature) will bind fewer water molecules from their surroundings than if each group were present alone. The resulting increase in lipophilicity is a function of both the distance between the two groups concerned and of their hydrophilic capacities, the former appar-

Refs. pg 103

TABLE III,3

OUTLIERS IN THE MULTIPLE REGRESSION ANALYSIS

The indication T. III,1 (no.) in the columns 4 and 5 (upper lines)
means that on the basis of the second regression analysis where proximity effects were taken into account, the structure concerned no longer had to be qualified as an outlier. Such a structure will be found
in Table III,1 with an indication of its original Table III,3 number.
The indication T. III,1 (no.) in the columns 4 and 5 (lower lines) indicates that the structure concerned had to be qualified as an outlier
on the basis of the second regression analysis. It was transferred from
Table III,1 to Table III,3 with an indication of its original Table
III,1 number.

no	Compound	obs. ref.	clc.I clc.III	res.I res.III	clc.II	res.II
1	$C_6H_5-CH_3$	2.11 [23]	2.61 2.61	−0.50 −0.50	2.63	−0.52
2	$C_6H_5-C_6H_5$	4.09 [24]	3.82 3.80	0.27 0.29	4.02	0.07
		3.16 [23]	3.82 3.80	−0.66 −0.64	4.02	−0.86
		4.04 [18]	T.III,1(68) 3.80	0.24	4.02	0.02
3	$C_6H_5-CH(OH)-C_6H_5$	2.03 [22]	2.63 2.60	−0.60 −0.57	3.16	−1.13
4	$(C_6H_5)_2CH-COOH$	2.06 [25]	3.09 3.04	−1.03 −0.98	3.65	−1.59
5	$(CH_3-CH_2)_3N$	1.44 [7]	1.86 T.III,1(80)	−0.42	1.57	−0.13
		1.45 [8]	1.86 T.III,1(80)	−0.41	1.57	−0.12
		1.15 [8]	1.86 1.57	−0.71 −0.42	1.57	−0.42
6	$(CH_3-CH_2-CH_2)_3N$	2.79 [7]	3.44 3.15	−0.65 −0.36	3.17	−0.38
		2.79 [8]	3.44 3.15	−0.65 −0.36	3.17	−0.38
7	$C_6H_5-(CH_2)_3-NH_2$	1.83 [16]	2.11 2.11	−0.28 −0.28	2.44	−0.61
8	$HO-CH_2-COOH$	−1.11 [6]	−1.89 T.III,1(74)	0.78	−1.33	0.22
9	$CH_3-CH(OH)-COOH$	−0.62 [6]	−1.47 T.III,1(76)	0.85	−0.83	0.21
10	$C_6H_5-CH(OH)-COOH$	0.56 [20]	−0.14 T.III,1(78)	0.70	0.60	−0.04

log P values

TABLE III,3 (continued)

11	$CH_3-CH_2-O-CH_2-O-CH_2-CH_3$	0.84 [7]	-0.04 0.88	T.III,1(79)	0.56	0.28
12	$HOOC-(CH_2)_2-COOH$	-0.59 [6]	-0.92 0.33	T.III,1(77)	-0.34	0.25
13	$C_6H_5-CH-CH-NH-CH_3$ OH CH$_3$	0.93 [7]	0.54 0.39	T.III,1(83)	1.27	-0.34
14	$HO-CH_2-CH_2-NH_2$	-1.31 [6]	-1.79 0.48	T.III,1(75)	-1.35	0.04
15		-0.42 [6]	-0.91 0.49	T.III,1(81)	0.06	-0.48
16	$(HO-CH-CH_2)_2NH$ CH$_3$	-0.82 [6]	-1.76 0.94	T.III,1(83)	-1.02	0.20
17	$HO-(CH_2)_2-NH-(CH_2)_2-OH$	-1.43 [6]	-2.60 1.17 -1.71[a] 0.28		-1.62	0.19
18	$C_6H_5-CH_2-N(CH_3)_2$	1.98 [8]	T.III,1(62) 1.71 0.27		2.33	-0.35
		1.91 [8]	T.III,1(62) 1.71 0.20		2.33	-0.28
19	$CH_3-(CH_2)_3-NH_2$	0.68 [6]	T.III,1(21) 0.91 -0.23		0.81	-0.13
20	$(CH_3)_3N$	0.27 [9]	T.III,1(12) -0.02 0.29		0.20	0.07
21	$CH_3-CH(OH)-CH(OH)-CH_3$	-0.92 [6]	T.III,1(19) -0.53[a] -0.39		-0.72	0.20
22	$CH_3-CH_2-O-(CH_2)_2-OH$	-0.54 [6]	T.III,1(20) -0.22[a] -0.32		-0.13	0.41

[a] Corrections for proximity effects included

ently being the more important.

More attention will be paid to the proximity effect in one of the following sections of this chapter.

Table III,4 lists those numbers of Table III,3 in which proximity effects are evident. Two sets with effects of 0.80 ± 0.08 and 0.46 ± 0.08 emerge, depending on whether the link between the two negative groups consists of one and two carbon atoms, respectively.

There is much evidence that attachment of a group with a negative character to a quaternary carbon atom lowers the partition value. As appears from Table III,5, this drop is fairly constant, and is approximately 0.47 for the first five compounds.

We do not wish to comment on the background of the phenomenon, es-

TABLE III,4

PROXIMITY EFFECTS

compound no. (Table III,3)	interacting groups		number of separating C's	proximity effect
8	HO	COOH	1	0.78
9	HO	COOH	1	0.85
10	HO	COOH	1	0.70
11	O	O	1	0.88
12	HOOC	COOH	2	0.33
13	HO	NH	2	0.39
14	HO	NH$_2$	2	0.48
15	O	O	2	0.49
16	HO	NH	2	2 × 0.47
17	HO	NH	2	2 × 0.58

Compounds 8–11: Av. = 0.80 (s.d.: 0.08)

Compounds 12–17: Av. = 0.46 (s.d.: 0.08)

pecially because the partition values of 3,3-dimethyl-2-butanol and 2-methyl-2-nitropropane are unclear. This matter will be dealt with in more detail in a following chapter.

Following the discovery of proximity effects, we decided to repeat the statistical treatment entirely, allowance now being made for those effects whenever necessary.

Two approaches are available in this repetitive procedure: (a) inclusion of the two proximity effects as unknown regressors in the total of f values, and (b) introduction of the mean values cited above as fixed data. The reason to choose (b) was that the limited material available might unbalance the data on applying procedure (a).

On the whole, the results of the second calculation showed an improvement because the significance levels of all f values were much higher and, strikingly, the f value of the tertiary N was now in the correct position relative to the secondary N.

Relevant figures are given in Tables III,1 and III,3 (columns "clc. III" and "res.III").Detailed information about the new improved set of f values is presented in Table III,2 (regr. anal. B).

The compound biphenyl is an unambiguous outlier now (the mean difference between computed and experimental partition value is +0.26). Its increased lipophilicity is probably due to a mesomeric interaction between the two phenyl rings. More attention will be paid to this subject in the following section.

The origin of the other outliers remains uncertain at present. Per-

TABLE III,5

STRUCTURES WITH A NEGATIVE GROUP ATTACHED TO A QUATERNARY C

compound	log P(obs)	ref.	log P(calc)	Δ
CH_3–$\underset{\underset{CH_3}{\mid}}{\overset{\overset{CH_3}{\mid}}{C}}$–OH	0.37	[3]	0.81	−0.44
CH_3–CH_2–$\underset{\underset{CH_3}{\mid}}{\overset{\overset{CH_3}{\mid}}{C}}$–OH	0.89	[3]	1.33	−0.44
CH_3–$\underset{\underset{CH_3}{\mid}}{\overset{\overset{CH_3}{\mid}}{C}}$–$NH_2$	0.40	[16]	0.87	−0.47
adamantane–OH	2.14	[16]	2.57	−0.43
$CH_2{=}CH{-}\underset{\underset{C{\equiv}CH}{\mid}}{\overset{\overset{C_2H_5}{\mid}}{C}}{-}O{-}\overset{\overset{O}{\parallel}}{C}{-}NH_2$	1.09	[31]	−1.66[a]	−0.57
CH_3–$\underset{\underset{H_3C}{\diagup}}{\overset{\overset{CH_3}{\mid}}{C}}$–$\underset{\underset{OH}{\diagdown}}{CH}$–$CH_3$	1.48	[29]	1.75	−0.27
CH_3–$\underset{\underset{CH_3}{\mid}}{\overset{\overset{CH_3}{\mid}}{C}}$–$NO_2$	1.01 1.17	[26] [28]	1.24	−0.23 −0.07

[a] Using $f(OOCNH_2) = -1.39$, $f(CH_2{=}CH) = 0.93$ and $f(CH{\equiv}C) = 0.73$ in addition to the necessary f values from Table III,2

haps some result from errors in partition measurements, while others are undoubtedly connected with the fact that not all factors that gov-ern the partitional behaviour of organic structures are fully understood. Table III,5 is illustrative of these assumptions.

Secondary Set of f Values

As stated above, a number of functional groups had to be excluded from statistical treatment because of the limited amount of material a-vailable, which did permit a set of secondary hydrophobic fragmental constants to be derived by using eqn. III-5:

Refs. pg 103

$$f(X) = \log P(SX) - \sum f(S) \qquad\qquad (III-5)$$

where $f(X)$ is the hydrophobic constant of the desired fragment that
forms part of structure SX and $\sum f(S)$ is the summation of the hydrophobic fragmental constants of the structure part S.

The accuracy with which these secondary f values are produced will
be lower than that of the primary constants calculated by means of the
MRA, firstly, because it is more dependent on individual log P determinations and secondly, because the whole of the error in the primary
constants of S is reflected in $f(X)$. It is assumed, however, that the
standard deviation in these secondary fragments will rarely exceed 0.15.

TABLE III,6

COMPILATION OF STRUCTURES USED FOR THE DERIVATION
OF SECONDARY ALIPHATIC f VALUES

fragment	compound	log P values			
		obs.	ref.	clc.	res.
F	See Table III,8, nos. 4 and 5				
Cl	See Table III,8, nos. 6,7 and 8				
Br	See Table III,8, nos. 9,10 and 11				
I	See Table III,8, nos. 12 and 13				
C = O	See Table III,8, nos. 20 – 26				
$CONH_2$	See Table III,8, nos. 37 – 40				
CN	See Table III,8, nos. 30 – 35				
NO_2	CH_3-NO_2	−0.33	[1]	−0.32	−0.01
		0.08	[6]		0.40*
	$CH_3-CH_2-NO_2$	0.18	[1]	0.11	0.07
	$CH_3-CH_2-CH_2-NO_2$	0.65	[1]	0.74	−0.09
	$C_6H_5-CH_2-CH_2-NO_2$	2.08	[27]	0.93	0.15
$CH_2{=}CH$	$CH_2{=}CH-CH_2-OH$	0.17	[24]	0.02	0.15
	$CH_2{=}CH-CH_2-NH_2$	0.03	[30]	0.08	−0.05
	$CH_2{=}CH-CH_2-NH-C_2H_5$	0.81	[30]	0.83	−0.02
	$CH_2{=}CH-CH_2-NH-\underline{n}.C_3H_7$	1.33	[30]	1.35	−0.02
	$CH_2{=}CH-CH_2-C_6H_5$	3.23	[19]	3.35	−0.12
	$CH_2{=}CH-CH_2-O-C_6H_5$	2.94	[13]	2.90[a]	0.04
C(quat.)	$(CH_3)_3C-CH_2-OH$	1.32	[29]	1.34	−0.02
		1.36	[1]		0.02
NH_2COO	$NH_2-COO-C_2H_5$	−0.15	[13]	−0.16	0.01

TABLE III,6 (continued)

	NH$_2$-COO-CH-CH=CH$_2$ C$_2$H$_5$	1.09	[31]	1.01	0.08
	NH$_2$-COO-CH$_2$ CH$_3$-C-n.C$_3$H$_7$ NH$_2$-COO-CH$_2$	0.70	[13]	0.88	-0.18
CF$_3$	CF$_3$-CF$_3$	2.00	[32]	2.04^b	-0.04
	CF$_3$-CH(CH$_3$)-OH	0.71	[29]	0.75	-0.04
	CF$_3$-CH$_2$-OH	0.41	[33]	0.34	0.07
		0.32	[10]		-0.02
	CF$_3$-CONH$_2$	0.12	[34]	-0.40	0.52*
	HCF$_3$	0.64	[36]	0.98	-0.34*
CCl$_3$	HCCl$_3$	1.97	[1]	1.99	-0.02
	CCl$_3$-CH$_2$-OH	1.35	[29]	1.34	0.01
	CCl$_3$-CONH$_2$	1.04	[24]	0.60	0.44*
SH	CH$_3$-CH$_2$-CH$_2$-CH$_2$-SH	2.28	[19]	2.28^c	
S	CH$_3$-CH$_2$-S-CH$_2$-CH$_3$	1.95	[1]	1.95^c	
SO	CH$_3$-SO-CH$_3$	-1.35	[35]	-1.35^c	

^{a} Using $f(O_{ar})$ = -0.45

^{b} Log P = 2 $f(CF_3)$ + p.e. 2

^{c} Only one data- point

* Outlier

Table III,6 presents a survey of the material employed in deriving the secondary aliphatic f values. Sixteen fragmental constants in all could be obtained by the method indicated. Three of these, SH, S and SO, refer to only one structure each, thus suggesting that caution should be exercised in the practical application of these f values.

Table III,7 combines all 16 secondary f values with the π values in so far as these have been reported in the literature or are easy to derive. The last column in the table shows that the differences between f and π are by no means constant.

Lipophilicity and Folding

One of the most striking findings on critical examination of the results of the computations, as presented in Table III,1, is that there is no longer a need for any folding correction. Diphenhydramine (A) can be taken as an example. The log P of this structure is composed as

Refs. pg 103

$$C_6H_5 \diagdown$$
$$\diagup CH-O-CH_2-CH_2-N\diagup^{CH_3}_{\diagdown CH_3} \quad (A)$$
$$C_6H_5 \diagup$$

follows:

$$\text{Log P} = 2\,f(C_6H_5) + f(CH) + f(O) + 2\,f(CH_2) + f(N) + 2\,f(CH_3) + \text{p.e.2}$$
$$= 3.792 + 0.236 + (-1.536) + 1.054 + (-2.133) + 1.404 + 0.46$$
$$= 3.28.$$

This result is in excellent agreement with the experimental values of 3.27 [7] and 3.40 [7]. The proximity correction for a 2C separation (p.e.2), applied in the above calculation, should not be regarded as a masked folding correction; it is necessary for all X-CH_2-CH_2-Y structures, even when folding can definitely be excluded.

The principle of folding was introduced by HANSCH and ANDERSON [1] on the basis of anomalies in the lipophilicity of a series of C_6H_5-CH_2-CH_2-CH_2-X structures with X = F, Cl, Br, I, $CONH_2$, CN, $COCH_3$, NH_2, OH, OCH_3, COOH and $COOCH_3$. The mean $(\pi - f)$ value appears to be 0.36 ± 0.11 (see Table III,7). This amount, plus the difference between log P(C_6H_6)

TABLE III,7

SECONDARY SET OF ALIPHATIC f VALUES

no	fragment	f	π	$\pi - f$
1	F	−0.51	−0.17	0.34
2	Cl	0.06	0.39	0.33
3	Br	0.24	0.60	0.36
4	I	0.59	1.00	0.41
5	$CONH_2$	−1.99	−1.71	0.28
6	CN	−1.13	−0.84	0.29
7	NO_2	−1.02	−0.85	0.17
8	C = O	−1.69	−1.21[a]	0.48
9	CH_2=CH	0.93	0.70[b]	−0.23
10	C(quat.)	0.15	0.10[c]	−0.05
11	NH_2COO	−1.39	−1.15[d]	0.24
12	CF_3	0.79	1.02[e]	0.23
13	CCl_3	1.79	2.01[f]	0.22
14	SH	0.00	0.28	0.28
15	S	−0.51	−0.05	0.46
16	SO	−2.75	−2.35[g]	0.40

NOTES: See next page.

Notes to Table III,7:

[a] $\pi(CO-CH_3) - \pi(CH_3)$; [b] $[\pi(CH_3) - 0.30] + \pi(CH_3)$

[c] $\pi(CH_3) - 2 \times 0.20$; [d] $\log P(NH_2COOC_2H_5) - 2\pi(CH_3)$

[e] $\log P(CF_3CH_2OH) - \pi(CH_3) - \pi(OH)$; [f] $\log P(CCl_3CH_2OH) - \pi(CH_3)$

[g] $\log P(CH_3SOCH_3) - 2\pi(CH_3)$. $- \pi(OH)$;

and $f(C_6H_5)$: $2.13 - 1.90 = 0.23$ gives a total of 0.59 ± 0.14 for the difference between the π and f summation.

This latter value equals almost exactly the difference between log P(obs) and log P(clc) for folded structures omitting the prescribed folding correction: 0.60 ± 0.05 (mean of the 12 structures mentioned above).

Detailed particulars on these 12 structures are given in Table III, 8, which also shows a number of congeneric structures where folding is presumed to be absent.

TABLE III,8

COMPARISON BETWEEN THE LIPOPHILICITY OF FOLDED AND UNFOLDED STRUCTURES

no	compound	log P values					
		obs.	ref.	clc.I	res.I	clc.II	res.II
1	$C_6H_5-CH_2-OH$	1.10	[14]	0.98	0.12	1.47	-0.37
2	$C_6H_5-(CH_2)_2-OH$	1.36	[16]	1.51	-0.15	1.97	-0.61
3	$C_6H_5-(CH_2)_3-OH$	1.88	[16]	2.04	-0.16	2.47	-0.59
4	$CH_3-(CH_2)_4-F$	2.33	[1]	2.30	0.03	2.33	0.00*
5	$C_6H_5-(CH_2)_3-F$	2.95	[16]	2.97	-0.02	3.46	-0.51
6	$CH_3-(CH_2)_3-Cl$	2.39	[1]	2.34	0.05	2.39	0.00*
7	$C_6H_5-(CH_2)_2-Cl$	2.95	[16]	3.01	-0.06	3.52	-0.57
8	$C_6H_5-(CH_2)_3-Cl$	3.55	[16]	3.54	0.01	4.02	-0.47
9	$CH_3-(CH_2)_2-Br$	2.10	[1]	2.00	0.10	2.10	0.00*
10	$C_6H_5-(CH_2)_2-Br$	3.09	[16]	3.19	-0.10	3.73	-0.64
11	$C_6H_5-(CH_2)_3-Br$	3.72	[16]	3.72	0.00	4.23	-0.51
12	CH_3-CH_2-I	2.00	[1]	1.82	0.18	2.00	0.00*
13	$C_6H_5-(CH_2)_3-I$	3.90	[15]	4.07	-0.17	4.63	-0.73
14	$C_6H_5-CH_2-COOH$	1.41	[14]	1.42	-0.01	1.96	-0.55
15	$C_6H_5-(CH_2)_2-COOH$	1.84	[16]	1.95	-0.11	2.46	-0.62
16	$C_6H_5-(CH_2)_3-COOH$	2.42	[16]	2.47	-0.05	2.96	-0.54
17	$C_6H_5-CH_2-COO-CH_3$	1.83	[16]	1.84	-0.01	2.36	-0.53
18	$C_6H_5-(CH_2)_2-COO-CH_3$	2.32	[16]	2.37	-0.05	2.86	-0.54
19	$C_6H_5-(CH_2)_3-COO-CH_3$	2.77	[16]	2.90	-0.13	3.36	-0.59

Refs. pg 103

TABLE III,8 (continued)

no	compound	obs.	ref.	clc.I	res.I	clc.II	res.II
20	$CH_3-CO-CH_3$	-0.24	[6]	-0.29	0.05	-0.21	-0.03
21	$CH_3-CH_2-CO-CH_3$	0.29	[1]	0.24	0.05	0.29	0.00*
		0.26	[6]		0.02		-0.03
22	$CH_3-(CH_2)_3-CO-CH_3$	1.38	[16]	1.30	0.08	1.29	0.09
23	cyclohexanone $C=O$	0.81	[26]	0.95	-0.14	1.20	-0.41
24	$CH_2=CH-(CH_2)_2-CO-CH_3$	1.02	[16]	1.00	0.02	0.99	0.03
25	$C_6H_5-CH_2-CO-CH_3$	1.44	[16]	1.44	0.00	1.92	-0.48
26	$C_6H_5-(CH_2)_3-CO-CH_3$	2.42	[16]	2.49	-0.07	2.92	-0.50
27	$\underline{C_6H_5-CH_2-NH_2}$	1.09	[16]	1.04	0.05	1.44	-0.35
28	$C_6H_5-(CH_2)_2-NH_2$	1.41	[16]	1.57	-0.16	1.94	-0.53
29	$C_6H_5-(CH_2)_3-NH_2$	1.83	[16]	2.10	-0.27	2.44	-0.61
30	$\underline{CH_3-CN}$	-0.34	[1]	-0.43	0.09	-0.34	0.00
31	CH_3-CH_2-CN	0.16	[1]	0.10	0.06	0.16	0.00
		0.04	[6]		-0.06		-0.12
32	$C_6H_5-CH_2-CN$	1.56	[16]	1.29	0.27	1.79	-0.23
33	$C_6H_5-(CH_2)_2-CN$	1.72	[16]	1.82	-0.10	2.29	-0.57
		1.70	[27]		-0.12		-0.59
		1.66	[84]		-0.16		-0.63
34	$C_6H_5-(CH_2)_3-CN$	2.21	[16]	2.35	-0.14	2.79	-0.58
35	$\underline{C_6H_5-(CH_2)_3-O-CH_3}$	2.70	[16]	2.64	0.06	3.16	-0.46
36	$\underline{CH_3-(CH_2)_2-CONH_2}$	-0.21	[1]	-0.23	0.02	-0.21	0.00*
37	$C_6H_5-CH_2-CONH_2$	0.45	[16]	0.43	0.02	0.92	-0.47
38	$C_6H_5-(CH_2)_2-CONH_2$	1.15	[27]	0.96	0.19	1.42	-0.27
		0.91	[8]		-0.05		-0.51
39	$\underline{C_6H_5-(CH_2)_3-CONH_2}$	1.41	[16]	1.49	-0.08	1.92	-0.51

*Only one structure was partitioned for the derivation of the relevant value of the functional group.
Folded structures are underlined.

The differences between the experimental partition values and the values calculated from f values are so small that there is no need to maintain the idea of a folding-induced lipophilicity decrease in any of the 12 structures. Only 3-phenyl-propylamine remains an outlier, for as yet unaccountable reasons (residual = -0.27).

AROMATIC STRUCTURES

Substituted Benzene Derivatives with at least one saturated
C Link between Ring and substituting functional Group;
Determination of unknown f Values

A number of monoalkyl-substituted benzene derivatives were included
in the preceding section, serving to obtain more insight into the partitional behaviour of C_6H_5 in structures where folding was not likely
to occur. Table III,9 shows the experimental partition values of five
monoalkyl-benzenes, compared with log P values from four different
modes of calculation. There appears to be reasonable agreement with the

TABLE III,9

LIPOPHILICITY DATA OF SOME SIMPLE ALKYL-SUBSTITUTED BENZENES
COMPARED WITH VALUES CALCULATED BY DIFFERENT ROUTES

I: using hydrophobic fragmental constants (f values).
II: using π values from the phenoxyacetic acid system [14].
III: using aliphatic π values with branching corrections.
IV: as III, but omitting branching corrections.

alkyl	obs.	ref.	clc. I	res. I	clc. II	res. II	clc. III	res. III	clc. IV	res. IV
CH_3	2.69	[14]	2.61	0.08	2.64	0.05	2.63	0.06	2.63	0.06
	2.73	[24]		0.12		0.09		0.10		0.10
	2.80	[15]		0.19		0.16		0.17		0.17
C_2H_5	3.15	[16]	3.13	0.02	3.10	0.05	3.13	0.02	3.13	0.02
$\underline{n}.C_3H_7$	3.68	[16]	3.66	0.02	3.56	0.12	3.63	0.05	3.63	0.05
	3.57	[7]		-0.09		0.01		-0.06		-0.06
$\underline{i}.C_3H_7$	3.66	[24]	3.56	0.10	3.43	0.23	3.43	0.23	3.63	0.03
	3.66	[17]		0.10		0.23		0.23		0.03
$\underline{t}.C_4H_9$	4.11	[24]	4.14	-0.03	3.81	0.30	3.73	0.38	4.13	-0.02
	4.11	[17]		-0.03		0.30		0.38		-0.02

values derived from f constants (clc.I). On application of the "phenoxyacetic acid" π values recommended by FUJITA et al. [14] as the system
of choice for structures in which special interactions between the substituents and the benzene nucleus are absent, both isopropylbenzene and
tert.-butylbenzene (see clc.II) deviate markedly from the experimental
log P values. Nearly identical deviations occur where CH_n π values are

applied and the necessary branching corrections are correctly taken into account. However, excellent agreement between calculated and experimental values is obtained if the branching corrections - ignoring the
advised rules - are neglected in the calculations (clc.IV).

In summary, it can be said that when the f system is used, aliphatic
and aromatic carbon fragments need not be differentiated, which con-
trasts with the π system, where any branching in the substituents may
lead to deviations that cannot be neglected.

Calculations of log P values for structures that belong to the
$C_6H_5-(CR_2)_nX$ type, where R = H or alkyl and X = some functional group,
will therefore proceed smoothly in the f system.

The C_6H_5 group is further useful in that it can be employed in determining unknown f values, use being made of its easy detectability by
means of UV spectrophotometry. Thus $f(X)$ can be established readily,
although X itself possesses no chromophoric properties that would ren-
der it suitable for UV detection, by measuring lipophilicity on one of
the following structures:

$$C_6H_5-CH_2-X \qquad C_6H_5-\underset{R_1}{\overset{R_2}{C}}-X \qquad C_6H_5-\overset{R}{C}\,\smile\,X$$

The last structure is intended to indicate that X may have a certain
complexity; for example, it may be incorporated in a ring structure.

The antiepileptic 3,5,5-trimethyloxazolidine-2,4-dione (trimetha-
dione, structure B) serves as an example.

(B)   (C)   (D)

This structure cannot be assayed by direct UV spectrophotometry.
This is, however, possible for the 3-methyl-5-phenylderivative (structure C). This compound shows UV absorption with λ_{max} = 262 nm and ϵ =
300, which mainly originates from its C_6H_5 chromophore. Its log P value
was found to be 1.06 [37], indicating that the oxazolidinedione fragment (see D) has an f value of 1.06 - $f(C_6H_5)$ - $f(H)$ = -1.04*. This

_____
* $f(H)$ is taken 0.20 (see one of the following headings).

value is now suitable for the calculation of log P of structure B in
the normal way:

$$\log P(B) = f(D) + 2 f(CH_3) = (-1.04) + 1.40 = 0.36$$

No log P measurements have been reported for 3,5,5-trimethyloxazol-
idine-2,4-dione. The result obtained by LIEN [38] by the use of π addition was

$$\log P(B) = \pi(OCO) + \pi(CH_3CON) + 2\,\pi(CH_3) + \pi(_NCH_3)^*$$
$$= (-1.14) + (-0.79) + 1.00 + 0.56 = -0.37$$

The result of the π calculation appears to be 0.73 lower than that of
the f calculation, which is partly attributable to the incorrect application of the basic π equation and partly to the ignorance of newly created proximities in building up the oxazolidinedione ring from the moieties OCO and CH$_3$CON. See also pg 126.

Monosubstituted Benzene Derivatives with the functional Group directly connected to the Ring; Di- and trisubstituted Benzene Derivatives

In a preliminary design for the calculation of aromatic f values of
functional groups, mono-, di-para- and di-meta-substituted derivatives
were combined into one group, it being assumed that the mono- and di-
-meta-derivatives would join together in one correct regression equation and that the observed deviations of the paraderivatives would permit the requirements for an adequate fitting to be established. Di-
-ortho-substituted benzene derivatives were all excluded, as it was
assumed that the accompanying steric factors would interfere considerably with the results of the regression analysis.

More than 100 partition values were involved in the first computation. The main features emerging from the regression equation were:

(a) for $f(C_6H_5)$, the value obtained was 1.88 (standard deviation 0.12),
     which is in good agreement with 1.896 (standard deviation 0.027)
     obtained for C$_6$H$_5$ attached to aliphatic hydrocarbon fragments (see
     one of the preceding sections);

(b) the value for $f(COOH)$ did not differ much from zero;

(c) meta- and paraderivatives did not show different behaviour; in
     both series the percentage of outliers was approximately equal;

(d) the set of f values obtained permitted a reasonably accurate check

---

* $\pi(_NCH_3)$ denotes a π value especially advised for CH$_3$ attached to an
  N of the type present in structures such as B [38].

of many experimentally derived partition values of di-<u>ortho</u>-substi-
tuted benzenes.

The following were deemed important in the final calculation:

(1) not to exclude apparently deviating values from the statistical
treatment and not to regard any partition value as an outlier until
the analysis had shown that double the standard error of estimate
was exceeded; as a consequence, di-<u>ortho</u>-substituted derivatives
were no longer omitted from the analysis;

(2) to reconcile the aromatic f values with the aliphatic f values by
simply introducing a fixed value of 1.90 for $f(C_6H_5)$;

(3) to introduce a fixed value of 0.00 for $f(COOH_{ar})$, in order to a-
void unduly poor statistical results for this fragment;

(4) to transfer the fragments CH_3, CH_2, CH and C, using their f values
listed in Tables III,2 and III,7, to the left-hand side of eqn.
III-3 before performing the calculations; this was done not only
in those instances where the fragment was not directly connected
with the benzene ring, for example C_6H_5-O-<u>CH</u>$_3$, but also when such
an attachment did occur, as in 2-<u>CH</u>$_3$-chlorobenzene; a similar
treatment was applied when a functional group was not directly
connected to the ring, for example 4-Cl-C_6H_4-CH_2-<u>OH</u>; in the latter
case, the summation of $f(CH_2)$ and $f(OH_{al})$ was transferred to the
left-hand side of eqn. III-3; once the computations had been completed, all those fragments were removed from the left-hand side
which had been accommodated there temporarily, the implication being that the log P(obs) and log P(calc) values can be compared
without any further processing.

(5) to include a number of pre-selected tri-substituted benzenederiv-
atives.

Of 253 different structures, 309 partition values — most taken from
the literature — were introduced into the final regression procedure.
These structures represent 14 mono-, 53 di-<u>para</u>-, 87 di-<u>meta</u>-, 87 di-
-<u>ortho</u>- and 12 tri-substituted benzene derivatives;these five groups
afforded 25, 62, 103, 107 and 12 partition values, respectively.

From this population, the f values for the following 15 fragments
were calculated: F, Cl, Br, I, OH, O, NH_2, NO_2, OCH_2COOH, C_6H_4, C_6H_3,
SO_2NH_2, (ar)OCONH(al), (ar)CO(al) and $CONH_2$.

On comparison of this fragment series with that included in a previous publication [39], it will be seen that the latter series has been

extended by the last five f values.

In the above fragment listing, a fragment such as $(ar)CO(ar)$ is excluded. It will be seen in one of the following sections that such functional groups, acting as a resonance transmitting link between two aromatic rings, have a distinct and deviating partitional behaviour. It stands to reason that we were unaware of these effects when starting the computations and we had to learn of them via trial and error calculations on allowed and forbidden manipulations.

The results of the multiple regression analysis are collected in Table III,10 and detailed information on all of the structures analyzed is given in Tables III,11 and III,12.

TABLE III,10

PRIMARY SET OF AROMATIC f VALUES SOLVED BY MULTIPLE
REGRESSION ANALYSIS

no.	fragment	f	standard deviation	$\underline{t}$
1	F	0.412	0.024	17.4
2	Cl	0.943	0.016	57.5
3	Br	1.168	0.024	48.5
4	I	1.460	0.028	51.9
5	OH	−0.359	0.016	−21.9
6	O	−0.454	0.021	−21.9
7	NH_2	−0.897	0.021	−43.5
8	NO_2	−0.077	0.017	−4.59
9	OCH_2COOH	−0.588	0.022	−26.3
10	$(ar)OCONH(al)$	−1.370	0.021	−64.2
11	$(ar)CO(al)$	−0.869	0.030	−28.8
12	SO_2NH_2	−1.506	0.029	−51.7
13	$CONH_2$	−1.120	0.025	−45.0
14	C_6H_3	1.440	0.037	38.7
15	C_6H_4	1.719	0.019	90.0
16	C_6H_5	1.90[a]		
17	COOH	0.00[a]		

$n = 234 (+15)$[b]; $s = 0.096$; $r = 0.996$;
$F = 2,167$;   constant term $= -0.036$

[a] Pre-fixed f values.
[b] 234 data-points with 15 zero-cards included for forcing.

Refs. pg 103

TABLE III,11

LOG P DATA OF ANALYZED BENZENE DERIVATIVES

I: Values based on f system.
II: Values based on π system.
The column "syst" codes the basic structures used for deriving
the eight sets of π values, as recommended by FUJITA et al. [14],
from the observed log P values.

PHA	= phenylacetic acid	PHL	= phenol
PHOA	= phenoxyacetic acid	AN	= aniline
BA	= benzoic acid	NB	= nitrobenzene
BZA	= benzyl alcohol	BN	= benzene

no	substituents	log P obs.	ref.	log P clc.I	res.I	syst.	log P clc.II	res.II
	Mono-substituted							
1	F	2.27	[14]	2.28	−0.01	BN		
2	Cl	2.84	[14]	2.81	0.03	BN		
3	Br	2.99	[14]	3.03	−0.04	BN		
4	I	3.25	[12]	3.32	−0.07	BN		
5	COOH	1.87	[14]	1.86	0.01	BN		
6	OH	1.46	[14]	1.51	−0.05	BN		
		1.48	[15]		−0.03			
		1.48	[40]		−0.03			
7	OCH_3	2.11	[14]	2.11	0.00	BN		
		2.04	[23]		−0.07			
8	OCH_2COOH	1.26	[14]	1.28	−0.02	BN		
		1.34	[83]		0.06			
9	NH_2	0.90	[14]	0.97	−0.07	BN		
		0.90	[41]		−0.07			
		0.98	[15]		0.01			
10	NO_2	1.86	[14]	1.79	0.06	BN		
		1.88	[15]		0.09			
11	$COCH_3$	1.58	[14]	1.70	−0.12	BN		
		1.73	[48]		0.03			
12	SO_2NH_2	0.31	[49]	0.36	−0.05			
		0.31	[14]		−0.05	BN		
13	$CONH_2$	0.64	[14]	0.74	−0.10	BN		
		0.65	[20]		−0.09			
14	$OOCNHCH_3$	1.16	[50]	1.19	−0.03			
		1.24	[1]		0.05			
	Ortho di-substituted							
15	Cl OCH_2COOH	2.02	[14]	2.04	−0.02	PHOA		

TABLE III,11 (continued)

16	Br	OCH$_2$COOH	2.10	[14]	2.26	-0.16	PHOA		
17	OH	"	0.85	[42]	0.74	0.11	PHOA		
18	F	OH	1.71	[14]	1.74	-0.03	PHL		
19	Cl	"	2.15	[14]	2.27	-0.12	PHL		
			2.19	[15]		-0.08			
20	Br	"	2.35	[14]	2.49	-0.14	PHL		
21	I	"	2.65	[14]	2.78	-0.13	PHL		
22	CH$_3$	"	1.95	[10]	2.03	-0.08		2.02	-0.07
23	OH	"	0.88	[12]	0.97	-0.09		0.79	0.09
			1.01	[15]		0.04			0.22
24	NH$_2$	"	0.52	[15]	0.43	0.09		0.23	0.29
			0.62	[12]		0.19			0.39
25	F	NH$_2$	1.26	[43]	1.20	0.06		1.04	0.22
26	Cl	"	1.90	[11]	1.73	0.17		1.61	0.29
			1.92	[15]		0.09			0.31
27	CH$_3$	"	1.29	[43]	1.49	-0.20		1.46	-0.17
			1.32	[15]		-0.17			-0.14
28	OCH$_3$	"	0.95	[43]	1.03	-0.08		0.88	0.07
29	CH$_3$	NO$_2$	2.30	[15]	2.31	-0.01		2.41	-0.11
30	NO$_2$	"	1.58	[15]	1.53	0.05		1.57	0.01
31	Cl	Cl	3.38	[15]	3.57	-0.19		3.55	-0.17
32	CH$_3$	"	3.42	[15]	3.33	0.09		3.40	0.02
33	"	CH$_3$	3.12	[19]	3.09	0.03		3.25	-0.13
34	NO$_2$	COCH$_3$	1.28	[48]	1.44	-0.16		1.30	-0.02
35	CH$_3$	SO$_2$NH$_2$	0.84	[49]	0.88	-0.04		0.87	-0.03
36	CH$_3$O	CONH$_2$	0.87	[46]	0.81	0.06		0.62	0.25
37	F	OOCNHCH$_3$	1.25	[50]	1.42	-0.17			
38	NO$_2$	"	1.02	[50]	0.94	0.08			

Meta di-substituted

39	F	OCH$_2$COOH	1.48	[14]	1.51	-0.04	PHOA		
40	Cl	"	2.03	[14]	2.04	-0.01	PHOA		
41	Br	"	2.22	[14]	2.26	-0.04	PHOA		
42	I	"	2.44	[14]	2.56	-0.12	PHOA		
43	CH$_3$	"	1.78	[14]	1.80	-0.02	PHOA		
44	OH	"	0.76	[14]	0.74	0.02	PHOA		
45	OCH$_3$	"	1.38	[14]	1.34	0.04	PHOA		
46	F	CH$_2$COOH	1.65	[14]	1.62	0.03	PHA		

Refs. pg 103

TABLE III,11 (continued)

no	substituents		log P obs.	ref.	log P clc.I	res.I	syst.	log P clc.II	res.II
47	Cl	CH$_2$COOH	2.09	[14]	2.15	−0.06	PHA		
48	Br	"	2.37	[14]	2.38	−0.01	PHA		
49	I	"	2.62	[14]	2.67	−0.05	PHA		
50	CH$_3$	"	1.95	[14]	1.91	0.04	PHA		
51	COOH	"	1.14	[14]	1.21	−0.07	PHA		
52	OH	"	0.85	[14]	0.85	0.00	PHA		
53	OCH$_3$	"	1.50	[14]	1.46	0.04	PHA		
54	F	COOH	2.15	[14]	2.10	0.05	BA		
55	Cl	"	2.68	[14]	2.63	0.05	BA		
56	Br	"	2.87	[14]	2.85	0.02	BA		
57	I	"	3.13	[14]	3.14	−0.01	BA		
58	CH$_3$	"	2.37	[14]	2.39	−0.02	BA		
59	COOH	"	1.66	[14]	1.68	−0.02	BA		
60	OH	"	1.50	[14]	1.32	0.18	BA		
61	OCH$_3$	"	2.02	[14]	1.93	0.09	BA		
62	CH$_3$	CH$_2$OH	1.60	[14]	1.47	0.13	BZA		
63	OH	"	0.49	[14]	0.41	0.08	BZA		
64	NH$_2$	"	−0.05	[14]	−0.12	0.07	BZA		
65	F	OH	1.93	[14]	1.74	0.19	PHL		
66	Br	"	2.63	[14]	2.49	0.14	PHL		
67	I	"	2.93	[14]	2.78	0.15	PHL		
			2.91	[20]		0.13			
68	CH$_3$	"	1.96	[14]	2.03	−0.07	PHL		
			1.95	[10]		−0.08			
			2.01	[15]		−0.02			
69	C$_2$H$_5$	"	2.40	[14]	2.55	−0.15	PHL		
70	OH	"	0.80	[14]	0.97	−0.17	PHL		
			0.77	[15]		−0.20			
71	OCH$_3$	"	1.58	[14]	1.57	0.01	PHL		
72	"	OCH$_3$	2.21	[19]	2.18	0.03		2.09	0.12
73	NH$_2$	"	0.93	[14]	1.03	−0.10	AN		
			0.93	[43]		−0.10			
74	F	NH$_2$	1.30	[14]	1.20	0.10	AN		
			1.30	[43]		0.10			
75	Cl	"	1.88	[14]	1.73	0.15	AN		

TABLE III,11 (continued)

	Cl	NH_2	1.90	[15]	1.73	0.17			
76	Br	"	2.10	[43]	1.95	0.15		1.76	0.34
77	CH_3	"	1.40	[14]	1.49	-0.09	AN		
			1.43	[15]		-0.06			
78	Cl	NO_2	2.46	[14]	2.55	-0.09	NB		
			2.41	[15]		-0.14			
79	Br	"	2.64	[14]	2.77	-0.15	NB		
80	CH_3	"	2.45	[14]	2.31	0.14	NB		
			2.40	[15]		0.09			
81	NO_2	"	1.49	[14]	1.53	-0.04	NB		
			1.49	[15]		-0.04			
82	Cl	Cl	3.38	[15]	3.57	-0.19		3.55	-0.17
83	CH_3	"	3.28	[15]	3.33	-0.05		3.40	-0.12
84	"	CH_3	3.20	[15]	3.09	0.11		3.25	-0.05
85	NO_2	$COCH_3$	1.42	[14]	1.44	-0.02	NB		
			1.28	[48]		-0.16			
86	$COCH_3$	OCH_2COOH	0.98	[14]	0.93	0.05	PHOA		
87	CH_3	SO_2NH_2	0.85	[49]	0.88	-0.03		0.87	-0.02
88	Cl	"	1.29	[49]	1.12	0.17		1.02	0.27
89	CH_3	$CONH_2$	1.18	[47]	1.26	-0.08		1.20	-0.02
90	CH_3O	"	0.94	[46]	0.81	0.13		0.62	0.32
91	OH	"	0.39	[47]	0.20	0.19		-0.03	0.42
92	F	"	0.91	[47]	0.97	-0.06		0.78	0.13
93	Cl	"	1.51	[47]	1.51	0.00		1.35	0.16
94	Br	"	1.65	[47]	1.73	-0.08		1.50	0.15
95	$(CH_3)_2N$	"	0.95	[47]	0.90	0.05		0.82	0.13
96	F	$OOCNHCH_3$	1.48	[50]	1.42	0.06			
97	Cl	"	2.03	[50]	1.96	0.07			
98	Br	"	2.25	[50]	2.18	0.07			
99	I	"	2.52	[50]	2.47	0.05			
100	CH_3	"	1.70	[50]	1.71	-0.01			
101	CH_3O	"	1.30	[50]	1.26	0.04			
102	C_2H_5	"	2.20	[50]	2.24	-0.04			
103	C_2H_5O	"	1.75	[50]	1.79	-0.04			
104	$\underline{i}-C_3H_7$	"	2.63	[50]	2.65	-0.02			
105	$\underline{n}\ C_3H_7$	"	2.64	[50]	2.76	-0.12			
106	$\underline{n}\ C_4H_9O$	"	2.96	[50]	2.85	0.11			

Refs. pg 103

TABLE III,11 (continued)

no	substituents		log P obs.	ref.	log P clc.I	res.I	syst.	log P clc.II	res.II
107	CH$_3$CO	OOCNHCH$_3$	0.90	[50]	0.84	0.06			
	Para di-substituted								
108	F	OCH$_2$COOH	1.41	[14]	1.51	−0.10	PHOA		
109	Cl	"	1.99	[14]	2.04	−0.05	PHOA		
110	Br	"	2.45	[14]	2.26	0.19	PHOA		
111	I	"	2.69	[14]	2.56	0.13	PHOA		
112	CH$_3$	"	1.86	[14]	1.80	0.06	PHOA		
113	OH	"	0.65	[14]	0.74	−0.09	PHOA		
114	OCH$_3$	"	1.23	[14]	1.34	−0.11	PHOA		
115	F	CH$_2$COOH	1.55	[14]	1.62	−0.07	PHA		
116	Cl	"	2.12	[14]	2.15	−0.03	PHA		
117	Br	"	2.31	[14]	2.38	−0.07	PHA		
118	I	"	2.72	[14]	2.67	0.05	PHA		
119	CH$_3$	"	1.86	[14]	1.91	−0.05	PHA		
120	CH$_3$O	"	1.42	[14]	1.46	−0.04	PHA		
121	F	COOH	2.07	[14]	2.10	−0.03	BA		
122	Cl	"	2.65	[14]	2.85	0.01	BA		
123	Br	"	2.86	[14]	2.85	0.01	BA		
124	I	"	3.02	[14]	3.14	−0.12	BA		
125	CH$_3$	"	2.27	[14]	2.39	−0.12	BA		
126	CH$_3$O	"	1.96	[14]	1.93	0.03	BA		
127	NH$_2$	"	0.68	[26]	0.79	−0.11		0.62	0.06
128	Cl	CH$_2$OH	1.94	[14]	1.71	0.23	BZA		
129	CH$_3$	"	1.58	[14]	1.47	0.11	BZA		
130	OH	"	0.25	[14]	0.41	−0.16	BZA		
131	CH$_3$O	"	1.10	[14]	1.02	0.08	BZA		
132	F	OH	1.77	[14]	1.74	0.03	PHL		
			1.81	[40]		0.07			
134	Cl	"	2.39	[14]	2.27	0.12	PHL		
			2.44	[15]		0.17			
			2.35	[40]		0.08			
135	Br	"	2.59	[14]	2.49	0.10	PHL		
			2.65	[40]		0.16			
136	I	"	2.91	[14]	2.78	0.13	PHL		
137	CH$_3$	"	1.94	[14]	2.03	−0.09	PHL		

TABLE III,11 (continued)

	CH_3	OH	1.92	[15]	2.03	-0.11			
			1.95	[40]		-0.08			
138	NH_2	OCH_3	0.95	[43]	1.03	-0.08		0.88	0.07
139	F	NH_2	1.15	[14]	1.20	-0.05	AN		
140	Cl	"	1.83	[15]	1.73	0.10		1.61	0.22
141	CH_3	"	1.39	[14]	1.49	-0.10	AN		
			1.41	[15]		-0.08			
			1.39	[41]		-0.10			
142	Cl	NO_2	2.39	[14]	2.55	-0.16	NB		
			2.41	[15]		-0.14			
143	CH_3	"	2.37	[14]	2.31	0.06	NB		
			2.42	[15]		0.11			
144	CH_3O	"	2.03	[14]	1.85	0.18	NB		
145	NO_2	"	1.46	[14]	1.53	-0.07	NB		
			1.49	[15]		-0.04			
146	Cl	Cl	3.39	[15]	3.57	-0.18		3.55	-0.16
147	CH_3	"	3.33	[15]	3.33	0.00		3.40	-0.07
148	"	CH_3	3.15	[15]	3.09	0.06		3.25	-0.10
149	NO_2	$COCH_3$	1.49	[14]	1.44	0.05	NB		
			1.53	[48]		0.09			
150	$COCH_3$	OCH_2COOH	0.87	[14]	0.93	-0.06		0.90	-0.03
151	CH_3	SO_2NH_2	0.82	[49]	0.88	-0.06		0.87	-0.05
152	Br	"	1.36	[49]	1.35	0.01		1.17	0.19
153	CH_3NH	"	0.08	[49]	-0.05	0.13			
154	CH_3O	"	0.47	[49]	0.42	0.05		0.29	0.18
155	NH_2	"	-0.72	[81]	-0.72	0.00		-0.92	0.20
			-0.72	[1]		0.00			0.20
			-0.83	[49]		-0.11			0.09
			-0.75	[80]		-0.03			0.17
156	CH_3O	$CONH_2$	0.86	[46]	0.81	0.05		0.62	0.24
157	CH_3	"	1.18	[19]	1.26	-0.08		1.20	-0.02
158	F	"	0.91	[47]	0.97	-0.06		0.78	0.13
159	Cl	"	1.55	[47]	1.51	0.04		1.35	0.20
			1.51	[19]		0.00			0.16
160	Br	"	1.76	[47]	1.73	0.03		1.50	0.26
161	OH	"	0.33	[47]	0.20	0.13		-0.03	0.36
162	C_2H_5O	"	1.30	[47]	1.34	-0.04			

Refs. pg 103

TABLE III,11 (continued)

no	substituents	log P obs.	ref.	log P clc.I	res.I	syst.	log P clc.II	res.II
163	i.C$_3$H$_7$ CONH$_2$	2.14	[47]	2.20	−0.06			
164	Br OOCNHCH$_3$	2.17	[50]	2.18	−0.01			
165	Cl "	2.01	[50]	1.96	0.05			
166	F "	1.28	[50]	1.42	−0.14			
167	I "	2.46	[50]	2.47	−0.01			
168	CH$_3$ "	1.66	[50]	1.71	−0.05			
169	CH$_3$O "	1.20	[50]	1.26	−0.06			
170	C$_2$H$_5$ "	2.23	[50]	2.24	−0.01			
171	C$_2$H$_5$O "	1.63	[50]	1.79	−0.16			
172	i.C$_3$H$_7$ "	2.80	[50]	2.65	0.15			
		2.57	[19]		−0.08			
173	n.C$_3$H$_7$ "	2.72	[50]	2.76	−0.04			
174	sec.C$_4$H$_9$ "	3.20	[50]	3.17	0.03			
175	n.C$_4$H$_9$O "	2.86	[50]	2.85	0.01			
176	CH$_3$CO "	1.01	[50]	0.84	0.17			
	Tri-substituted							
177	1,3,5-tri NO$_2$-benzene	1.18	[19]	1.17	0.01			
178	1,3-di CH$_3$, 2-NO$_2$-benzene	2.95	[26]	2.73	0.22			
179	3,5-di CH$_3$-phenol	2.35	[51]	2.45	−0.10			
180	2,6-di CH$_3$-phenol	2.36	[51]	2.45	−0.09			
181	1,3,5-tri CH$_3$-benzene	3.42	[19]	3.50	−0.08			
182	3-Br,4-Cl-phenoxy-acetic acid	2.75	[30]	2.93	−0.18			
183	3-Cl,5-F-phenoxy-acetic acid	2.20	[30]	2.17	0.03			
184	3-I,4-Cl-phenoxy-acetic acid	3.10	[30]	3.22	−0.12			
185	3-NO$_2$,4-Cl-phenoxy-acetic acid	1.85	[30]	1.68	0.17			
186	2,4-di Cl-phenoxy-acetic acid	2.81	[14]	2.70	0.11			
187	3,4-di Cl-phenoxy-acetic acid	2.81	[13]	2.70	0.11			
188	3-CN,4-Cl " "	1.56	[30]	1.53	0.03			

TABLE III, 12

OUTLIERS IN THE MULTIPLE REGRESSION ANALYSIS

no	substituents		log P obs.	ref.	log P clc.I	res.I	syst.	log P clc.II	res.II
	Ortho di-substituted								
1	F	OCH$_2$COOH	1.26	[14]	1.51	-0.25	PHOA		
2	I	"	2.19	[14]	2.56	-0.37	PHOA		
3	CH$_3$CO	"	1.25	[14]	0.90	0.35	PHOA		
4	Cl	COOH	1.98	[26]	2.63	-0.65		2.56	-0.58
5	I	"	2.40	[26]	3.14	-0.74		2.97	-0.57
6	OH	"	2.26	[1]	1.32	0.94		1.18	1.08
			2.21	[7]		0.86			1.03
7	"	CH$_2$OH	0.73	[22]	0.41	0.32		0.43	0.30
8	NO$_2$	OH	1.79	[14]	1.25	0.54		1.18	0.61
			1.73	[15]		0.48			0.55
9	Br	NH$_2$	2.29	[43]	1.95	0.34		1.76	0.53
10	I	"	3.34	[43]	2.25	1.09		2.02	1.32
11	NH$_2$	"	0.15	[15]	-0.11	0.26		-0.33	0.48
12	NO$_2$	"	1.44	[43]	0.71	0.73		0.62	0.82
			1.83	[1]		1.12			1.21
			1.79	[15]		1.08			1.17
13	Cl	NO$_2$	2.24	[15]	2.55	-0.31		2.56	-0.32
14	Br	Br	3.64	[19]	4.02	-0.38		3.85	-0.21
15	CH$_3$	CH$_3$	2.77	[15]	3.09	-0.32		3.25	-0.48
16	Cl	SO$_2$NH$_2$	0.74	[48]	1.09	-0.35		1.02	-0.28
17	NO$_2$	"	0.34	[48]	0.08	0.26		0.03	0.31
18	CH$_3$	OOCNHCH$_3$	1.46	[49]	1.71	-0.25			
19	Cl	"	1.64	[49]	1.94	-0.30			
20	Br	"	1.77	[49]	2.19	-0.42			
21	I	"	1.96	[49]	2.48	-0.52			
22	CN	"	1.11	[49]	0.78	0.33			
23	$\underline{i}$.C$_3$H$_7$	"	2.31	[49]	2.65	-0.34			
24	$\underline{n}$.C$_3$H$_7$	"	2.40	[49]	2.76	-0.36			
25	$\underline{i}$.C$_3$H$_7$O	"	1.52	[49]	2.20	-0.68			
26	$\underline{sec}$.C$_4$H$_9$	"	2.78	[49]	3.17	-0.39			
27	$\underline{tert}$.C$_4$H$_9$	"	2.65	[49]	3.27	-0.62			
28	C$_2$H$_5$O	"	1.24	[49]	1.79	-0.55			
29	CH$_3$O	"	0.81	[49]	1.25	-0.44			

Refs. pg 103

TABLE III,12 (continued)

no	substituents		log P obs.	ref.	log P clc.I	res.I	syst.	log P clc.II	res.II
30	C$_2$H$_5$	OOCNHCH$_3$	1.93	[49]	2.24	−0.31			
	Meta di-substituted								
31	NO$_2$	COOH	1.83	[14]	1.61	0.22	BA		
32	Cl	CH$_2$OH	1.94	[14]	1.71	0.23	BZA		
33	NO$_2$	"	1.21	[14]	0.69	0.52		0.82	0.39
34	Cl	OH	2.50	[14]	2.27	0.23	PHL		
			2.47	[15]		0.20			
35	NH$_2$	"	0.17	[14]	0.43	−0.26	PHL		
			0.15	[15]		−0.28			
36	NO$_2$	"	2.00	[14]	1.25	0.75	PHL		
			2.00	[15]		0.75			
37	I	NH$_2$	2.98	[43]	2.25	0.73		2.02	0.96
38	NO$_2$	"	1.37	[14]	0.71	0.66	AN		
			1.37	[15]		0.66			
39	CH$_3$O	NO$_2$	2.16	[14]	1.85	0.31	NB		
40	Br	Br	3.75	[19]	4.02	−0.27			
41	NO$_2$	SO$_2$NH$_2$	0.55	[48]	0.08	0.47		0.03	0.52
42	"	OOCNHCH$_3$	1.39	[49]	0.92	0.47			
43	CF$_3$	"	2.73	[49]	2.26	0.47			
44	i.C$_3$H$_7$O	"	1.96	[49]	2.20	−0.24			
45	NO$_2$	CONH$_2$	0.77	[46]	0.48	0.29		0.36	0.41
46	OH	COCH$_3$	1.39	[14]	1.14	0.25		0.91	0.42
	Para di-substituted								
47	OH	COOH	1.58	[14]	1.32	0.26	BA		
48	NO$_2$	"	1.89	[14]	1.61	0.28	BA		
49	Cl	CH$_2$OH	1.96	[14]	1.71	0.25	BZA		
50	NO$_2$	"	1.26	[14]	0.69	0.56	BZA		
51	OH	OH	0.59	[14]	0.97	−0.38	PHL		
			0.50	[15]		−0.47			
52	CH$_3$O	"	1.34	[14]	1.57	−0.23	PHL		
53	NH$_2$	"	0.04	[15]	0.43	−0.39	PHL		
54	NO$_2$	"	1.91	[14]	1.25	0.66	PHL		
			1.91	[15]		0.66			
55	Br	NH$_2$	2.26	[43]	1.95	0.31			
56	I	"	3.34	[43]	2.25	1.09			

TABLE III,12 (continued)

57	NO_2	NH_2	1.39	[14]	0.71	0.68	AN	
58	Cl	SO_2NH_2	0.84	[48]	1.09	−0.25	1.02	−0.18
59	CN	"	0.23	[48]	−0.07	0.30	−0.26	0.49
60	CH_3CO	"	0.20	[48]	−0.02	0.22	−0.24	0.44
61	NO_2	"	0.64	[48]	0.08	0.56	0.03	0.61
62	"	$OOCNHCH_3$	1.47	[49]	0.92	0.55		
63	"	$CONH_2$	0.82	[19]	0.48	0.34	0.36	0.46
64	OH	$COCH_3$	1.35	[14]	1.14	0.21	0.91	0.44

The standard error of estimate (s = 0.093, the correlation coefficient (r = 0.996) and the over-all goodness of fit expressed by means of
the F value (2,167) are all indicative of the excellent quality of the
final regression equation. In order to adjust the constant term to an
acceptable level — minimizing it to a value that is negligible for cal-
culative purposes but still above the mean standard deviation of the
regressor values — , forcing through the introduction of 15 zero-cards
was applied. This procedure lowered the constant term to −0.036, which
remains small enough to be ignored.

The standard deviations of all 15 f values are sufficiently low and
the high t values in the Student test invariably show that all of the
f values determined satisfy a severe criterion of significance comparable with that in our set of f values for aliphatic structures (see
Table III,2).

Included in the total number of outliers in the above regression a-
nalysis are 74 partition values belonging to 64 different structures
distributed as follows: 20 values (18 compounds) in the para-, 20 values (16 compouds) in the meta- and 34 values (30 compounds) in the
ortho-series.

In most instances it is still uncertain what makes a structure behave as an outlier but it is assumed that suppression or reinforcement
of mesomeric and/or inductive effects by (or in) the substituting group
is operative. Steric factors and hydrogen bonding are, of course, known
to play an important role in ortho-substituted structures. What exactly happens in such a case cannot be described in a simple way. There
are clear indications that in some instances there are superpositions
of factors that increase and factors that decrease the lipophilic be-
havioural pattern.

Refs. pg 103

Secondary Set of Aromatic f Values

Also in the aromatic series are a number of functional groups that belong to structures that are only occasionally involved in partition experiments. Therefore, they were not included in multiple regression analysis but used to derive a supplementary set of secondary fragmental constants. Table III,13 presents a survey of such structures; the fragmental constants are to be found in Table III,14.

TABLE III,13

COMPILATION OF STRUCTURES USED FOR DERIVING SECONDARY AROMATIC f VALUES

fragment	compound	log P values			
		obs.	ref.	clc.	res.
COO	$C_6H_5-COO-CH_3$	2.12	[14]	2.17	−0.05
	$C_6H_5-COO-C_2H_5$	2.64	[46]	2.70	−0.06
	$4-OH-C_6H_4-COO-CH_3$	1.96	[19]	1.63	0.33*
	$4-OH-C_6H_4-COO-C_2H_5$	2.47	[24]	2.15	0.32*
	$4-OH-C_6H_4-COO-\underline{n}.C_3H_7$	3.04	[24]	2.68	0.36*
	$(C_5H_4N)-3-COO-C_2H_5$	1.36	[43]	1.34	0.02
		1.32	[44]		−0.02
	$(C_5H_4N)-4-COO-C_2H_5$	1.43	[44]	1.34	0.09
	$(C_5H_4N)-3-COO-\underline{n}.C_3H_7$	1.94	[43]	1.86	0.08
	$(C_5H_4N)-3-COO-\underline{n}.C_4H_9$	2.27	[43]	2.39	−0.12
NH	$C_6H_5-NH-CH_3$	1.82	[42]	1.67	0.15
		1.66	[40]		−0.01
		1.66	[13]		−0.01
	$C_6H_5-NH-C_2H_5$	2.16	[40]	2.20	−0.04
		2.26	[42]		0.06
	$2-CH_3-C_6H_4-NH-CH_3$	2.16	[40]	2.20	−0.04
	$4-CH_3-C_6H_4-NH-CH_3$	2.15	[40]	2.20	−0.05
	$C_6H_5-NH-\underline{n}.C_3H_7$	2.45	[42]	2.72	−0.27*
N	$C_6H_5-N(CH_3)_2$	2.31	[40]	2.24	0.07
		2.31	[14]		0.07
		2.62	[23]		0.38*
	$3-OH-C_6H_4-N(CH_3)_2$	1.56	[14]	1.70	−0.14
		1.57	[39]		−0.13
	$2-CH_3-C_6H_4-N(CH_3)_2$	2.81	[40]	2.78	0.03
	$4-CH_3-C_6H_4-N(CH_3)_2$	2.81	[40]	2.78	0.03
	$C_6H_5-N(C_2H_5)_2$	3.31	[40]	3.29	0.02

TABLE III,13 (continued)

	$4-HOCH_2-C_6H_4-N(C_2H_5)_2$	2.29	[30]	2.27	0.02
CF_3	$3-CF_3-C_6H_4-COOH$	2.95	[14]	2.97	-0.02
	$3-CF_3-C_6H_4-CH_2COOH$	2.62	[14]	2.50	0.12
	$3-CF_3-C_6H_4-OCH_2COOH$	2.36	[14]	2.38	-0.02
	$4-CF_3-C_6H_4-CONH_2$	1.71	[46]	1.75	-0.04
CN	C_6H_5-CN	1.56	[14]	1.67	-0.11
	$3-CN-C_6H_4-COOH$	1.48	[14]	1.50	-0.02
	$4-CN-C_6H_4-COOH$	1.56	[14]	1.50	0.06
	$3-CN-C_6H_4-OCH_2COOH$	0.93	[14]	0.89	0.04
	$4-CN-C_6H_4-OCH_2COOH$	0.94	[14]	0.89	0.05
SH	C_6H_5-SH	2.52	[19]	2.52[a]	
S	$C_6H_5-S-CH_3$	2.74	[19]	2.71	0.03
	$3-CH_3-S-C_6H_4-OCH_2COOH$	1.90	[14]	1.93	-0.03
SO	$C_6H_5-SO-CH_3$	0.55	[19]	0.55[a]	
SO_2	$C_6H_5-SO_2-CH_3$	0.47	[1]	0.62	-0.15
		0.50	[24]		-0.12
	$3-CH_3-SO_2-C_6H_4-CH_2COOH$	0.06	[14]	-0.15	0.21
	$3-CH_3-SO_2-C_6H_4-OCH_2COOH$	0.01	[14]	-0.03	0.04

* Outliers
a Only one data-point available

TABLE III,14

SECONDARY SET OF AROMATIC f VALUES

no	fragment	f
1	COO	-0.43
2	NH	-0.93
3	N	-1.06
4	CF_3	1.25
5	CN	-0.23
6	SH	0.62
7	S	0.11
8	SO	-2.05
9	SO_2	-1.87

At present, 26 fragments are available in the series of aromatic substituents; 17 of the primary set and 9 of the secondary set. It is obvious that the material can be extended as one thinks fit, and we shall do so for some special groups in following sections.

Refs. pg 103

When sufficient log P values become available, it is possible to transfer an item originally classed amongst the secondary fragments to the primary set by renewed computations, the implication being an enhanced statistical reliability.

It may be useful to conclude this section by a block diagram representing all of the sequences in the procedure for the determination of hydrophobic fragmental constants and their interrelations (Fig.III,1).

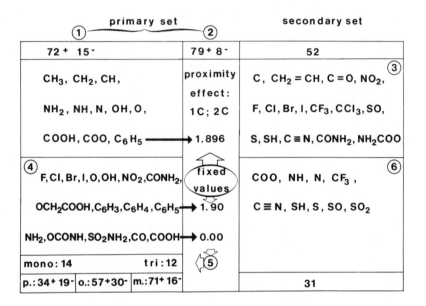

Fig.III,1: Development of the *f* System

Step 1: Fragment calculations on a series of 87 aliphatic structures, providing the following primary set: CH_3, CH_2, CH, NH_2, NH, N, C_6H_5, OH, O, COOH and COO. Of these structures, 72 fitted into statistical analysis and the remaining 15 had to be qualified as outliers.

Step 2: Following the evaluation of the proximity effect, p.e. (1C) = 0.80 and p.e. (2C) = 0.46, step 1 had to be repeated. This raised the number of structures that could be included in the statistical treatment to 79, the remaining 8 being definitely discarded as outliers.

Step 3: Fragment calculations on a series of 52 aliphatic structures, providing the following secondary set: F, Cl, Br, I, $CONH_2$, CN, NO_2, C=O, CH_2=CH, C, NH_2COO, CF_3, CCl_3, SH, S and SO.

Step 4: Orientational fragment calculations on aromatic structures to obtain more information about possible differences between <u>ortho</u>, <u>meta</u> and <u>para</u> compounds and to ascertain whether the aromatic fragment set and the aliphatic sets could be linked up.

<u>Step 5</u>: Fixing of $f(C_6H_5)$ as 1.90, with in view the value of 1.88 obtained by step 4, and the value of 1.896 found for C_6H_5 in step 2; fixing of $f(COOH)$ as 0.00. Final fragment calculations were performed on 236 aromatic structures using for CH_3, CH_2, CH and C the f values found in steps 2 and 3, irrespective of whether or not these fragments were connected directly to the aromatic ring. The following set of <u>primary</u> aromatic f values was obtained: F, Cl, Br, I, OH, O, NH_2, OCH_2COOH, C_6H_4, C_6H_3, NO_2, (ar)OCONH(al), (ar)C=O(al), SO_2NH_2 and $CONH_2$. Some 171 structures appeared suitable for statistical treatment, and the other 65 had to be dismissed as outliers.

<u>Step 6</u>: Fragment calculations on a series of aromatic structures to provide the following secondary set: COO, NH, N, CF_3, CN, S, SH, SO and SO_2.

Differences between Aliphatic and Aromatic f Values

LEO <u>et al.</u> [5] have already pointed to the differences between aliphatic and aromatic structures in lipophilic behaviour. Inherent in the definition of the substituent constant and the consequential set-up of the π system, these differences were fully accounted for by the substituents.

In our design for the calculations of hydrophobic fragmental constants, all C_6H_5 were <u>a priori</u> equalized with each other, no matter what was the character of the substituent on the benzene ring (+M, −M, +I, −I*). The same applies to C_6H_4 and C_6H_3 units and to all aromatic fragments discussed later. In principle, it is possible to differentiate between the various phenyl and phenylene types, but the result would be a rather complicated and unwieldy system of fragmental constants.

The lipophilic behaviour of the major functional groups is presented in Table III,15, which shows that the (ar - al) differences, except in two cases (CF_3 and NH_2) all indicate increased lipophilicity of the aromatic structure. For a number of groups, this increase may hardly be regarded as significant, but for groups such as COOH, O, N, OH and NO_2 significance is clear. Electron delocalization might be responsible for this discrepency, but in that case the behaviour of the NH_2 and CF_3

---

* M represents mesomeric and I inductive effects.

Refs. pg 103

TABLE III,15

COMPARISON OF AROMATIC AND ALIPHATIC PARTITION VALUES FOR A SERIES

OF FUNCTIONAL GROUPS IN THE π SYSTEM AND THE f SYSTEM

function	π(Ar)	π(Al)	π(Ar-Al)	f(Ar)	f(Al)	f(Ar-Al)
CF_3	0.88	1.02	−0.14	1.25	0.79	0.46
NH_2	−1.23	−1.19	−0.04	−0.90	−1.38	0.48
SH	0.39	0.28	0.11	0.62	0.00	0.62
I	1.12	1.00	0.12	1.47	0.59	0.88
S	0.11[a]	−0.05[a]	0.16	0.11	−0.51	0.62
C=O	−1.05[a]	−1.21[a]	0.16	−0.89	−1.69	0.80
$CONH_2$	−1.49	−1.71	0.22	−1.13	−1.99	0.86
COO	−0.51[a]	−0.77[a]	0.26	−0.43	−1.28	0.85
Br	0.86	0.60	0.26	1.18	0.24	0.94
CN	−0.57	−0.84	0.27	−0.23	−1.13	0.90
F	0.14	−0.17	0.31	0.42	−0.51	0.93
Cl	0.71	0.39	0.32	0.93	0.06	0.87
COOH	−0.32	−0.67	0.35	0.00	−1.00	1.00
O	−0.52[a]	−0.97[a]	0.45	−0.46	−1.54	1.08
N	−0.82[a]	−1.30[a]	0.48	−1.06	−2.13	1.07
OH	−0.67	−1.16	0.49	−0.36	−1.44	1.08
NO_2	−0.28	−0.85	0.57	−0.09	−1.02	0.93

[a] Derived from $\pi(X - CH_3) - \pi(CH_3)$

groups is peculiar.

LEO et al. [5] assumed that the relatively low aromatic $\pi(NH_2)$ results from the better hydrogen bonding of the two hydrogen atoms, thus
increasing the affinity of the aromatic NH_2 group for the water phase.
Removal of these hydrogen atoms will offset this effect, the result being that the behaviour of a tertiary N approaches normal (see Table
III,15). The characteristics of OH and −O−(al), however, are not consistent with this reasoning.

The f(ar-al) differences are much larger than the π(ar-al). It is
true that no negative values occur but again the functions CF_3 and NH_2
tend to deviate because of their extremely low value for f(ar-al).

The mean value of f(ar-al) for all groups listed in Table III,15 is
0.86, with a standard deviation of 0.22. This Δf will be discussed in
more detail in subsequent parts of this monograph.

Differences between Aromatic π and f Values

The differences between aromatic π and f values are much more consistent than those found between the aliphatic values. Table III,16
compares the π and f values of some of the commonest substituents. For
the six functions listed, the mean difference between f and π is 0.25,
with a standard deviation of 0.08. As the relationship between π and f
is expressed by

$$f(X) \;=\; \pi(X) \;+\; f(H) \tag{III-6}$$

the difference 0.25 must agree with the $f(H)$ value. As will be seen in
the next section, $f(H)$ is indeed of this magnitude.

The fluctuations in $f - \pi$ as expressed by the standard deviation
are due to inaccuracies in π rather than in f (cf., Table III,10).

TABLE III,16

DIFFERENCES BETWEEN f AND π VALUES OF A FEW MAJOR AROMATIC SUBSTITUENTS

substituent	f	π^a	$f - \pi$
NO_2	−0.09	−0.28	0.19
NH_2	−0.90	−1.23	0.33
OH	−0.36	−0.67	0.31
COOH	0.00	−0.32	0.32
Cl	0.93	0.71	0.22
CH_3	0.70	0.56	0.14

[a] π values belonging to the $C_6H_5OCH_2COOH$ system.

The π system, originally derived for aromatics, consists of eight
different systems, as seen in the preceding chapter (FUJITA et al.
[14]). Variations in the π values are greatest when two groups with
strong mutual interactions are placed together on the aromatic ring;
examples are the NO_2 − NH_2 and NO_2 − OH combinations. FUJITA et al. [14]
attempted to obtain more insight into these variations by including
additional electronic parameters in their considerations. Sometimes
this was successful but in other instances these attempts failed completely:

$$\Delta\pi \;=\; \pi(\text{phenol}) - \pi(\text{benzene}) \;=\; 0.823\,\sigma + 0.061 \tag{III-7}$$
$$n = 24; \quad r = 0.954; \quad s = 0.097$$

$$\Delta\pi \;=\; \pi(\text{benzylalcohol}) - \pi(\text{benzene}) = 0.469\,\sigma + 0.041 \tag{III-8}$$
$$n = 11; \quad r = 0.937; \quad s = 0.086$$

Refs. pg 103

$$\Delta\pi \;=\; \pi(\text{phenoxyacetic ac.}) \;-\; \pi(\text{benzene}) \;=\; 0.307\ \sigma \;-\; 0.004 \quad (\text{III-9})$$
$$n = 22; \quad r = 0.754; \quad s = 0.100$$

$$\Delta\pi \;=\; \pi(\text{nitrobenzene}) \;-\; \pi(\text{benzene}) \;=\; -0.509\ \sigma \;+\; 0.282 \quad (\text{III-10})$$
$$n = 20; \quad r' = 0.676; \quad s = 0.250$$

where $\Delta\pi$ is the difference between the π constants in the two systems and σ is the HAMMETT constant, which provides a measure of the elec-tronic interaction of the substituting group.

TABLE III, 17

LOG P VALUES OF SOME BENZENE DERIVATIVES WITH STRONGLY INTERACTING SUBSTITUENTS

substituents			log P (obs)[*]	log P(clc) (π)	log P(clc) (f)	residuals (π)	residuals (f)
NO_2	NH_2	m	1.37	0.62	0.73	0.75	0.64
		p	1.39			0.77	0.66
NO_2	OH	m	2.00	1.18	1.27	0.82	0.73
		p	1.91			0.73	0.64
NO_2	COOH	m	1.83	1.53	1.63	0.30	0.20
		p	1.89			0.36	0.26
NO_2	Cl	m	2.44	2.56	2.56	-0.12	-0.12
		p	2.40			-0.16	-0.16
NO_2	NO_2	m	1.49	1.57	1.55	-0.08	-0.06
		p	1.48			-0.09	-0.07
NH_2	NH_2	m	nm	-0.13	-0.08		
		p	nm				
NH_2	OH	m	0.16	0.23	0.46	-0.07	-0.30
		p	0.14			-0.09	-0.32
NH_2	COOH	m	nm	0.58	0.82		
		p	0.68			0.10	-0.14
NH_2	Cl	m	1.89	1.61	1.75	0.28	0.14
		p	1.83			0.22	0.08
OH	COOH	m	1.50	1.14	1.36	0.36	0.14
		p	1.58			0.44	0.22
OH	Cl	m	2.48	2.17	2.29	0.31	0.19
		p	2.39			0.22	0.10
Cl	COOH	m	2.68	2.52	2.65	0.16	0.03
		p	2.65			0.13	0.00

[*] Values taken from Tables III,11 and III,12 and averaged where more than one value was available.
nm = not measured.

It will be seen that a good correlation is obtained for phenol (eqn. III-7) and benzyl alcohol systems (eqn. III,8); for the phenoxyacetic

acid and nitrobenzene systems (eqns. III,9 and III,10, respectively) the
correlation is very poor.

In addition, Table III,17 presents some log P values of benzene derivatives that contain groups which are known to display strong interactions. It is remarkable that, on the whole, the meta – para log P differences can be neglected whenever the log P values are well'established. Since the σ_m and σ_p values of the groups are distinctly different,
the question arises of how this can be reconciled with the fact that
the meta- and para- log P values are scarcely different. Also, the
SWAIN and LUPTON parameters $\mathcal{F}$ and $\mathcal{R}$ are of no avail. They serve to express the field and resonance effects of a given substituent [52,53]
and are associated with σ as follows:

$$\sigma = a\mathcal{F} + b\mathcal{R} \qquad\qquad (III-11)$$

where a and b are constants to be determined in a regression analysis.

TABLE III,18

ELECTRONIC CONSTANTS OF SOME AROMATIC SUBSTITUENTS

substituent	σ_m	σ_p	$\mathcal{F}$	$\mathcal{R}$
F	0.34	0.06	0.43	−0.34
Cl	0.37	0.23	0.41	−0.15
Br	0.39	0.23	0.44	−0.17
I	0.35	0.18	0.40	−0.19
NH_2	−0.16	−0.66	0.02	−0.68
NO_2	0.71	0.78	0.67	0.16
OH	0.12	−0.37	0.29	−0.64
CH_3	−0.07	−0.17	−0.04	−0.13
COOH	0.37	0.45	0.33	0.15
$COOCH_3$	0.37	0.45	0.33	0.15
$CONH_2$	0.28	0.36	0.24	0.14

In most of the subsequent studies by HANSCH et al. [54] and by others (TUTE [55]), exclusive use was made of π values derived from the
phenoxyacetic acid series, as these values are considered to be a good
compromise between electron-withdrawing and electron-releasing factors.

Table III,18 gives the most frequently used electronic constants of
a few of the most important aromatic substituents.

Condensed Ring Systems and Heterocyclic Structures

Partially hydrogenated condensed Ring Systems

The log P values of ring systems of this type can be easily derived from the fragment values obtained in the previous sections.

A few examples are:

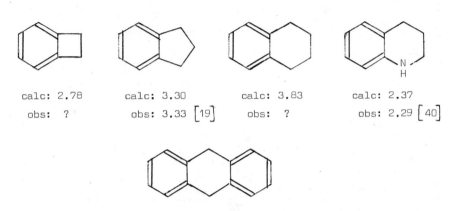

calc: 2.78   calc: 3.30   calc: 3.83   calc: 2.37

obs: ?   obs: 3.33 [19]   obs: ?   obs: 2.29 [40]

calc: 4.50

obs: 4.25 [19]

Where comparisons can be made between the derived and experimental partition values, there is fairly good agreement.

Condensed Ring Systems with Benzene and Pyridine Units

The concept that lipophilicity may be broken down into strictly additive fragmental f values, as successfully applied in the preceding sections, is so attractive that we were tempted to investigate whether it might be applicable to even smaller fragments of the benzene ring and perhaps even to other aromatic structures, condensed ring systems, etc. In this connection, the structures recorded in Table III, 19 were tried in a preliminary test.

The problem in such comparative tests is that partition values that are first regarded for seemingly good reasons as serious measuring values appear later to be "outliers".

If the value of 1.56 for benzene is immediately disregarded as being an oulier, the following equation can be put forward to derive a value for the fragment CH_{ar} from the remaining benzene partition values:

$$f(CH_{ar}) = 1/6 \log P(C_6H_6) = 1/6 \times 2.14 = 0.36$$

TABLE III,19

PARTITION VALUES OF SOME INTERRELATED AROMATICS

aromatic	log P values		
benzene	2.13 [14]	2.15 [15]	1.56 [23]
pyridine	0.64 [46] 1.04 [56]	0.66 [58]	0.65 [16]
naphthalene	3.59 [59] 3.45 [18]	3.37 [57] 3.30 [47]	3.01 [23]
anthracene	4.45 [57]		
phenanthrene	4.46 [57]		
quinoline	2.03 [41]	2.03 [16]	2.06 [23]
iso quinoline	2.08 [1]		

On comparison of the values of $\log P(C_6H_6)$ and $f(C_6H_5)$, i.e., 2.14 and 1.90, with $f(H_{al}) = 0.21^*$, it is reasonable to assume that $f(H_{ar}) \approx f(H_{al}) = 0.21$, which means that there is no significant difference between $f(C_{ar})$ and $f(C_{al})$:

$$f(C_{ar}) = f(CH_{ar}) - f(H) = 0.36 - 0.21 = 0.15$$

and

$$f(C_{al}) = 0.15 \text{ (see Table III,7)}$$

This aspect will be considered further in the following section; in the following paragraphs the above data for $f(CH_{ar})$, $f(C_{ar})$ and $f(H_{ar})$ will be used to ascertain how the structures listed in Table III,19 are related to their lipophilicities.

The following partition coefficients can be derived for naphthalene, anthracene, phenanthrene, quinoline and iso quinoline:

Naphthalene

$$\log P = 8 \, f(CH_{ar}) + 2 \, f(C_{ar})$$
$$= 2.88 + 0.30 = 3.18$$

Anthracene and phenanthrene

$$\log P = 10 \, f(CH_{ar}) + 4 \, f(C_{ar})$$
$$= 3.60 + 0.60 = 4.20$$

Quinoline and iso quinoline

$$\log P = \log P(\text{pyridine}) + 2 \, f(CH_{ar}) + 2 \, f(C_{ar})$$

* This value of $f(H_{al}) = 0.21$ can be derived by relating $f(CH_3)$, $f(CH_2)$ and $f(CH)$ mutually. The averaged differences, 0.17, 0.29 and 0.38, then refer to 1, 1 and 2 hydrogen atoms, respectively.

Refs. pg 103

$$= 0.65 + 0.72 + 0.30 = 1.67,$$

it being understood that 0.65 for log P(pyridine) represents a correct value and that the value 1.04 is incorrect.

We should point out that the principle of additivity implies that isomeric differences between anthracene and phenanthrene and between quinoline and _iso_ quinoline have no influence on lipophilicity (in accordance with the experimental results). The last calculation also shows how the principle of "redistribution" is applicable to fragmentation (fragments may be transferred from one part of a molecule to another part), so that fragments that are not present as such in the structure are allowed to occur in the log P calculation.

Comparison of the above calculated values with the observed log P values in Table III,19 indicates that condensation of aromatic rings always leads to _higher_ log P values than simple fragment addition suggests.

This increase in lipophilicity is not surprising per se in that a similar effect was noted in the preceding section (see Table III,15) for substitution of a functional group on a benzene ring. The lipophilicity then appears to be about 0.85 unit higher than when the same group is included in an aliphatic structure.

An attempt to correct this effect was made by the introduction of an additional parameter into the regression analysis of a number of structures (see Table III,20). This could, as it turned out, be achieved in several ways, two of which are discussed below:

(a) The structures may be fragmented into CH, H, N and C_c, where C_c is a common carbon atom in any condensed system. It should be noted that fragmentation may require the introduction of (- H) as a formal fragment, as illustrated below.

The unit C_6H_4 (No. 3) is introduced with a lipophilicity of 1.72, its fragmentation being 6 f(CH) - 2 f(H); 2-methylpyridine (No. 6) is introduced with a lipophilicity of 0.41 (log P - f(CH$_3$) = 1.11 - 0.70) and fragmented into 5 f(CH) + 1 f(N) - 1 f(H).

The results of the regression analysis, carried out with Nos. 1 - 25 in Table III,20 (33 log P values in all), were as follows:

$$f \text{(CH)} = 0.345; \quad \text{standard deviation: } 0.018; \quad \underline{t} = 19.4$$
$$f \text{(H)} = 0.184; \quad \text{standard deviation: } 0.026; \quad \underline{t} = 7.11$$
$$f \text{(N)} = -1.003; \quad \text{standard deviation: } 0.036; \quad \underline{t} = -28.2$$
$$f \text{(C}_c\text{)} = 0.305; \quad \text{standard deviation: } 0.026; \quad \underline{t} = 11.7$$

Constant term = -0.000; n = 33 (+1); r = 0.996; s = 0.098; F = 1,123

TABLE III,20

STATISTICAL TREATMENT OF SOME AROMATIC COMPOUNDS

(three "Φ" fragments included)

no	compound or fragment	log P obs.	ref.	log P(clc) I	res.	II	res.
1	benzene	2.13	[14]	2.07	0.06	2.07	0.06
		2.15	[15]		0.08		0.08
2	C_6H_5	1.90	a	1.89	0.01	1.88	0.02
3	C_6H_4	1.72	b	1.72	0.00	1.70	0.02
4	C_6H_3	1.44	b	1.54	-0.10	1.51	-0.07
5	pyridine	0.65	[16]	0.71	-0.06	0.72	-0.07
		0.66	[54]		-0.05		-0.06
		0.64	[46]		-0.07		-0.08
6	2-methylpyridine	1.11	[19]	1.23	-0.12	1.24	-0.13
7	3-methylpyridine	1.20	[22]	1.23	-0.03	1.24	-0.04
8	4-methylpyridine	1.22	[22]	1.23	-0.01	1.24	-0.02
9	naphthalene	3.37	[53]	3.37	0.00	3.37	0.00
		3.45	[18]		0.08		0.08
		3.30	[47]		-0.07		-0.07
		3.59	[55]		0.22		0.22
10	α-naphthylamine	2.25	[41]	2.29	-0.04	2.29	-0.04
11	α-naphthol	2.98	[1]	2.83	0.15	2.83	0.15
12	β-naphthol	2.84	[1]		0.01		0.01
13	quinoline	2.03	[41]	2.01	0.02	2.03	0.00
		2.03	[16]		0.02		0.00
		2.06	[23]		0.05		0.03
14	iso quinoline	2.08	[1]	2.01	0.07	2.03	0.05
15	2-methylquinoline	2.59	[41]	2.53	0.06	2.54	0.05
16	6-methylquinoline	2.57	[10]	2.53	0.04	2.54	0.03
17	7-methylquinoline	2.47	[10]	2.53	-0.06	2.54	-0.07
18	8-methylquinoline	2.60	[10]	2.53	0.07	2.54	0.06
19	6-chloroquinoline	2.73	[10]	2.77	-0.04	2.78	-0.05
20	5-nitroquinoline	1.86	[10]	1.75	0.11	1.76	0.10
21	6-nitroquinoline	1.84	[10]	1.75	0.09	1.76	0.08
22	7-nitroquinoline	1.82	[10]	1.75	0.07	1.76	0.06
23	anthracene	4.45	[53]	4.67	-0.22	4.67	-0.22
24	phenanthrene	4.46	[53]	4.67	-0.21	4.67	-0.21
25	acridine	3.40	[53]	3.31	0.09	3.33	0.07

Refs. pg 103

TABLE III,20 (continued)

no	compound or fragment	log P obs.	ref.	log P(clc)			
				I	res.	II	res.
26	4-phenylpyridine	2.45	[26]	2.68	−0.23	*	
27	biphenyl	3.95	[60]	4.04	−0.09	*	
		4.09	[24]		0.05	*	
		4.04	[18]		0.00	*	
		4.17	[59]		0.13	*	
Outliers:							
1	pyridine	1.04	[56]	0.71	0.33	0.72	0.32
2	naphthalene	3.01	[23]	3.37	−0.36	3.37	−0.36
3	8-nitroquinoline	1.40	[10]	1.75	−0.35	1.76	−0.36
4	8-chloroquinoline	2.33	[11]	2.77	−0.44	2.78	−0.45

Regression (a)				Regression (b)			
fragment	f	st.dev.	$\underline{t}$	fragment	f	st.dev.	$\underline{t}$
CH	0.336	0.014	23.3	C_6H_3	1.489	0.026	57.2
H	0.176	0.026	6.84	H	0.188	0.022	8.69
N	−1.025	0.036	−28.3	C_5H_4N	0.518	0.061	8.49
C_c	0.315	0.024	13.3				
	n = 38 (+1)				n = 34 (+1)		
	r = 0.996				r = 0.996		
	s = 0.103				r = 0.096		
	F = 1,323				F = 1,546		
constant term: 0.050				constant term: 0.020			

[a] values from Table III,10.
* These structures do not lend themselves to the desired type of fragmentation.

The effect of forcing is striking in this particular regression a-nalysis. When the zero card is omitted, the parameter N is refused, and the ultimate outcome is:

f(CH) = 1.348; standard deviation: 0.035; $\underline{t}$ = 28.9
f(H) = 0.184; standard deviation: 0.026; $\underline{t}$ = 7.11
f(C_c) = −0.698; standard deviation: 0.038; $\underline{t}$ = −18.1
Constant term = −6.016; n = 33; r = 0.996; s = 0.098; F = 1,340

By introducing only one zero card, the constant term is eliminated and the parameter N is included.

Pyridine (log P = 1.04), naphthalene (log P = 3.01), 8-nitro-quin-oline and 8-chloro-quinoline were removed as outliers. Strictly, the structures naphthalene (log P = 3.59), anthracene and phenanthrene with

residuals of +0.22, -0.22 and -0.21, respectively, ought to have been
excluded also, but as these residuals are marginal, the structures in
question were retained in the data set.

On comparison of the lipophilicity of a structure such as biphenyl
with that of a condensed aromatic system, it will be seen that the interconnecting carbon atoms of biphenyl behave more or less as the common carbon atoms in a condensed aromatic structure. The regression a-
nalysis was therefore repeated with the series of Table III,20, but now
including 4-phenyl-pyridine and biphenyl (the latter with four data
points). Biphenyl was fragmented into 10 f(CH) + 2 f(C_c) and 4-phenyl-
-pyridine into 9 f(CH) + f(N) + 2 f(C_c).

The results of these calculations, presented in Table III,20, are
not appreciably different from those of the previous analysis. This permits the conclusion that the two interconnecting carbon atoms in an
Ar - Ar structure are able to enhance the lipophilicity in a manner
that is not significantly different from the lipophilicity effect of
the pair of carbon atoms representing the common element in a condensed
aromatic structure. This type of carbon atoms is indicated by clarendon
dots in the following formulae:

and ($C^\bullet$) is used to denote a fragment of this type.

(b) From the results discussed above (a), it follows that f($C^\bullet$) is a-
bout double f(C_{ar}). This difference may serve as a basis of a regression analysis whose attraction is, that the excessively large number of
CH fragments is reduced drastically enough to permit more congruity in
the fragment pattern. The procedure to be adopted is as follows:

(1) C_6H_3, C_5H_4N and H are introduced as formal basic fragments into
the regression analysis, even if they are not present as indicated
(principle of "redistribution"). The additivity concept is, in fact, on
severe trial now but the ultimate outcome appears to confirm its applicability.

(2) The fragmental equation is so balanced that it gives an f value to
the common carbon atoms in a condensed aromatic system that is twice as
large as that for the other aromatic carbon atoms. The practical realization of this procedure can be illustrated by a few examples:

$$\log P(\text{naphthalene}) = 2\,f(C_6H_3) + 2\,f(H)$$
$$\log P(\text{quinoline}) = f(C_6H_3) + f(C_5H_4N)$$
$$\log P(\text{acridine}) = 2\,f(C_6H_3) + f(C_5H_4N) - f(H)$$

In the first and second examples two carbon atoms and in the third example four carbon atoms are counted twice. In all three, no other fragments but those mentioned under (1) are used. Special attention is drawn to the fact that the acridine fragmentation requires the introduction of an $f(H)$ with a minus sign in order to provide the correct number of H atoms.

The results of a regression analysis, carried out along these lines, are recorded in Table III,20 (regression (b)). There is no essential difference between regressions (a) and (b) but perhaps the statistics of (b) are slightly better on an average. The outliers of (a) and (b) are also similar.

Comparison of Aliphatic and Aromatic CH Fragments

As already indicated, there is little, if any, difference between the f values of an H atom in an aliphatic structure and an H atom bound to an aromatic system. A satisfactory average for $f(H_{al})$ is 0.21, while $f(H_{ar})$ is at most 0.03 lower. The difference between $f(C_{al})$ and $f(C_{ar})$ is also hardly significant; compare the value of log P(benzene) determined experimentally to be 2.13 and the calculated value of:

$$6 \times f(C_{al} + H_{al}) = 6 \times (0.15 + 0.21) = 2.16$$

This remarkable agreement induced us to perform a statistical analysis with a number of CH fragments and compounds of different origin with $f(C)$ and $f(H)$ introduced as the sole regressors in the computation. The results are recorded in Table III,21.

The fragment with the largest deviation is CH. It is reasonable to assume that the sequence from CH_3 via the two fragments with one and two fewer H atoms, respectively, will eventually lead to the quaternary C with gradually diminishing lipophilicity values. The fact that this pattern did not emerge from our regression analysis (Section on aliphatic structures: primary set of f values) may be due to the frequent occurrence of structures of the $>$CH–X type, where X is a negative group. This should probably be interpreted to mean that in the analysis part of the effect of X, i.e., a decrease in lipophilicity, is transferred automatically to the CH fragment value. It may be concluded that the

TABLE III,21

STATISTICAL ANALYSIS OF A SELECTED SERIES OF C_xH_y FRAGMENTS
OF DIFFERENT ORIGINS

I = introduced lipophilicities; II = estimated lipophilicities.

no	compound or fragment	lipophilicities		
		I	II	res.
1	CH_3-CH_2	1.23[a]	1.27	−0.04
2	$CH_2=CH$	0.93	0.87	0.06
3	$CH\equiv C$	0.51	0.47	0.04
4	C_6H_6	2.13	2.10	0.03
5	C_6H_5	1.90	1.90	0.00
6	C_6H_4	1.72	1.70	0.02
7	C_6H_3	1.44	1.50	−0.06
8	CH_3	0.70	0.71	−0.01
9	CH_2	0.53	0.52	0.01
10	CH	0.24	0.32	−0.08
11	C	0.15	0.12	0.03

fragment	f	st.dev.	$\underline{t}$
C	0.157	0.009	17.8
H	0.199	0.011	18.2

n = 11

r = 0.997

s = 0.048

F = 996

constant term = −0.040

[a] Derived from a series of suited structures

lipophilic behaviour of the constituent Cs and Hs is not related to the
type of bond between the carbon atoms involved (single, double, triple
and aromatic) and in any event it is outside the range of detectability.

Extension of the f System to other Heteroaromatic Structures; Introduction of [N], [NH], [S] and [O]

The medicinal chemist's attention is being increasingly directed towards heteroatomic structures. There are, however, too few reliable log
P values to consider a procedure as indicated for pyridine and its condensed analogues. The material at hand does lend itself to the treat-

Refs. pg 103

ment discussed for the secondary aliphatic and aromatic fragments.

Relevant data are collected in Table III,22, and refer to structures derived from pyrrole, imidazole, thiophene and furan. The f values calculated by means of these data are given in Table III,23, which also includes some fragments that may be useful in computing log P values for aromatic structures.

TABLE III,22

STRUCTURES USED IN DERIVING f CONSTANTS OF SOME IMPORTANT
HETEROCYCLIC STRUCTURES

no	compound	log P value			
		obs.	ref.	clc.	res.
1	pyrrole	0.75	[1]	0.72	0.03
2	indole	2.00	[22]	2.06	−0.06
		2.14	[1]		0.08
		2.25	[61]		0.19
3	5-bromoindole	3.00	[23]	3.03	−0.03
4	2-methylindole	2.53	[62]	2.57	−0.04
5	3-methylindole	2.60	[23]	2.57	0.03
6	5-methylindole	2.68	[23]	2.57	0.11
7	5-methoxyindole	2.06	[62]	2.11	−0.05
8	indole-3-acetic acid	1.41	[13]	1.39	0.02
9	carbazole	3.29	[23]	3.40	−0.11
		3.72	[19]		0.32*
10	diphenylamine	3.34	[22]	3.49	−0.15
		3.22	[23]		−0.27*
		3.50	[19]		0.01
		3.72	[63]		0.23*
11	thiophene	1.81	[16]	1.79	0.02
12	2-phenylthiophene	3.74	[47]	3.80	−0.06
13	thiophene-2-carboxylic acid ethylester	2.33	[45]	2.39	−0.06
14	benzothiophene	3.12	[13]	3.13	−0.01
		3.09	[23]		−0.04
15	diphenylsulfide	3.47	[22]	4.48	−1.01*a
		4.45	[19]		−0.03a
16	furan-2-carboxamide	−0.11	[44]	0.24	−0.35*
17	benzofuran	2.67	[13]	2.70	−0.03

TABLE III,22 (continued)

18	dibenzofuran	4.12	[13]	4.04	0.08
19	diphenylether	4.21	[24]	4.25	-0.04[a]
		4.36	[22]		0.11[a]
20	imidazole	-0.08	[60]	0.06	-0.14
		0.04	[63]		-0.02
21	2-phenylimidazole	1.87	[63]	2.07	-0.20
22	benzimidazole	1.46	[60]	1.40	0.06
		1.34	[13]		-0.06
		1.50	[64]		0.10
		1.20	[23]		-0.20
23	2-trifluoromethyl-benzimidazole	2.67	[64]	2.45	0.22[*]
		2.58	[60]		0.13
24	2-trifluoromethyl-5-bromo-benzimidazole	3.57	[64]	3.42	0.15
25	5,6-dimethylbenzimidazole	2.35	[64]	2.40	-0.05

[*] Outliers
[a] Calculated with the inclusion of 2 $C^{\bullet}$ (see Table III,25)

TABLE III,23

FRAGMENTAL CONSTANTS RECOMMENDED FOR THE CALCULATION OF

PARTITION VALUES OF AROMATIC RINGS

no	fragment	f
1	H	0.199
2	C_{ar}	0.157
3	$(CH)_{ar}$	0.356
4	$C^{\bullet}$	0.314
5	C_6H_5	1.896
6	C_6H_4	1.719
7	C_6H_3	1.440
8	$C_{10}H_7$ (naphthalenyl)	3.17
9	C_5H_4N (pyridinyl)	0.52
10	C_4H_3O (furanyl)	1.36
11	C_4H_3S (thienyl)	1.59
12	C_4H_4N (pyrrolyl)	0.52 [a]
13	C_9H_5N (quinolinyl)	1.81
14	$C_3H_3N_2$ (2- or 4-imidazolyl)	-0.14

Refs. pg 103

TABLE III,23 (continued)

no	fragment	f
15	C_4H_4N (1-pyrrolyl)	0.24*
16	$C_3H_3N_2$ (1-imidazolyl)	-0.42*
17	[NH]	-0.70
18	[N]	-1.06
19	[O]	0.14
20	[S]	0.37

* See one of the following sections of this chapter.
a Recommended for all substitutions except 1 (no. 15).

In order to enhance the usefulness of the f system, the f values of
a few heteroatoms were determined from heterocyclic systems that are of
importance in SAR studies. Using $f(CH_{ar})$ and $f(C_{ar})$ as a basis, values
for hetero-N (pyridine), hetero-NH (pyrrole), hetero-S (thiophene) and
hetero-O (furan) were obtained. In order to prevent these fragments being mistaken for fragments already included in the preceding tables, we
decided to use braces, i.e., [N], [NH], [S] and [O].

Cross-conjugated Systems

Cross-conjugated systems consist of conjugation partners separated
by two single bonds via what may be called a transmissive link. Examples are 1,1-diphenylethene (E), benzophenone (F), diphenyl ether (G)
and diphenylamine (H).

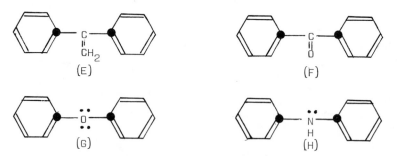

In structures E and F, the polarizable double bonds act as the
transmissive link, and in structures G and H lone pairs of electrons
fulfil this function. Polar resonance hybrids of the type indicated by
I and J, where through-conjugation occurs from ring to ring, will be
involved here. The extent to which the cross-conjugation accentuates an
existing interaction in the non-cross-conjugated halves can be deduced

(I)   (J)

from the UV absorptions of the structures concerned. Table III,24 presents a compilation of the UV absorption data for some compounds as far
as their conjugative absorption band is concerned.

TABLE III,24

UV DATA FOR THE CONJUGATIVE ABSORPTION BANDS OF SOME
CROSS-CONJUGATED STRUCTURES IN COMPARISON WITH DATA
FOR RELATED NON-CROSS-CONJUGATED ANALOGUES

compound	λ(nm)	ϵ	ref.
C_6H_5-CO-H	246	13,200	[65]
$C_6H_5-CO-CH_3$	242	12,400	[65]
$C_6H_5-CO-C_6H_5$	253	18,000	[66]
$C_6H_5-O-CH_3$	220	7,100	[67]
$C_6H_5-O-C_6H_5$	225	11,300	[68]
$C_6H_5-S-CH_3$	254	9,300	[69]
$C_6H_5-S-C_6H_5$	250	12,600	[70]
$C_6H_5-NH-CH_3$	240	10,000	[71]
$C_6H_5-NH-C_6H_5$	285*	20,000	[72]

* Broad maximum covering both conjugative and benzenoid absorption
and hence not well suited for comparison with the corresponding
band of $C_6H_5-NH-CH_3$.
Solvent: ethanol.

On the assumption that in benzene and alkylbenzenes the conjugative
absorption band is to be found between 200 and 210 nm, it may be concluded that the effect of cross-conjugation on UV absorption is much
smaller than the first attachment of C=O, O or S to a benzene ring.

On consideration of the energy aspects of the resonance, a similar
trend is noted: in acetophenone, the resonance energy is 7 kcal/mole in
addition to that in the benzene ring; in benzophenone, however, the resonance energy is not 14 (= 2 × 7) but only 10 kcal/mole in addition to
that in two benzene rings [73].

From these data, we may expect an increased lipophilicity in cross-
-conjugated systems. This increase will, however, remain significantly

TABLE III,25

APPLICATION OF C$^\bullet$ IN THE CALCULATION OF LOG P VALUES FOR CROSS-
-CONJUGATED SYSTEMS AND FOR STRUCTURES WITH AN ELONGATED
CONJUGATION PATTERN

no	compound	log P values			
		obs.	ref.	clc.	res.
1		3.18	[19]	3.24	-0.06
2		1.88	[37]	1.86	0.02
3		1.88	[37]	1.86	0.02
4		1.98 1.98	[37] [74]	1.86	0.12 0.12
5		2.61	[75]	2.62	-0.01
6		1.37	[75]	1.32	0.05
7		4.21 4.36	[24] [22]	4.25	-0.04 0.11
8		4.45	[19]	4.48	-0.03
9		3.34 3.50	[22] [19]	3.49	-0.15 0.01
10	$CH_3-CH=CH - COOH$	0.72 0.72	[51] [44]	0.74 0.7	-0.02 -0.02
11		2.95	[19]	3.14	-0.19

TABLE III,25 (continued)

No.	Structure				
12	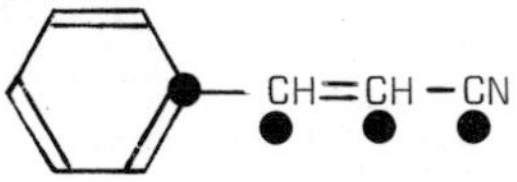 C6H5—CH=CH—CN	1.96	[27]	2.12	−0.16
13	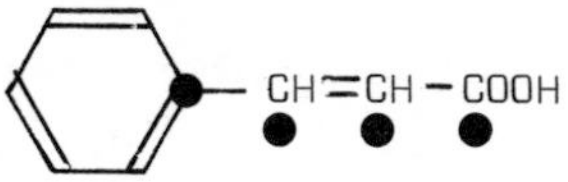 C6H5—CH=CH—COOH	2.13	[13]	2.25	−0.12
14	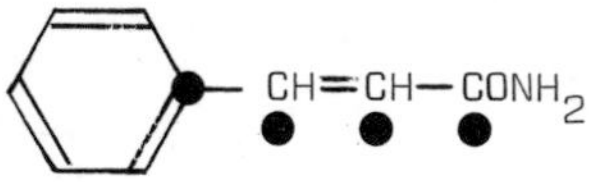 C6H5—CH=CH—CONH2	1.41	[27]	1.26	0.15
15	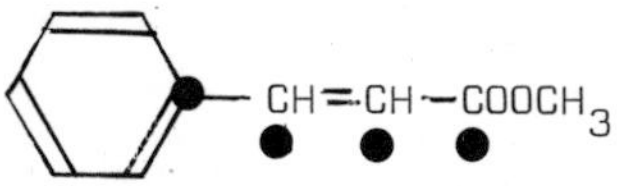 C6H5—CH=CH—COOCH3	2.62	[27]	2.67	−0.05
16	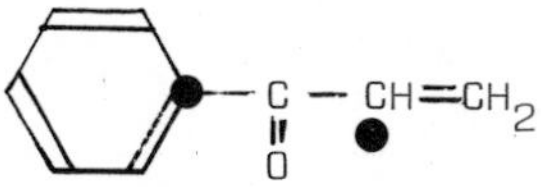 C6H5—CO—CH=CH2	1.88	[76]	2.27	−0.39*
17	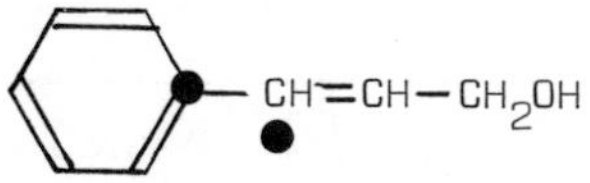 C6H5—CH=CH—CH2OH	1.95	[12]	2.03	−0.08

* Outlier.

below the value of 0.85 observed on attachment of a functional group to a benzene ring. Our practical findings are consistent with this view but, surprisingly, a virtually constant Δlog P value of 0.28 was found throughout. Again, the $f(C^\bullet)$ symbol is used to denote the carbon atoms with clarendon dots in formulae E, F, G and H. Details are given in Table III,25.

Further study has shown that a correct calculation of lipophilicity is also feasible for structures with an elongated conjugation pattern by means of $f(C^\bullet)$ values. The calculation of the partition values of crotonic and cinnamic acids is exemplified below:

$$\text{Log P(crotonic acid)} = f(CH_3) + f(CH_2=CH) - f(H)$$
$$+ 2\left[f(C^\bullet) - f(C_{ar})\right] + f(COOH_{al}) = 0.74$$

$$\text{Log P(cinnamic acid)} = f(C_6H_5) + f(CH_2=CH) - f(H)$$
$$+ 4\left[f(C^\bullet) - f(C_{ar})\right] + f(COOH_{al}) = 2.25$$

The values reported in the literature are 0.72 [44], 0.72 [51] and 2.13 [13], respectively. For other examples, see Table III,25.

Refs. pg 103

Lipophilicity of H attached to an electron-withdrawing Centre

As indicated in a previous section, there is no detectable difference between $f(H_{al})$ and $f(H_{ar})$ and the value of the fragment can be
taken to be 0.20. This value can be employed in calculations of various
kinds. It appears imperative, however, that the hydrogen atom for which
f = 0.20 should never be attached to an electron-withdrawing centre.
These centres include C=O, COOH, COOR and CON. While the electron-withdrawing capacities of these centres differ substantially as indicated
by their σ^* values, all H atoms attached to these centres can be included in the calculations with f = 0.47. This is evident from the examples given in Table III,26.

TABLE III,26

APPLICATION OF $f(H)$ = 0.47 IN STRUCTURES WITH H ATOMS
ATTACHED TO AN ELECTRON-WITHDRAWING CENTRE

no	compound	log P values			
		obs.	ref.	clc.	res.
1	H-COOH	-0.54	[6]	-0.53	-0.01
2	H-COO-n.C_3H_7	0.83	[12]	0.95	-0.12
3	C_6H_5-CO-H	1.48	[19]	1.50	-0.02
4	n.C_3H_7-$CONH_2$	-0.21	[1]	-0.24	0.03
5	C_6H_5-CH_2-$CONH_2$	0.45	[16]	0.46	-0.01
6	CH_3-CONH-CH_3	-1.05	[19]	-1.06	0.01
7	H-CON$(CH_3)_2$	-1.01	[47]	-1.06	0.05
8	CH_3-CON$(CH_3)_2$	-0.77	[19]	-0.82	0.05
9	C_6H_5-$CONH_2$	0.64	[14]	0.78	-0.14
		0.65	[63]		-0.13
10	C_6H_5-CONH-CH_3	1.18	[47]	1.01	0.17
		0.86	[47]		-0.15
11	C_6H_5-NHCO-H	1.15	[12]	1.15	0.00
		1.12	[77]		-0.03
		1.26	[63]		0.11
12	C_6H_5-NHCO-CH_3	1.16	[14]	1.38	-0.22
		1.36	[63]		-0.02
13	C_6H_5-CONH-C_6H_5	2.62	[12]	2.52	0.10
		2.70	[63]		0.18

When calculating the log P values of carbonamide structures, use was made of the values f = -2.93 for an aliphatic CON, f = -2.06 for CON with an aromatic ring connected to C and f = -1.69 for attachment of the aromatic ring to N. As can be seen in Table III,26, it is irrelevant whether H is attached to N ar to C=O; in all instances f can be taken as 0.47.

This method of calculation is not applicable, however, to a number of carbonamide-like structures such as N-methylformanilide, N-methyl-acetanilide, N,N-dimethylbenzamide and N,N-diethylbenzamide. The log P values of these structures are found to be much too low, as will be discussed in Chapter V.

Benzanilide (K) is a structure in which the NCO unit functions as a transmissive link between the two phenyl rings. Therefore, the log P

 calculation should be based on the formula shown on the left. The question remains, however, whether the

value f(CON) = -2.06 or -1.69 should be used. With the former value, the final result is log P = 2.52 and with the latter value, log P = 2.89. The observed lipophilicities are 2.62 [24] and 2.70 [63], respectively, and it is difficult to choose between the two f values although it seems that f(CON) = -1.69 will represent the situation more accurately than f(CON) = -2.06 (see also Table V,2).

A similar effect can be expected with the COO group. Table III,27 shows that for phenyl benzoate a value of -0.48, as derived from methyl benzoate, is suitable for f(COO), whereas a value of -1.11 (from phenyl acetate) would result in a far too hydrophilic structure for phenyl benzoate.

TABLE III,27

PARTITION VALUES OF DIFFERENT TYPES OF COO GROUP

	log P(obs)	ref.	f(COO)
(al)COO(al)		Table III,2	-1.26
$C_6H_5-COO-CH_3$	2.12	[14]	-0.48
$CH_3-COO-C_6H_5$	1.49	[14]	-1.11
$C_6H_5-COO-C_6H_5$	3.59	[12]	-0.52
	3.58	[63]	-0.53

It is tempting to check the value f(H) = 0.47 with some pairs of functional groups that differ in only one H atom and satisfy the cri-

Refs. pg 103

terion that this H atom should be attached to a centre with electron-
-withdrawing capacity. Data available to us are recorded in Table III,
28.

TABLE III,28

COMPARISON OF f VALUES BELONGING TO PAIRS OF FUNCTIONAL
GROUPS THAT DIFFER FROM EACH OTHER BY ONE H ATOM

functional group	al	f	ar
COOH	−1.00		0.00 [a]
COO	−1.28		−0.43
OH	−1.44		−0.36
O	−1.54		−0.45
NH$_2$	−1.38		−0.90
NH	−1.86		−0.93
N	−2.13		−1.06
SH	0.00		0.62
S	−0.51		0.11

[a] Combinations enclosed by heavy lines differ within
the range of the error by an amount of 0.47.

It appears that the f differences, within the limits of error, are
about 0.47 in only four of the ten cases. It may therefore be concluded
that this matter leaves several questions outstanding, which will receive further attention later.

PROXIMITY EFFECT

In a previous section, mention was made of the proximity effect: in
the presence of certain functional groups separated from each other by
one or two carbon atoms, enhanced lipophilicity is observed. This increase is 0.80 ± 0.08 for a 1C separation and 0.46 ± 0.08 for a 2C separation. For 3C and higher separations, the proximity effect disappears
or lies below the limit of detectability $(\Delta 0.20)$.

Our initial view was that the proximity effect would be due to the
fact that hydrophilic groups in the closest possible connection would
lose part of their hydration mantle or perhaps even share it to a
certain extent. For structures such as X−CH$_2$−Y, this assumption presents
no problems as there is no doubt that X and Y lie within each other's
hydration peripheries . With a structure such as X−CH$_2$−CH$_2$−Y, an e-

clipsed conformation would be essential in order to bring X and Y close
enough together, which is at variance with actual practice, where a
staggered conformation is common (see Fig.III,2).

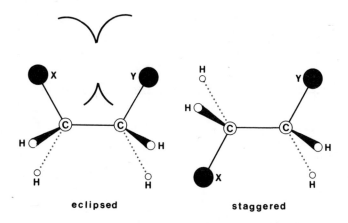

eclipsed   **staggered**

Fig.III,2: Eclipsed and Staggered Conformations of
X-CH₂-CH₂-Y (the eclipsed conformation with an indication of the intersecting hydration peripheries

   By coupling the proximity effect exclusively to the hydrophylic nature of the group in question, the question arises as to what exactly
is the criterion for hydro- and lipophilicity?

   The definition of a partition coefficient and the subsequent choice
of a standard solvent pair (octanol - water) make it plausible that the
change between hydro- and lipophilicity occurs at log P = 0. Hence any
structure for which this is true (log P = 0) is equally well (or poorly)
soluble in octanol and in water; if log P > 0 the structure is lipophilic and if log P < 0 it is hydrophilic. By extension of this criterion to hydrophobic constants, any functional group would have to be
designated as lipophilic if $f > 0$ and as hydrophilic if $f < 0$. This has
remarkable implications, e.g., of the halogens, F ($f = -0.51$) is fairly
hydrophilic, Cl ($f = 0.06$) is a borderline case and Br ($f = 0.24$ and I
($f = 0.59$) are clearly lipophilic. The above considerations hold for
the octanol - water system with the halogens accommodated in an aliphatic structure. If accommodated in an aromatic structure, all of the halogens are found to be lipophilic in the octanol - water system. As will
be seen in Part II, the use of solvent systems other than octanol - wa -
ter results in appreciable changes in the given pattern.

   A better definition of the concept hydro-lipophilicity will be at-

Refs. pg 103

tempted in Part II but, in anticipation, the proximity effect will be
considered an implication of the electronegativity of certain substituents. The structures X-CH$_2$-Y and X-CH$_2$-CH$_2$-Y, where X and/or Y is a
halogen, thus appear to possess a lipophilicity pattern that permits
identification by means of the proximity effects described above. In
this connection, reference is made to Nos. 1 - 9 in Table III,29.

TABLE III,29

SELECTION OF STRUCTURES FOR DEMONSTRATING PROXIMITY CORRECTIONS

no	compound	log P values			
		obs.	ref.	clc.	res.
1	Br-CH$_2$-COOH	0.41	[6]	0.57	-0.16
2	Cl-CH$_2$-CONH$_2$	-0.53	[12]	-0.60	0.07
3	Br-CH$_2$-CONH$_2$	-0.52	[34]	-0.42	-0.10
4	F-CH$_2$-CONH$_2$	-1.05	[34]	-1.18	0.13
5	I-CH$_2$-CONH$_2$	-0.19	[34]	-0.07	-0.12
6	F-CH$_2$-CH$_2$-OH	-0.92	[29]	-0.44	-0.48*
7	Cl-CH$_2$-CH$_2$-OH	0.03	[29]	0.13	-0.10
8	Br-CH$_2$-CH$_2$-OH	0.23	[29]	0.31	-0.08
9	Br-CH$_2$-CONH-CH$_2$-CH$_3$	0.34	[76]	0.34	0.00
10	C$_6$H$_5$-O-CH$_2$-COOH	1.26	[14]	1.32	-0.06
11	C$_6$H$_5$-O-CH$_2$-CON(CH$_3$)$_2$	0.77	[78]	0.79	-0.02
12	C$_6$H$_5$-NH-CH$_2$-COOH	0.62	[79]	0.84	-0.22*
13	C$_6$H$_5$-O-CH$_2$-CH$_2$-OH	1.16	[12]	1.06	0.10
14		2.08	[19]	2.04	0.04
15		1.05	[1]	0.91	0.14

* Outliers.

Except for 2-fluoroethanol, all of the structures listed can be calculated correctly. It is notable that in those benzene derivatives substituted with an electronegative group in _ortho_ positions the proximity
effect does not operate, although there is certainly a 2C separation
(reference is made to the collected material in Table III,11).

Phenoxyacetic acid, which, together with many of its derivatives, was examined in the regression analysis discussed in a previous section, also appears to have a markedly deviating behaviour. In the $-OCH_2COOH$ group there is a 1C separation, which, in principle, should call for a proximity effect of 0.80; in practice the value is not more than 0.34. This unduly low value should also be employed in calculating log P for N,N-dimethylphenoxyacetamide (No. 11 in Table III,29). For log P for N-phenylglycine (No. 12), even this value of 0.34 appears not to be low enough.

These considerations could lead to the conclusion that for unknown reasons the proximity effects in all structures of the type $C_6H_5-X-CH_2-$ $-Y$ are perhaps decreased from 0.80 to 0.34 and that, in agreement with this, those of $C_6H_5-X-CH_2-CH_2-Y$ can be neglected instead of being assigned a value of 0.46. Further support for such a conclusion is provided by log P for 2-phenoxyethanol, while no proximity effect needs to be used in the calculation of this value.

It is curious, however, that in a correct description of the partitional behaviour of structures of the 1,2-methylenedioxybenzene type, the proximity correction cannot be dispensed with (see Table III,29, Nos. 14 and 15). This also holds for 2-phenylindandione (L):

$$Log\ P = f(C_6H_5) + f(CH) + f(C_6H_4) +$$
$$2\ f(C=O) + pe\ (1)$$
$$= 1.90 + 0.24 + 1.72 + (-1.74) + 0.80$$
$$= 2.92$$

Observed value [82]: 2.90

The manner in which the proximity effect should be calculated most conveniently in practice for structures in which several electronegative centres with separations of one and/or two carbon atoms are present, remains to be established.

While there is no general basis for comparison and sufficiently reliable measurements are lacking, the following guidelines are recommended.

In M, the two X atoms indicated by ✔ are separated by 1 C atom, for which 1 x pe (1) should be assigned. In N, these two X atoms are combined to X_2 and this X_2 combination is again separated from the remaining X by 1 C atom. Therefore, 1 x pe (1) needs to be assigned again. The X_3 combination O is considered as a whole relative to what is attached to the remaining C valences: either pe (1), pe (2) or zero should be assigned, depending on the substitution pattern.

Refs. pg 103

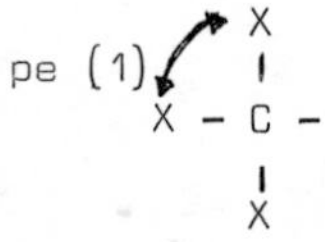

(M) 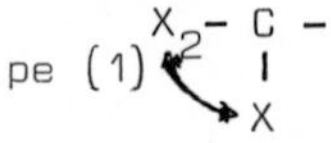   (N) 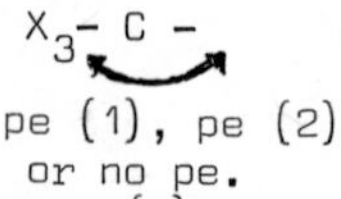   (O)

General example:

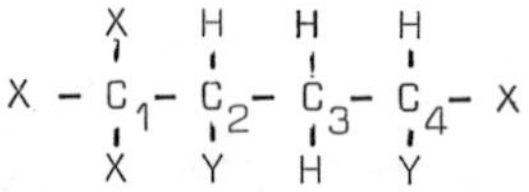

$$X - \underset{\substack{| \\ X}}{\overset{\substack{X \\ |}}{C_1}} - \underset{\substack{| \\ Y}}{\overset{\substack{H \\ |}}{C_2}} - \underset{\substack{| \\ H}}{\overset{\substack{H \\ |}}{C_3}} - \underset{\substack{| \\ Y}}{\overset{\substack{H \\ |}}{C_4}} - X$$

where X and Y are electronegative groups.

$$\text{Log } P = f(C) + 2\,f(CH) + f(CH_2) + 4\,f(X) + 2\,f(Y)$$

$$+ \ 2 \text{ pe } (1) \qquad \text{(effect around } C_1\text{)}$$
$$+ \ 1 \text{ pe } (1) \qquad \text{(effect around } C_4\text{)}$$
$$+ \ 1 \text{ pe } (2) \qquad \text{(effect over } C_1\text{-} C_2\text{)}$$

When the value $f(CX_3)$ is available, the calculation is simplified to:

$$\log P = f(CX_3) + 2\,f(CH) + f(CH_2) + f(X) + 2\,f(Y)$$

$$+ \ 1 \text{ pe } (1) \qquad \text{(effect around } C_4\text{)}$$
$$+ \ 1 \text{ pe } (2) \qquad \text{(effect over } C_1\text{-} C_2\text{)}$$

Some practical calculations are as follows:

1. Chloroform: $CHCl_3$
    $$\text{Log } P = f(CH) + 3\,f(Cl) + 2 \text{ pe } (1)$$
    $$= 0.24 + 0.18 + 1.60 = 2.02$$
    Observed value [1]: 1.97

2. Carbon tetrachloride: CCl_4
    $$\text{Log } P = f(C) + 4\,f(Cl) + 3 \text{ pe } (1)$$
    $$= 0.15 + 0.24 + 2.40 = 2.79$$
    Observed value [83]: 2.83

3. Malic acid: $HOOC-CH_2-CH(OH)-COOH$
    $$\text{Log } P = f(CH_2) + f(CH) + 2\,f(COOH) + f(OH) + 1 \text{ pe } (1) + 1 \text{ pe } (2)$$
    $$= 0.53 + 0.24 + (-2.00) + (-1.44) + 0.80 + 0.46 = -1.41$$
    Observed value [6]: -1.26

4. Citric acid: $HOOC-CH_2-C(OH)(COOH)-CH_2-COOH$
    $$\text{Log } P = 2\,f(CH_2) + f(C) + 3\,f(COOH) + f(OH) + 1 \text{ pe } (1) + 2 \text{ pe } (2)$$
    $$= 1.05 + 0.15 + (-3.00) + (-1.44) + 0.80 + 0.92 = -1.52$$
    Observed value [6]: -1.72

5. <u>Glyceryl monobutyrate</u>: $HO-CH_2-CH(OH)-CH_2-OOC-\underline{n}.C_3H_7$

$\begin{aligned}
\text{Log } P &= f(CH_3) + 4\ f(CH_2) + f(CH) + 2\ f(OH) + f(COO) + 2\ pe\ (2) \\
&= 0.70 + 2.11 + 0.24 + (-2.88) + (-1.28) + 0.92 = -0.19
\end{aligned}$

Observed value [11] : -0.17

6. <u>Diethyl tartrate</u>: $H_5C_2-OOC-CH(OH)-CH(OH)-COO-C_2H_5$

$\begin{aligned}
\text{Log } P &= 2\ f(CH_3) + 2\ f(CH_2) + 2\ f(CH) + 2\ f(OH) + 2\ f(COO) \\
&\quad + 2\ pe\ (1) + 1\ pe\ (2) \\
&= 1.40 + 1.05 + 0.47 + (-2.88) + (-2.56) + 1.60 + 0.46 = -0.46
\end{aligned}$

Observed value [24]: -0.29

7. <u>Trifluromethoxybenzene</u>: $CF_3-O-C_6H_5$

$\begin{aligned}
\text{Log } P &= f(CF_3) + f(O_{ar}) + f(C_6H_5) + 1\ pe\ (1) \\
&= 0.79 + (-0.45) + 1.90 + 0.80 = 3.04
\end{aligned}$

Observed value [24]: 3.17

8. <u>Trifluromethylthiobenzene</u>: $CF_3-S-C_6H_5$

$\begin{aligned}
\text{Log } P &= f(CF_3) + f(S_{ar}) + f(C_6H_5) + 1\ pe\ (1) \\
&= 0.79 + 0.11 + 1.90 + 0.80 = 3.60
\end{aligned}$

Observed value [19]: 3.57

The generally obtained agreement between the observed and estimated values is such that we feel that we have presented the correct method of calculation. Additional material is needed, however, especially if one takes into account the lipophilic behaviour of structures of the P type with X and Y representing negative groups. They need no more than 1 pe (2) correction to bring the calculated log P value into agreement with the experimental value. An example is dioxane; see Table III,1.

$$X\overset{\displaystyle CH_2-CH_2}{\underset{\displaystyle CH_2-CH_2}{\diagdown\diagup}}Y \quad (P)$$

REFERENCES

1 C. Hansch and S.M. Anderson, J. Org. Chem., 32(1967)2583.

2 S.S. Davis, J. Pharm. Pharmac., 25(1973).

3 S.S. Davis, J. Pharm. Pharmac., 25(1973)293.

4 G.G. Nys and R.F. Rekker, Chim. Thérap., 8(1973)521.

5 C. Hansch and A. Leo, Pomona College Medicinal Chemistry Project, Issue 6, January 1975.

6 R. Collander, Acta Chem. Scand., 5(1951)774.

7 S. Anderson and C. Hansch, unpublished analysis (see ref. [5]).

8 T. Fujita, unpublished analysis (see ref. [5]).

9 K.B. Sandell, Naturwissenschaften, 49(1962)12.

10 W.J. Dunn III and C. Hansch, unpublished analysis (see ref. [5]).

11 W.R. Glave and C. Hansch, unpublished analysis (see ref. [5]).

12 C. Church and C. Hansch, unpublished analysis (see ref. [5]).

13 S. Anderson, unpublished analysis (see ref. [5]).

14 T. Fujita, J. Iwasa and C. Hansch, J. Amer. Chem. Soc.,
    86(1964)5175.

15 M. Tichy and K. Bocek, communicated to Hansch (see ref. [5]).

16 J. Iwasa, T. Fujita and C. Hansch, J. Med. Chem., 8(1965)150.

17 V. Lee, M.S. Thesis, San Jose State College, 1967.

18 M. Gorin and C. Hansch, unpublished analysis (see ref. [5]).

19 D. Nikaitani and C. Hansch, unpublished analysis (see ref. [5]).

20 G.G. Nys and R.F. Rekker, unpublished analysis.

21 T. Bultsma, Free University Amsterdam, unpublished analysis.

22 M. Tute, unpublished analysis (see ref. [5]).

23 K.S. Rogers and A. Cammarata, J. Med. Chem., 12(1969)692.

24 C. Church, unpublished analysis (see ref. [5]).

25 K. Bowden and R.C. Young, J. Med. Chem., 13(1970)225.

26 D. Soderberg and C. Hansch, unpublished analysis (see ref. [5]).

27 C.E. Lough, Suffield Memorandum No. 17/71.

28 D. Mackintosh and C. Hansch, unpublished analysis (see ref. [5]).

29 E.O. Dillingham, R.W. Mast, G.E. Bass and J. Autian,
    J. Pharm. Sc. 62(1973)22.

30 N. Kurihara and T. Fujita, Pest. Biochem. and Physiol., 2(1973)383.

31 C. Hansch, A.R. Steward, S.M. Anderson and D.L. Bentley,
    J. Med. Chem., 11(1967)1.

32 A. Vittoria, C. Silipo and C. Hansch, unpublished analysis
    (see ref. [5]).

33 F. Helmer, unpublished analysis (see ref. [5]).

34 R. Kerely and C. Hansch, unpublished analysis (see ref. [5]).

35 J. Clayton and C. Hansch, unpublished analysis (see ref. [5]).

36 P. Jow and C. Hansch, unpublished analysis (see ref. [5]).

37 R.F. Rekker and H.M. de Kort, unpublished analysis.

38 E.J. Lien, J. Med. Chem., 13(1970)1189.

39 G.G. Nys and R.F. Rekker, Eur. J. Med. Chem., 9(1974)361.

    R.F. Rekker and G.G. Nys, Relations Structure – Activité, Séminaire
    Paris – 25/26 Mars 1974; Edité par la Société de Chimie Thérapeu-
    tique, Paris 1975.

40 S. Roth and P. Seeman, Biochem. Biophys. Acta, 255(1972)207.

41 O.E. Schultz, C. Jung and K.E. Moeller, Z. Naturforsch.,
    25(1970)1024.

42 T. Fujita and C. Hansch, unpublished analysis (see ref. [5]).

43 Y. Ichikawa, T. Yamano and H. Fujishima, Biochem. Biophys. Acta,
    171(1969)32.

44 H. Benthe (see ref. [5]).

45 E. Coats and C. Hansch, unpublished analysis (see ref. [5]).

46 M. Tute, communicated to Hansch (see ref. [5]).

47 K.H. Kim and C. Hansch, unpublished analysis (see ref. [5]).

48 J.D. Chapman, J.A. Raleigh, I. Borsa, R.G. Webb and R. Whitehouse,
    Int. J. Radiat. Biol., 17(1972)475.

49 N. Kakeya, N. Yata, A. Kamada and M. Aoki, Chem. Pharm. Bull.,
    17(1969)2558.

50 K. Kamishita, communicated to Hansch (see ref. [5]).

51 E. Kutter, unpublished results (see ref. [5]).

52 C.G. swain and E.C. Lupton, J. Amer. Chem. Soc., 90(1968)4328.

53 C. Hansch, A. Leo, S.H. Unger, K.H. Kim, D. Nikaitani and E.J. Lien,
    J. Med. Chem., 16(1973)1207.

54 C. Hansch, in E.J. Ariëns (Editor), Drug Design, Vol. I, Academic
    Press, New York and London, 1971.

55 M.S. Tute, in N.J. Harper and A.B. Simmonds (Editors), Advances in
    Drug Research, Vol. VI, Academic Press, London and New York, 1971.

56 P.J. Gehring, T.R. Torkelson and F. Oyen, Toxic. and Appl. Phar-
    mac., 11(1967)361.

57 C. Hansch and T. Fujita, J. Amer. Chem. Soc., 86(1964)1616.

58 I. Higbee, communicated to Hansch (see ref. [5]).

59 W.B. Neely, unpublished results (see ref. [5]).

60 J. Tollenaere, communicated to Hansch (see ref. [5]).

61 P. Harrass, Arch. Int. Pharmacol. & Therap. 11(1903)431.

62 J.P. Liberti and K.S. Rogers, Biochem. Biophys. Acta,
    222(1970)90.

63 R.F. Rekker, H.M. de Kort and W.F.C. Vlaardingerbroek,
    unpublished results.

64 K. Buchel and W. Draber, Progr. in Photosynth. Rev., 3(1969)1777.

65 R.P. Mariella and R.R. Raube, J. Amer. Chem. Soc., 74(1952)521.

66 R.F. Rekker and W.Th. Nauta, Rec. Trav. chim., Pays Bas,
    73(1954)969.

67 H.E. Ungnade, J. Amer. Chem. Soc., 75(1953)432.

68 A. Cerniani, R. Passerini and G. Righi, Bull. sci. chim. ind.
    Bologna, 12(1954)114.

69 K. Bowden and E.A. Braude, J. Chem. Soc., (1952)1068.

70 L. Láng (Editor), Absorption Spectra in the Ultraviolet and Visi-
    ble Region, Vol. IV. Publishing House of the Hungarian Academy of

Sciences, Budapest.

71 R. Huisgen and H.J. Koch, Liebigs Ann. Chem., 591(1955)200.

72 W.A. Schroeder, P.E. Wilcox, K.N. Trueblood and A.O. Dekker,
    Anal. Chem., 23(1951)1740.

73 L. Pauling and J. Sherman, J. Chem. Phys., 1(1933)606.

74 W.J. Dunn III, communicated to Hansch (see ref. [5]).

75 M.P. Breen, E.M. Bojanowski, R.J. Cipolle, W.J. Dunn III, E. Frank
    and J.E. Gearien, J. Pharm. Sci., 62(1973)847.

76 E.J. Lien, unpublished analysis (see ref. [5]).

77 J. Fastrez and A.R. Fersht, Biochem., 12(1973)1067.

78 C. Silipo and C. Hansch, unpublished analysis (see ref. [5]).

79 P.P. Maloney and C. Hansch, unpublished analysis (see ref. [5]).

80 K. Kistao, K. Kubo, T. Mori,
    Chem. Pharm. Bull. Jap., 21(1973)2417.

81 T. Morishita, M. Yamazaki, N. Yata and A. Kamada, Chem. Pharm.
    Bull. Jap., 21(1973)2309.

82 H. Whitehouse and J. Leader, Biochem. Pharmacol., 16(1967)537.

CHAPTER IV

A "MAGIC" CONSTANT CONNECTING ALL (?) PARTITION VALUES

A FREQUENTLY OCCURRING CONSTANT IN THE LIPOPHILICITY STUDIES
DESCRIBED SO FAR

In the calculations of the f constants as described in Chapter III, it sometimes appeared necessary to use correction terms. When properly listed, all of these corrections appear to share a greatest common divisor of 0.28. They are collected in Table IV,1 and a short survey is added of the main reasons that led to their introduction.

TABLE IV,1

THE CONSTANT 0.28

source	value
proximity effect	
(2c)	2 × 0.23
(1c)	3 × 0.27
H attached to a negative group	Δ = 0.27
condensed aromatic units	0.31
Ar − Ar conjugation	0.28
cross-conjugation	0.28
aromatic − aliphatic differences	3 × 0.28
incorrectly presumed folding	2 × 0.29^5
abnormal proximity correction for	
phenoxyacetic acid	Δ = 2 × 0.23

Proximity Effect

The reason for introducing the proximity effect was to correct the excessive log P of those structures in which two electronegative groups were separated by one or two carbon atoms. In the former instance, the proximity effect is 0.80 (standard deviation 0.08), and in the latter instance 0.46 (standard deviation 0.08). These figures can be interpreted as 3 × 0.27 and 2 × 0.23, respectively, and in the light of Table IV,1, within the limits of accuracy, as three times and double the same

Refs. pg 131

amount, respectively.

H attached to a negative Group

The hydrophobic fragmental constant of H always appears to have the
same value of approx. 0.20, no matter whether it is aliphatic or aromatic in origin. However, when H is attached to a group that has an obviously electronegative character, it is necessary to use a value of
0.47, which implies an increase of 0.27 over a "normal" H.

Condensed aromatic Units

During the study of the lipophilic behaviour of condensed aromatic
systems, such as naphthalene and quinoline, a reasonable agreement between observed and estimated partition values was not possible unless,
for each pair of common carbon atoms in these systems, an extra lipophilicity of 0.31 was taken into account, which implied that the common
carbon atoms of the condensed system had been assigned double lipophilicity over a "normal" aromatic carbon atom.

Ar – Ar Conjugation

In structures such as biphenyl, in which two aromatic systems are
conjugated, a correction for enhanced lipophilicity is also needed. In
this instance the correction is 0.28. As with the condensed systems
mentioned above, the two through-connected carbon atoms of the Ar – Ar
system account for this phenomenon. Although no lipophilicity measure-

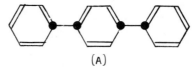

(A)

ments are available as controls, a correction of 2 × 0.28 may be proposed for a structure such as terphenyl (A). The following partition
coefficient can thus be predicted:

$$\text{Log } P = 2 \ f(C_6H_5) + f(C_6H_4) + (2 \times 0.28) = 6.08$$

Cross-Conjugation

In order to reconcile experiment with calculation, an extra lipophilicity of 0.28 should be included for each cross-conjugation. Again,
the carbon atoms, which provide the through-conjugation via a transmissive link, are thought to be responsible.

Aromatic – aliphatic Differences

While the standard deviation is not inconsiderable (0.22), the difference between f(ar) and f(al), which was calculated as 0.84 for 21
functional groups, should, no doubt, be regarded as three times the value 0.28. The problem of these aromatic – aliphatic differences will be
discussed at length in the next section.

Incorrectly presumed Folding

In Chapter III, the possibility of a folding that ultimately results
in decreased lipophilicity was dismissed and the difference in partition values resulting from $\Sigma \pi$ and Σf calculations was determined to
be 0.59 ± 0.14. This result is, in fact, approximately double the value of 0.28.

Abnormal Proximity Correction for Phenoxyacetic Acid
and its Derivatives

Instead of the expected amount of 0.80, due to a 1C separation in
the group $O-CH_2-COOH$, the necessary correction appeared to be 0.34. The
difference between the expected and observed corrections is 2 × 0.23,
which does not differ much from 2 × 0.28.

The frequent recurrence of the value 0.28, or a multiple of it,
suggests that this figure could have major significance in the partitioning process.

CLOSER INSPECTION OF THE AROMATIC – ALIPHATIC DIFFERENCES BETWEEN
FUNCTIONAL GROUPS; THE "MAGIC" CONSTANT; THE "KEY NUMBER"

As already stated, the mean value of 0.84 for $\Delta f = f$(ar) – f(al) is
known with a standard deviation that is much too high. This prompted us
to subject these Δf values to scrutiny.

It is notable that the series can be divided into three sub-series:
eleven Δf values with no appreciable differences, the mean being 0.87
with a standard deviation of 0.07; six Δf values that are hardly different, the mean being 0.55 with a standard deviation of 0.09; and four Δf
values with a mean of 1.06 with a standard deviation of 0.04. The three
standard deviations are such that this sub-division of Δf values is certainly justified.

This sub-division is indicated by the headings "-1", Δf and "+1" in

TABLE IV,2

$f(ar) - f(al)$ DIFFERENCES

function	$f(al)$	$f(ar)$	"–1"	Δf	"+1"
NH_2	−1.38	−0.90	0.48 →	0.76	
NH	−1.86	−0.93		0.93	
N	−2.13	−1.06		0.79 ←	1.07
F	−0.51	0.41		0.92	
Cl	0.06	0.94		0.88	
Br	0.24	1.17		0.93	
I	0.59	1.46		0.87	
O	−1.54	−0.45		0.81 ←	1.09
OH	−1.44	−0.36		0.80 ←	1.08
COO	−1.28	−0.43		0.85	
OCH_2COOH	−1.21	−0.59	0.62 →	0.90	
CN	−1.13	−0.23		0.90	
$CONH_2$	−1.99	−1.12		0.87	
SH	0.00	0.62	0.62 →	0.90	
S	−0.51	0.11	0.62 →	0.90	
SO	−2.75	−2.05		0.70	
CO	−1.69	−0.87		0.82	
NO_2	−1.02	−0.08		0.94	
NH_2COO	−1.39	−0.90	0.49 →	0.77	
COOH	−1.00	0.00		0.72 ←	1.00
CF_3	0.79	1.25	0.46 →	0.74	

the last three columns of Table IV,2. It should be noted that the differences between $\Delta f - \Delta f"+1"$ and $\Delta f"-1" - \Delta f$, respectively, again agree with the value 0.28 within the range of deviation; in other words, <u>it appears as if the lipophilicity difference of 0.28 between an aliphatic and an aromatic structure whose functional groups are identical operates as a kind of quantifiable factor</u>. The lipophilicity difference between aromatic and aliphatic values in a given functional group then tends to be 3 × 0.28, but in some instances it may be one 0.28 unit higher or lower. Using a simple minimalization procedure for the values listed in Table IV,2, it was found that the series as a whole is consistent with a basic value of exactly 0.28: increasing $\Delta f"-1"$ and decreasing $\Delta f"+1"$ by 0.28 results in the lowest possible standard deviation of the combined Δf series: Δf(mean) = 0.84 (standard deviation 0.07).

The postulated concept of quantifiability will be further verified
in the following sections. For the number that indicates how many times
the factor 0.28 is operating in a given instance, the designation "key
number" is proposed, whereas the factor 0.28 itself will be denoted as
the "magic" constant, symbolized by c_M.

The proximity effect is apparently governed by a key number of 2 or
3, depending on the number of carbon atoms between the two negative
groups; for a correct description of f(ar) − f(al) differences, a key
number of 3 ± 1 is necessary; the incorrect postulation of folding must
be associated with the masking of two key numbers, while all of the remaining examples from Table IV,1 are connected with the operation of
one key number.

PARTITION VALUES OF ALIPHATIC AND AROMATIC STRUCTURES
ON A GENERAL DENOMINATOR

It will be clear that when the factor c_M is really as constant as
suggested in the previous sections, it is useful to look for a procedure that relates all structures considered so far, i.e., all aliphatic
and aromatic structures listed in Chapter III, to a general denominator.
In this connection, a regression analysis was carried out with most of
the partition values tabulated so far. The following were taken into
account:

(1) Fragmentation was confined to CH_3, CH_2, CH, C_6H_5, C_6H_4, C_6H_3
(the last three fragments representing a benzene ring that is substituted once, twice and three times, respectively), CH_2=CH, H, CF_3, C=O,
COO, COOH, OCH_2COOH, O, OH, F, Cl, Br, I, NH_2, NH, N, NO_2, SO_2NH_2, CON,
NHCOO, $C_3H_3N_2$ (imidazolyl) and C_5H_4N (pyridinyl).

(2) Used in an aliphatic structure, a fragment receives a key number of zero, and on application in an aromatic structure, the key number indicated in column 3 in Table IV,3 is allotted. Most of these key
numbers have simply been copied from Table IV,2.

(3) The factor c_M functions as an individual parameter (see No. 30
in Table IV,3).

(4) For the presence of proximity effects, hydrogen atoms attached
to negative groups, condensed aromatic units, Ar − Ar conjugations and
cross-conjugations, the appropriate key number is assigned according to
the indications in the last paragraph of the previous section.

Refs. pg 131

TABLE IV,3

COMPILATION OF f VALUES SOLVED BY MULTIPLE REGRESSION
ANALYSIS USING A FIXED SET OF KEY NUMBERS

No.	fragment	key number	f	stand. dev.	$\underline{t}$
1	CH_3	0	0.702	0.013	53.5
2	CH_2	0	0.530	0.005	103.
3	CH	0	0.235	0.017	14.0
4	C_6H_5	0	1.886	0.018	107.
5	C_6H_4	0	1.688	0.019	89.3
6	C_6H_3	0	1.431	0.042	34.4
7	$CH_2=CH$	0	0.935	0.030	31.4
8	H	0 or 1 [a]	0.175	0.016	11.1
9	CN	3	-1.066	0.034	-31.6
10	CF_3	2	0.757	0.030	24.9
11	C=O	3	-1.703	0.027	-62.4
12	COO	3	-1.292	0.032	-40.1
13	COOH	3	-0.954	0.021	-45.3
14	OCH_2COOH	2	-1.155	0.027	-42.8
15	O	4	-1.581	0.024	-65.9
16	OH	4	-1.491	0.022	-67.5
17	F	3	-0.462	0.029	-15.9
18	Cl	3	0.061	0.021	2.88
19	Br	3	0.270	0.027	10.1
20	I	3	0.587	0.032	18.2
21	NH_2	2	-1.428	0.024	-60.3
22	NH	3	-1.825	0.029	-62.2
23	N	4	-2.160	0.037	-58.0
24	NO_2	3	-0.939	0.024	-39.3
25	SO_2NH_2	b	-1.530	0.036	-43.0
26	CON	3 or 4 [c]	-2.894	0.040	-72.7
27	NHCOO	2 [d]	-1.943	0.022	-88.3
28	$C_3H_3N_2$ [e]	0	-0.119	0.040	-3.00
29	C_5H_4N [f]	0	0.526	0.028	18.6
30	c_M		0.287	0.005	56.6
31	p.e. (1C)	3			
32	p.e. (2C)	2			

TABLE IV,3 (continued)

n =	509 (+5)
r =	0.995
s =	0.107
F =	1,740
constant term =	-0.009

(a) Depending on whether H is attached to a "neutral" C (key number = 0) or to an electronegative centre, for instance N, C=O (key number = 1);

(b) Only $f(SO_2NH_2)_{ar}$ is known. The tabulated value -1.530 is used with key number = 0 in aromatic structures;

(c) Key number = 3 if an aromatic substituent is attached to C of CON (amides) and key number = 4 if the aromatic substituent is attached to N of CON (anilides);

(d) If relevant, with the aromatic substituent attached to O of NHCOO; for structures with an aromatic unit attached to N, the key number 4 is recommended.

(e) Imidazolyl group;

(f) Pyridinyl group.

Note: The tabulated key numbers are applicable when the substituent is used on an aromatic system. In aliphatic substitutions, all key numbers are zero.

(5) On the introduction of condensed aromatic systems and their derivatives, two treatments are possible. The simplest is to consider the aromatic rings in terms of separate parameters. In that event, however, it would be impossible to pay due regard to the extra lipophilicity encountered in condensed aromatic systems, as one key number per condensation would be masked and hence not be related to the general c_M value. It was therefore decided to follow a different procedure and to fragment these condensed structures into the items discussed under (1), allowance being made for the inevitable consequence of broken H and C_6H_5 numbers, etc. Some practical parameters are as follows:

(a) Naphthalene:

$$\text{Log P} = 1.667\, f(C_6H_5) - 0.333\, f(H) + 1\, c_M$$

(b) Naphthalenyl:

$$f = 1.667\, f(C_6H_5) - 1.333\, f(H) + 1\, c_M$$

(c) Quinolinyl:

$$f = 1\, f(C_5H_4N) + 0.667\, f(C_6H_5) - 1.333\, f(H) + 1\, c_M$$

After completion of the regression analysis, the condensed fragments (b and c) can be re-composed from their constituent parts and used, if desired, as secondary f values.

Refs. pg 131

(6) The hydrogen atoms emerging from the procedure recommended under (5) are introduced into the regression analysis with a key number of zero.

(7) The SO_2NH_2 group does not occur as an aliphatic substituent in the available collection of partition values, so that the aromatic SO_2NH_2 fragment had to be included in the regression analysis with a key number of zero.

(8) For each structure, the total of key numbers is introduced as an independent parameter.

Example: 3-methoxybenzamide (B)

$$\text{Log P} = f(C_6H_4) + f(CH_3) + f(CON) + f(O) + 2f(H)$$
$$+ \left[kn(O) + kn(CON) + 2\ kn(H) \right] c_M$$

which implies a total key number of 9.

In order that there may be no misunderstanding, it should be noted that the constant c_M is <u>not introduced as a numerical value but as its total frequency of occurrence</u> in a given structure. The real numerical value of c_M will therefore appear as a regressor in the computation.

The analysis has a two-fold purpose: firstly, it should answer the question of whether the factor c_M is really as constant as claimed; this should be evidenced by a sufficiently low standard deviation and, in addition, by the requirement that the various fragmental constants remain satisfactory, i.e., do not undergo statistical deterioration relative to the regressions listed in Chapter III; and secondly, it should obviate the need for a number of secondary fragmental constants (of the sixteen secondary aliphatic and nine secondary aromatic fragmental constants, no more than five and four, respectively, have remained).

The results recorded in Table IV,3 make it sufficiently clear how successful the operation has been. The constant c_M appears with a value of 0.287 and a standard deviation of 0.005, and it is plain that the factor c_M is operating with an astonishing constancy in a variety of aspects connected with the phenomenon of lipophilicity.

In conclusion, the calculation of some log P and f values, instanced above, are presented below:

$$\text{Log P (3-methoxybenzamide)} = 1.688 + 0.702 + (-2.894) + (-1.581)$$
$$+ (2 \times 0.175) + (9 \times 0.287) = 0.85$$

The experiment afforded a value of 0.94 [1], while in Chapter III a value of 0.81 was calculated.

$$\text{Log P (naphthalene)} = (1.667 \times 1.886) - (0.333 \times 0.175) + 0.287$$
$$= 3.37$$

The observed values are: 3.37 [2] and 3.45 [3], while in Chapter III a value of 3.37 was calculated.

$$f \text{(naphthalenyl)} = (1.667 \times 1.886) - (1.333 \times 0.175) + 0.287 = 3.20$$

The value calculated in Chapter III was 3.17.

$$f \text{(quinolinyl)} = 0.526 + (0.667 \times 1.886) - (1.333 \times 0.175) + 0.287$$
$$= 1.84$$

The value calculated in Chapter III was 1.81.

EQUIDISTANT ALIGNMENT OF PARTITION VALUES; THE "NEAREST HOLE" THEOREM; LIPOPHILICITY OF INERT GASES

Once the constant c_M was adopted, we considered what could be at its basis, and the question arose of whether it could not be assigned a role even more frequently. In addition, accumulated evidence suggests that there must be some regular pattern in our series of f constants. We wish to demonstrate this with the f values of the halogens, and refer to Table IV,4, where their f values are given.

On inspection of the column of Δf values, where the differences between two consecutive halogens are listed, it can be seen that the aliphatic f values do not present an encouraging picture as the values 0.57, 0.18 and 0.35 do not seem to be related. However, when the aromatic f values and the "combined al/ar f values" from the previous section are also taken into consideration, the solution is obvious provided that Δf(F/Cl) is given a double interval compared with the remaining Δf values. In this fashion, the mean Δf 0.275 (for A) and 0.26 (for B and C) are obtained, which are values that, within the limits of accuracy, are identical with the magic constant c_M. The best fit, i.e., that with the lowest standard deviation, is to be found in the last column of Table IV,4. The f values given here show excellent agreement with the f values in column 2, except perhaps for those of Br, which deviate somewhat *.

TABLE IV,4

INDICATIONS OF EQUIDISTANCY IN HALOGEN LIPOPHILICITY

halogen	f	st.dev.	Δf	f after alignment	

A. Aliphatic structures (Table III,7)

halogen	f	st.dev.	Δf	f after alignment	
F	−0.51	a		−0.52	
			0.57		
Cl	0.06	a		0.03	
			0.18		$\Delta f = 0.27^5$
Br	0.24	a		0.30	
			0.35		
I	0.59	a		0.58	

B. Aromatic structures (Table III,10)

halogen	f	st.dev.	Δf	f after alignment	
F	0.412	0.024		0.41	
			0.53		
Cl	0.943	0.016		0.93	
			0.23		$\Delta f = 0.26$
Br	1.168	0.024		1.19	
			0.29		
I	1.460	0.028		1.45	

C. Al/Ar combined structures (Table IV,3)

halogen	f	st.dev.	Δf	f after alignment	
F	−0.462	0.029		−0.47	
			0.52		
Cl	0.061	0.021		0.05	
			0.21		$\Delta f = 0.26$
Br	0.270	0.027		0.31	
			0.32		
I	0.587	0.032		0.57	

[a] Secondary f values, no standard deviations determined

Other f values can be manipulated along similar lines and the magic constant appears to operate in all instances. For this reason, it was decided to treat all available f values as follows.

A pattern of constant intervals:

```
    ×    ×    ×    ×    ×    ×
    A    B    C    D    E    F  . . . . .
```

(AB = BC = CD = ...) was constructed with the locations of A, B, C, etc., being so chosen that our collection of f values (taken mainly from Table IV,3) achieves an optimal fit with the points A, B, C, etc., with minimum deviations. This optimal fit could be represented by the following equation:

* (previous page) The doubling of the F–Cl interval relative to the others is less surprising when one considers the van der Waals radii [4] of the halogens: F, 1.35 Å; Cl, 1.80 Å; Br, 1.95 Å; I, 2.15 Å. The difference between the radii of F and Cl is clear.

TABLE IV,5

EQUIDISTANT ALIGNMENT OF f AND log P VALUES

n	k	examples
-9	-3.12	
-8	-2.85	CON (-2.89)
-7	-2.59	
-6	-2.32	
-5	-2.06	N (-2.16) NHCOO (-1.94) SO_{ar} (-2.05)
-4	-1.80	CO (-1.70) NH (-1.83) $SO_{2,ar}$ (-1.87)
-3	-1.53	O (-1.58) OH (-1.49) NH_2 (-1.43) SO_2NH_2 (-1.53)
-2	-1.27	COO (-1.29)
-1	-1.00	CN (-1.07) COOH (-0.95) NO_2 (-0.94)
0	-0.74	
1	-0.48	F (-0.46) S_{al} (-0.51)
2	-0.21	
3	0.05	Cl (0.06) S_{ar} (0.11) SH_{al} (0.00)
4	0.32	Br (0.27)
5	0.58	I (0.59) C_5H_4N (0.53) SH_{ar} (0.62)
6	0.84	CF_3 (0.76) pyrrole (0.75)
7	1.11	
8	1.37	C_6H_3 (1.43)
9	1.64	C_6H_4 (1.69)
10	1.90	C_6H_5 (1.89) thiophene (1.81)
11	2.16	C_6H_6 (2.14) indole (2.13)
12	2.43	
13	2.69	toluene (2.71)
14	2.96	α-naphthol (2.98)
15	3.22	
16	3.48	naphthalene (3.43)
17	3.75	
18	4.01	biphenyl (4.04)
19	4.28	
20	4.54	anthracene (4.45) phenanthrene (4.46)

$$k = -0.740 \pm 0.264 \, n \qquad\qquad (IV-1)$$

where n = 0, 1, 2, 3 . . . . . The value -0.740 is arbitrary; it would
also have been possible to take, for example, -1.004 or -0.476 as the

Refs. pg 131

intercept value. In both regressors the standard deviation is ca. 0.025.

Table IV,5 shows the application of eqn. IV-1. In addition to primary fragmental constants, a supplementary number of secondary values from the tables in Chapter III are incorporated. Complete structures can also be accommodated with their partition values. Examples are benzene, toluene and biphenyl, and it can be increasingly postulated that many, if not all, lipophilicity values can be aligned this way, i.e., can be written as a multiple of the constant c_M and thus be considered as quantifiable values.

We wish to emphasize that the constant c_M, (perhaps owing to its magic nature?) should be treated with care. The accuracy with which a log P value can be determined in practice (and especially the relatively high and low partition values should be taken into consideration) is such that two log P values which, in effect, differ by 0.28 can actually fall into each other's range of accuracy; it is even possible that the measurements will show the higher of the two log P values to be the lower.

By comparing the log P values of structures alone, we would never have found the existence of the factor c_M. That we did succeed via the hydrophobic fragmental constants is connected with the considerably greater accuracy in determining these constants compared with the average accuracy of log P determinations; also this high accuracy is above all the result of the fairly drastic procedure applied for eliminating outliers from a log P data set. Thus, the standard deviation of all fragments listed in Table IV,3 can be averaged as 0.027. Such a small value implies that on combining *f* values to give a log P, the regular pattern that emerges from Table IV,5 will remain intact for a fairly long period: at least ten fragments can be put together before one needs to be afraid of the summed inaccuracies exceeding the 0.28 threshold. In this connection, we draw attention to Table III,11, which includes tri-substituted benzene derivatives. They are reproduced below and we have added those log P values from Table IV,5 that fit optimally.

1,3,5-trinitrobenzene	1.18(1.17);	"fit":1.11
1,3-dimethyl-2-nitrobenzene	2.95(2.73);	2.69
3,5-dimethylphenol	2.35(2.45);	2.43
2,6-dimethylphenol	2.36(2.45);	2.43
1,3,5-trimethylbenzene	3.42(3.50);	3.48
3-bromo-4-chlorophenoxyacetic acid	2.75(2.93);	2.96

3-chloro-5-fluorophenoxyacetic acid	2.20(2.17);	2.16
3-iodo-4-chlorophenoxyacetic acid	3.10(3.22);	3.22
3-nitro-4-chlorophenoxyacetic acid	1.85(1.68);	1.64
2,4-dichlorophenoxyacetic acid	2.81(2.70);	2.69
3,4-dichlorophenoxyacetic acid	2.81(2.70);	2.69
3-cyano-4-chlorophenoxyacetic acid	1.56(1.53);	1.64

There is striking agreement which, in our opinion, cannot be purely coincidental.

CH_3, CH_2, CH and C fragments were excluded from eqn. IV-1. When their f values are correlated with the set-up in Table IV,5, the most acceptable fit can be proposed as follows:

$$CH_3 \text{ with } = 0.70 \text{ on } 0.84$$
$$CH_2 \text{ with } = 0.53 \text{ on } 0.58$$
$$CH \text{ with } = 0.24 \text{ on } 0.32$$
$$C \text{ with } = 0.15 \text{ on } 0.05$$

This result is not particularly good. On the one hand, the series 0.84, 0.58, 0.32, 0.05 looks attractive in that the inconstancy of H, from which the f values of CH_3, CH_2, CH and C suffer and which was discussed in some detail in Chapter III, has disappeared. On the other hand, the values 0.84, 0.58, 0.32, 0.05, when used for the calculation of the par-tition value of an "average" structure, are definitely not correct. This is plain in the presence of branchings: "normal" f values provide $\Sigma f = 1.64$ for, e.g., an i. C_3H_7 group, whereas the alignments via Table III,5 would have afforded $\Sigma f = 2.00$ $[(2 \times 0.84) + 0.32]$.

We believe that the irregular pattern of the calculated f values of the hydrocarbon fragments is due to the smoothing effect of the regres-sion procedure on the introduced collection of partition values, which forces the fragments CH_3, CH_2, CH and C to fit with the others optimal-ly. In practice, the use of the values 0.70, 0.53, 0.24 and 0.15 will nearly always be associated with an "average" group of other f values and lead to correct log P values provided that this group is as "aver-age" as the original data set from which the f values were derived. When used more or less exclusively, the f set 0.70 . . . 0.15 may give rise to marked deviations because they lack correcting "satellites", as already appears from the log P calculation for methane:

$$\log P(CH_4) = f(CH_3) + f(H) = 0.702 + 0.175 = 0.88$$

The value obtained is 1.09 $[5]$, the difference being 0.21.

Refs. pg 131

With ethane, the result is even poorer:

$$\log P(CH_3-CH_3) \;=\; 2\,f(CH_3) = 2 \times 0.702 = 1.40$$

The experimental value is 1.81 [5], the difference being 0.41.

Fuller reference will be made to the partition coefficients of methane, ethane, etc., in a subsequent section.

In Chapter II, LANGMUIR's concept was considered, indicating that the major factor in solution and partition is the energy necessary to make a hole in the solvent in which a solute is to be accommodated [6]. It seems obvious to think of an action primarily originating from the solute that is accommodating itself, rather than of an accentuation of the solvent that provides the solute with accommodation. In the former instance, the solute volume is an important or even the chief factor in the dissolution or partitioning process; in the latter instance, solvent structuring and its possibilities regarding hole formation are essential.

The observed quantifiability of the partition coefficient has, however, prompted the question of whether, in achieving the partition equilibrium, solvent structuring would have to be given much more emphasis than solute volume.

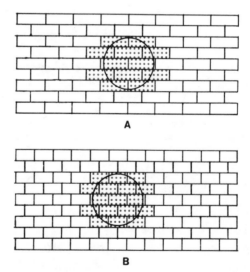

Fig.IV,1: The partition process visualized as the passage of a ball through a brick wall. The dotted bricks need removal to ensure free passage. A and B denote the application of two different brick sizes.

The model in Fig. IV,1 can assist in imagining the situation. For a ball to be pushed through a brick wall, more bricks have to be removed than actually correspond with the balldiameter, while the individual bricks in the masonry have to be considered as indivisible (quantifiable) units. A similar situation may be postulated in a solvent, and is certainly not alien to water, which FRANK and WEN [7] believe to be made up of flickering clusters of hydrogen-bonded molecules; and would it indeed, be so unusual when, for a hole that is released with the nearest fit to a solute looking for accommodation, only quantifiable small unit clusters are displaced from an "ice-like" structuring?

It should be recognized that the above considerations have by no means solved any part of the problem. They indicate the direction in which we started thinking about the problem. An additional complication is that partition cannot, in effect, be modelled by a single-hole procedure, as it is connected with a "hole to hole" transfer of the solute.

However, this line of thought can hardly be avoided, and becomes still more forcible on inspection of the lipophilicities of the inert gases (see Table IV,6).

TABLE IV,6

PARTITION VALUES OF INERT GASES [8]

inert gas	log P	log P (est)	
	(obs)	I	II
helium	0.28	0.32	0.32
neon	0.28	0.32	0.58
argon	0.74	0.84	0.84
krypton	0.89	0.84	1.11
xenon	1.28	1.37	1.37
radon	?	1.37	1.64

I According to the "nearest" fitting hole
II According to the "nearest" succeeding hole

With the inert gases, it is notable that the partition values of helium and neon are identical while those of neon and argon differ by 0.46, those of argon and krypton differ by only 0.15 and those of krypton and xenon differ by as much as 0.39. As the partition values refer to one series of measurements from one institute, there is no reason to question the reliability of the data.

The remarkable partitional behaviour outlined here can be more

Refs. pg 131

readily understood only in the light of the above nearest "hole" theorem[*]. The identity of the log P values of helium and neon then supposedly arises from the circumstance that the nearest "hole" for helium is, in effect, too large for this gas and even suitable for accommodating neon, which is much wider in diameter. We are even inclined to suggest that the "hole" created for accommodating helium and neon might be identical with the minimally removable unitcluster. By extension, it may be said that the argon molecule, which is much larger than the neon molecule, must show a rise in lipophilicity, and it may be that it misses the "hole" originally intended for neon, thus causing the gap in partition between neon and argon. As a consequence, it is likely that the log P values of argon and krypton may become identical; this appears to be true within the standard deviations accepted. The next log P, that of xenon is higher than that of krypton, just as the log P value of argon rises compared with that of neon ; in other words, xenon misses the "hole" intended for krypton. Fig. IV,2 gives a schematic visualization of the lipophilic features of the inert gases.

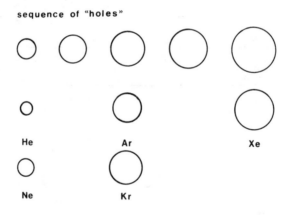

Fig.IV,2: Schematic presentation of the lipophilic behaviour of the inert gases.

The above considerations are illustrated by the data in the column log P(est) in Table IV,6. The values indicated by II are taken from Table IV,5; starting with 0.32, which represents the best fit with the log P value observed for helium (0.28), they extend to 1.37 and indicate

---

[*] Within this context, the term "hole" will be used to denote the essentially more complex "hole to hole" transfer of solute.

predictions when the inert gases would show a continuously increasing, although quantifiable, series of partition values. The data denoted by I indicate the nearest fits for the observed values, allowing for the discontinuously increasing partition values.

The partition coefficient of radon can now be predicted as 1.37.

APPLICATION OF f IN THE CALCULATION OF PARTITION VALUES OF "SMALL" MOLECULES

With the help of methane and ethane, it was demonstrated in the preceding section how an exclusive use of the f values of hydrocarbon fragments in the absence of any correcting "satellites" can lead to markedly deviating log P values. A significant amount of information, largely relating to "small" molecules, is collected in Table IV,7. This table presents the results of three different log P estimations:

(a) The calculation indicated by I utilizes "normal" f values, as represented in Table IV,3. On inspection of Table IV,7, unduly low estimates can be seen; e.g., in the series 1 - 14. Nos. 15 - 18, where the unit C=C features, are estimated correctly, and this is also true for all members following No. 20. They belong to combinations of CH_3 with a negative element or a negative group which function as correcting "satellites".

(b) Estimates denoted by II are based on f values taken from Table IV,5 plus the values 0.84, 0.58, 0.32 and 0.05 for CH_3, CH_2, CH and C, respectively, 0.84 for CH_2=CH and 0.32 for CH≡C. As for H, there was some difficulty. The value $f(H) = 0.175$, as revealed by the regression analysis of Table IV,3, should correspond to either 0.32 or 0.05 on the alignment given by Table IV,5. The value of our choice was 0.32, in spite of the larger difference with 0.175.

On the whole, the estimated values are in excellent agreement with the observed ones. Significant deviations were noted for tetramethylmethane (No. 10), −0.30, ethyne (No. 12), −0.27, butadiene (No. 18), +0.31, propyne (No. 19), −0.22, methyl acetate (No. 28), −0.23 and ethyl acetate (No. 35), −0.26 and −0.36. With one exception, there is always a decrease in the observed log P compared with the estimated value of ca. 0.27, i.e., by one key number, while in the exceptional case the log P is increased compared with the estimated value by one key number.

Refs. pg 131

TABLE IV,7

CALCULATION OF PARTITION VALUES IN A SERIES OF
"SMALL" MOLECULES

no.	compound	obs.	log P calc. I	calc. II	calc. III
1	H_2	0.45	0.35 (0.10)	0.64 (-0.19)	0.32 / 0.58 (0.13 / -0.13)
2	CH_4	1.09	0.88 (0.21)	1.16 (-0.07)	1.11 (-0.02)
3	CH_3-CH_3	1.81	1.40 (0.41)	1.68 (0.13)	1.64 (0.17)
4	CH_2-CH_2 \ / CH_2	1.72	1.59 (0.13)	1.74 (-0.02)	1.64 (0.08)
5	$CH_3-CH_2-CH_3$	2.36	1.93 (0.43)	2.26 (0.10)	2.43 (-0.07)
6	$CH_3-(CH_2)_2-CH_3$	2.89	2.46 (0.43)	2.84 (0.05)	2.96 (-0.07)
7	$CH_3-(CH_2)_3-CH_3$	3.39	2.99 (0.40)	3.42 (-0.03)	3.48 (-0.09)
8	CH_2-CH_2 / \ CH_2 CH_2 \ / CH_2-CH_2	3.44	3.18 (0.26)	3.48 (-0.04)	3.48 (-0.04)
9	$(CH_3)_2CH-CH_3$	2.76	2.34 (0.42)	2.84 (-0.08)	2.69 (0.07)
10	$C(CH_3)_4$	3.11	2.95 (0.16)	3.41 (-0.30)	3.22 (-0.11)
11	$CH_2=CH_2$	1.13	1.11 (0.02)	1.16 (-0.03)	1.11 (0.02)
12	$CH\equiv CH$	0.37	0.76 (-0.40)	0.64 (-0.27)	0.32 (0.05)
13	$CH_3-CH=CH_2$	1.77	1.64 (0.13)	1.68 (0.09)	1.64 (0.13)
14	$CH_3-CH_2-CH=CH_2$	2.40	2.17 (0.23)	2.26 (0.14)	2.43 (-0.03)
15	$(CH_3)_2C=CH_2$	2.34	2.16 (0.18)	2.34 (0.10)	2.43 (-0.09)
16	$CH_3-CH=CH-CH_3$ (cis)	2.33	2.16 (0.17)	2.34 (-0.01)	2.43 (-0.10)
17	$CH_3-CH=CH-CH_3$ (trans)	2.31	2.16 (0.15)	2.34 (-0.03)	2.43 (-0.12)
18	$CH_2=CH-CH=CH_2$	1.99	1.87 (0.12)	1.68 (0.31)	1.90 (0.09)

TABLE IV,7 (continued)

19	$CH_3-C\equiv CH$	0.94	1.28 (-0.34)	1.16 (-0.22)	0.84 (0.10)
20	CH_3-F	0.51	0.24 (0.27)	0.36 (0.15)	0.58 (-0.07)
21	CH_3-Cl	0.91	0.76 (0.15)	0.89 (0.02)	0.84 (0.07)
22	CH_3-Br	1.19	0.97 (0.22)	1.16 (0.03)	1.11 (0.08)
23	CH_3-I	1.51	1.29 (0.22)	1.42 (0.09)	1.37 (0.14)
		1.69	1.29 (0.40)	1.42 (0.27)	1.37 (0.32)
24	CH_3-OH	-0.64	-0.79 (0.15)	-0.69 (0.05)	-0.74 (0.10)
		-0.82	-0.79 (-0.03)	-0.69 (-0.13)	-0.74 (-0.08)
		-0.77	-0.79 (0.02)	-0.69 (0.08)	-0.74 (-0.03)
25	CH_3-NH_2	-0.57	-0.73 (0.16)	-0.69 (0.12)	-0.48 (-0.09)
26	CH_3-COOH	-0.17	-0.16 (-0.01)	-0.11 (-0.06)	-0.21 (0.04)
		-0.31	-0.16 (-0.15)	-0.11 (-0.20)	-0.21 (-0.10)
27	CH_3-O-CH_3	0.10	-0.17 (0.27)	0.15 (-0.05)	0.05 (0.05)
28	$CH_3-COO-CH_3$	0.18	0.11 (0.07)	0.41 (-0.23)	0.32 (-0.14)
29	$CH_3-CO-CH_3$	-0.24	-0.30 (0.06)	-0.12 (-0.12)	-0.21 (-0.03)
30	CH_3-CH_2-Cl	1.43	1.29 (0.14)	1.47 (-0.04)	1.37 (0.06)
31	CH_3-CH_2-Br	1.61	1.50 (0.11)	1.74 (-0.13)	1.64 (-0.03)
32	CH_3-CH_2-I	2.00	1.82 (0.18)	2.00 (0.00)	1.90 (0.10)
33	CH_3-CH_2-OH	-0.31	-0.26 (-0.05)	-0.11 (-0.20)	-0.21 (-0.10)
		-0.32	-0.26 (-0.06)	-0.11 (-0.21)	-0.21 (-0.11)
34	$CH_3-CH_2-NH_2$	-0.13	-0.15 (0.02)	-0.11 (-0.02)	-0.21 (0.08)
35	$CH_3-COO-CH_2-CH_3$	0.73	0.65 (0.08)	0.99 (-0.26)	0.84 (-0.11)

Refs. pg 131

TABLE IV,7 (continued)

0.63	0.65	0.99	0.58
	(-0.02)	(-0.36)	(0.05)

Note: Values in parenthesis are residuals

(c) Estimates denoted by III are taken directly from Table IV,5 and are, therefore, alignments with optimal fits for log P(obs). Apart from the deviating structures mentioned under (b), they agree well with the log P estimates denoted by II.

We would like to stress, that the f values from our collections (cf., those from Table IV,3) cannot be applied in instances where fragments occur without correcting "satellites" (such as CH_2 in cycloalkanes). In such case one should take recourse to the data in Table IV,5. In combinations with a correctly chosen partner, this drawback is largely removed and, therefore, with small molecules of the type CH_3-X, CH_3-CH_2-X or CH_3-X-CH_3, where X represents a negative unit such as NH_2 or halogen, our log P calculations based on f values produce acceptable results.

LIPOPHILIC BEHAVIOUR OF SOME COMPLEX FUNCTIONAL GROUPS

When discussing the principles of the fragmentation of structures in the first section of Chapter III, we pointed out that in these fragmentations a group such as COO had to remain intact and must not be split up into C=O and O.

It may be said that any complex functional group in which two or more negative structural portions are directly connected needs to be considered in its entirety in the fragmentation procedure, as increases in the partition value, even higher than those observed for 1C separation, could be expected. Once the proximity effects were recognized as double and triple the constant c_M and, more particularly, after the discovery of the far-reaching consequences of this constant (the nearest "hole" theorem), the question arose as to what extent the difference $f(XY) - [f(X) + f(Y)]$, where X and Y represent negative groups,

(a) would show constancy, as was observed for the proximity effects p.e. (1) and p.e. (2);

(b) could be likewise expressed as a multiple of the magic constant and thus be written as

TABLE IV,8

LIPOPHILICITIES OF SOME COMPLEX FUNCTIONAL GROUPS

functional group		via f addition	lipophilicity obs.	Δ	kn
$-\overset{\text{O}}{\underset{\text{\textbardbl}}{C}}-OH$	$-\overset{\text{O}}{\underset{\text{\textbardbl}}{C}}-$ $-OH$	$\left.\begin{array}{l}-1.70\\[1em]-1.49\end{array}\right\}$ -3.19	-0.95	2.24	8 (0.01)
$-\overset{\text{O}}{\underset{\text{\textbardbl}}{C}}-O-$	$-\overset{\text{O}}{\underset{\text{\textbardbl}}{C}}-$ $-O-$	$\left.\begin{array}{l}-1.70\\[1em]-1.58\end{array}\right\}$ -3.28	-1.29	1.99	7 (0.04)
$-\overset{\text{O}}{\underset{\text{\textbardbl}}{C}}-NH_2$	$-\overset{\text{O}}{\underset{\text{\textbardbl}}{C}}-$ $-NH_2$	$\left.\begin{array}{l}-1.70\\[1em]-1.43\end{array}\right\}$ -3.13	-2.03	1.10	4 (-0.02)
$NH_2-\overset{\text{O}}{\underset{\text{\textbardbl}}{C}}-O-$	$-NH_2$ $-\overset{\text{O}}{\underset{\text{\textbardbl}}{C}}-$ $-O-$	$\left.\begin{array}{l}-1.43\\-1.70\\-1.58\end{array}\right\}$ -4.71	-1.48	3.23	12 (-0.12)
$NH_2-\overset{\text{O}}{\underset{\text{\textbardbl}}{C}}-NH-$	$-NH_2$ $-\overset{\text{O}}{\underset{\text{\textbardbl}}{C}}-$ $-NH-$	$\left.\begin{array}{l}-1.43\\-1.70\\-1.83\end{array}\right\}$ -4.96	-1.83	3.13	11 (0.05)
$-\overset{\text{O}}{\underset{\text{\textbardbl}}{C}}-NH-\overset{\text{O}}{\underset{\text{\textbardbl}}{C}}-NH_2$	$2 \times -\overset{\text{O}}{\underset{\text{\textbardbl}}{C}}-$ $-NH-$ $-NH_2$	$\left.\begin{array}{l}-3.40\\-1.83\\-1.43\end{array}\right\}$ -6.66	-1.67	4.99	18 (-0.03)
$-\overset{\text{O}}{\underset{\text{\textbardbl}}{C}}-NH-\overset{\text{O}}{\underset{\text{\textbardbl}}{C}}-NH-\overset{\text{O}}{\underset{\text{\textbardbl}}{C}}-$	$3 \times -\overset{\text{O}}{\underset{\text{\textbardbl}}{C}}-$ $2 \times -NH-$	$\left.\begin{array}{l}-5.10\\[1em]-3.66\end{array}\right\}$ -8.76	-2.10	6.66	24 (-0.04)
	same	-8.76	-2.81	5.95	21 (0.09)
	$-O-$ $2 \times -\overset{\text{O}}{\underset{\text{\textbardbl}}{C}}-$ $-NH-$	$\left.\begin{array}{l}-1.58\\-3.40\\-1.83\end{array}\right\}$ -6.81	-2.26	4.55	16 (0.09)

$$f(XY) - \left[f(X) + f(Y)\right] = kn \times c_M \qquad (IV-2)$$

The results of this examination are presented in Table IV,8. The f values of nine complex functional groups were calculated by addition of the f values of their constituent parts and compared with the observed values as listed in Table IV,3 or with the partition values of a few selected structures (for details on the latter structures see Part II).

The differences between the two f values (Δ) are given in the last but one column of Table IV,8. It appears that they can be expressed as $kn \times c_M$, using the value 0.28 for c_M. The last column shows these kn values (key numbers) and, in parenthesis, the differences $\Delta - (kn \times c_M)$. Except for the slightly deviating group $NH_2-CO-O-$, the differences have strikingly low values and thus the significance of the function of the constant c_M in producing the lipophilic behaviour of a complex functional group is plain. Attention may also be drawn to the remarkable difference between an open diacylurea structure and a cyclic analogue, as seen in the barbiturates. It should be noted here that prior to establishing the value $f = -2.81$, the correction for p.e. (1) was applied, so that the difference of three key numbers between an open and a closed form of the diacylurea pattern is accounted for entirely by the changes brought into play by the cyclization (alterations in the hydratation pattern?; changes in the molecular shape, resulting in the missing of three "holes"?).

CALCULATION OF PARTITION COEEFICIENTS WITH f VALUES DERIVED FROM "SMALL" MOLECULES

Recently, LEO et al. proposed a method for calculating partition values that deviates somewhat from the design outlined above. It involves the use of f values derived from the partition coefficients of a series of small molecules and the application of the appropriate corrections to reconcile the Σf values with the measured log P values of larger molecules [5].

Table IV,9 surveys the most important modified fragmental constants. The f values in the upper part of this table were obtained by the use of the partition values of H_2, CH_4 and CH_3-CH_3:

$$f(H) = \tfrac{1}{2} \log P(H_2) = \tfrac{1}{2} \times 0.45 = 0.23$$

TABLE IV,9

LEO et al.'s MODIFIED FRAGMENTAL CONSTANTS

fragment	f	
H	0.23	
CH_3	0.89	
CH_2	0.66	
CH	0.43	
C	0.20	
b (single bond in chain)	−0.12	
b̲ (single bond in rings)	−0.09	
cbr (chain branching)	−0.13	
gbr (group branching)	−0.22	
F	−0.38*	0.37*
Cl	0.06	0.94
Br	0.20	1.09
I	0.60	1.35
OH	−1.64	−0.40
COOH	−1.09	−0.03
COO	−1.49	−0.56
O	−1.81	−0.57
S	−0.79	0.03
NH_2	−1.54	−1.00
NH	−2.11	−1.03
N	−2.16	−1.17
NO_2	−1.26	−0.02
CN	−1.28	−0.34
CO	−1.90	−0.32
$CONH_2$	−2.18	−1.26
C_6H_5	1.90	

* Left hand column: aliphatic constants
  Right hand column: aromatic constants

$$f(CH_3) = \tfrac{1}{2}\left[\tfrac{1}{2}\log P(CH_3-CH_3) + (\log P(CH_4) - f(H)\right]$$
$$= \tfrac{1}{2}\left[0.91 + 0.86\right] = 0.89$$

$$f(CH_2) = f(CH_3) - f(H) = 0.66$$

$$f(CH) = f(CH_2) - f(H) = 0.43$$

Refs. pg 131

$$f(C) = f(CH) - f(H) = 0.20$$

On application of these f values in the calculation of log P values of simple carbon chains or rings, the first corrections are found. They are included in the middle part of Table IV,9, and their use can be seen in the following examples:

(1) n. Butane

Log P $= 2\ f(CH_3) + 2\ f(CH_2) + 2\ f_b$

$= 1.78 + 1.32 + (-0.24) = 2.86$

Observed value [5]: 2.89.

(2) Cyclopentane

Log P $= 5\ f(CH_2) + 4\ f_\underline{b}$

$= 3.30 + (-0.36) = 2.94$

Observed value [5]: 3.00.

(3) Isobutane

Log P $= 3\ f(CH_3) + f(CH)\ 2\ f_b + f_{cbr}$

$= 2.67 + 0.43 + (-0.24) + (-0.13) = 2.73$

Observed value [5]: 2.76.

It appears that acceptable results cannot be obtained unless the number of f_b and $f_\underline{b}$ corrections (representing single-bond corrections between fragments in chains and rings, respectively) is always lower by one than the number demanded on structural grounds. LEO et al. assumed that "every single bond after the first makes a negative contribution to hydrophobicity (perhaps a volume reduction through flexibility)".

In the above evaluation of the modified f system, the authors started from H_2 and CH_4 and the appearance of the corrections indicated by f_b, $f_\underline{b}$ and f_{cbr} is strongly connected with the wish to arrive at a uniformly constant $f(H)$ value. This was achieved by deriving $f(CH_2)$, $f(CH)$ and $f(C)$ from $f(CH_3)$ by simple substraction of 1, 2 and 3 H values, respectively. An alternative route would have been to start with propane, butane and pentane and to derive first $f(CH_2)$ (0.53), than $f(CH_3)$ (0.91) and finally to end with isobutane, neopentane, methane and H_2. This procedure would have yielded $f(CH) = 0.03$, $f(C) = -0.53$ and variable H values ranging from 0.18 to 0.56.

The f values in the lower part of Table IV,9 are mainly determined from small molecules by means of f values and corrections mentioned above. The derivation of the aromatic f series is based on $f(C_6H_5)$, which can be obtained as follows:

$$f(C_6H_5) \;=\; \log P(C_6H_6) - f(H) \;=\; 2.13 - 0.23 = 1.90$$

In its current form, the modified f system does not allow for some corrections that, in our opinion, do have significance. The Ar — Ar conjugation correction is a good example. LEO et al. claimed that "the most direct way to obtain $f(C_6H_5)$ for an 'undisturbed' phenyl is from $\frac{1}{2}$ log P $(C_6H_5-C_6H_5) = \frac{1}{2} \times 4.04 = 2.02$". But, as any log P of more than 4 units is difficult to establish with sufficient experimental accuracy, they decided to take the value of 1.90, as derived above, for $f(C_6H_5)$.

By doing so, it is assumed a priori that Ar — Ar conjugation has no effect on partitional behaviour. It may be true that the determination of log P values above 4 presents some difficulties, but it is a fact that not less than five partition coefficients of biphenyl are available : 3.95 [9], 4.17 [10], 4.09 [11], 3.16 [12] and 4.04 [3]. The outlying value of 3.16 belongs to a series of measurements which includes benzene with log P = 1.56 instead of 2.13. This justifies the removal of 3.16 from the above collection and it further permits the conclusion that the real log P value of biphenyl is probably expressed more appropriately by our calculation, resulting in log P = 4.08, than by a value of 3.80, which is the consequence of the concept of LEO et al.

The modified f system cannot be judged conclusively until more material becomes available.

REFERENCES

1   M. Tute, communicated to Hansch (see ref. [13]).

2   C. Hansch and T. Fujita, J. Amer. Chem. Soc., 86(1964) 1616.

3   M. Gorin and C. Hansch, unpublished analysis (see ref. [13]).

4   Handbook of Chemistry and Physics (Editor R.C. West) 49th Ed.
    1968, The Chemical Rubber Comp., Ohio.

5   A. Leo, P.Y.C. Jow, C. Silipo and C. Hansch, J. Med. Chem.,
    18(1975) 865.

6   I. Langmuir, in H.N. Holmes (Editor), Colloid Symposium
    Monograph III, Chemical Catalog Co., 1925.

7   H.S. Frank and W.Y. Wen, Discuss. Faraday Soc., 24(1957) 133.

8   C. Hansch, in Relations Structure — Activité, Séminaire Paris,
    25-26 Mars 1974, Edité par la Société de Chimie Thérapeutique.

9   J. Tollenaere, communicated to Hansch (see ref. [13]).

10 W.B. Neely, unpublished results (see ref. [13]).

11 C. Church, unpublished results (see ref. [13]).

12 K. Rogers and A. Cammarata, J. Med. Chem., 12(1969)692.

13 A. Leo, C. Hansch and D. Elkins, Chem. Rev., 71(1971)525.
    See also the more extended listing of log P values published
    by C. Hansch, Pomona College, Claremont, Cal.

CHAPTER V

SOME SPECIAL APPLICATIONS OF THE HYDROPHOBIC FRAGMENTAL CONSTANT

STERIC FACTORS AND LIPOPHILICITY

Correlations between E_s and Lipophilicity

The partition coefficient is, as indicated in Chapter II, dependent on the parachor. As the latter should be regarded as a volume-related quantity (actually it is the molecular volume of a liquid with a surface tension equal to unity), a correlation between log P or f and certain steric parameters is clearly not out of question.

In eqn. V-1, the steric TAFT parameters, E_s [1], of six substituents, namely H (1.24), CH_3 (0.00), $\underline{i}.C_3H_7$ (-0.47), $\underline{tert}.C_4H_9$ (-1.54), C_6H_5 (-2.58) and $C(CH_2CH_3)_3$ (-3.8), are correlated with their f values:

$$E_s = -1.34\,f + 1.16 \qquad\qquad (V-1)$$
$$r = 0.936; \quad s = 0.72$$

The parameter E_s provides a measure of the width of the substituent at the periphery of its site of attachment to the parent structure, and as the six substituents form an ascending series as regards their widths, a value of r of 0.936 is not surprising.

The C_6H_5 group combines a relatively large E_s value with a moderate f value. By omitting this group, the correlation is improved considerably:

$$E_s = -0.91\,f + 1.73 \qquad\qquad (V-2)$$
$$r = 0.988; \quad s = 0.33$$

Deterioration occurs on introduction of an $\underline{n}$. butyl group (which combines a relatively low steric factor, $E_s = -0.39$, with a high f value) or an OH group (which has a steric factor of 0.69 combined with a negative f value). The r values change to 0.943 and 0.923, respectively, and the accompanying s values are 0.63 and 0.70.

The substituent series H, F, Cl, Br and I, with E_s = 1.24, 0.78, 0.27, 0.08 and -0.16, respectively, also show a clear E_s - f correlation; cf., eqn. V-3:

Refs. pg 173

$$E_s = -0.72\,f + 0.35 \qquad (V-3)$$
$$r = 0.955; \quad s = 0.12$$

By omitting H, which fits the series least, the following equation is obtained:

$$E_s = -0.87\,f + 0.33 \qquad (V-4)$$
$$r = 0.997; \quad s = 0.03$$

It is not easy to destroy a correlation between E_s and f, as can be seen from eqn. V-5, where the six substituents of eqn. V-1 are combined with the halogens:

$$E_s = -1.12\,f + 0.62 \qquad (V-5)$$
$$r = 0.931; \quad s = 0.57$$

It is obvious that in practical SAR research, where E_s and f (or log P) are often used concurrently as parameters, a correlation between these two parameters should be taken into account. Only the introduction of clearly hydrophilic groups such as OH, COOH, NH_2 and $CONH_2$ into the series of substituents will enable the investigator to destroy the correlation between E_s and f at an acceptable level.

Steric Effects in 2-Alkyltriazinones

An interesting example of the influence which the steric parameter is supposed to have on the partition value is shown by a series of 2-alkyltriazinones (A). DRABER et al. [2] examined the lipophilic behaviour

(A)

of nine of these structures, and LEO et al. [3] correlated the observed log P values with a set of log P values calculated on the basis of the relevant π values (eqn. V-6):

$$\log P(obs) = 1.094 \log P(clc) + 0.172 \qquad (V-6)$$
$$n = 9; \quad r = 0.971; \quad s = 0.222; \quad F = 117; \quad \underline{t} = 10.8$$

Although satisfactory, the regression has the disadvantage of a large residual for the tert. butyl derivative (0.50), which is the main reason for the rather large standard error of estimate.

TABLE V,1

LIPOPHILICITY OF 2-ALKYLTRIAZINONES

R	log P(obs)	E_s	log P(calc) (a)	(b)	(c)
CH_3	-0.16	0.00	-0.16	-0.16	-0.21
C_2H_5	0.46	-0.07	0.34	0.37	0.32
$\underline{n}.C_3H_7$	0.93	-0.36	0.84	0.89	0.84
$\underline{i}.C_3H_7$	1.01	-0.47	0.64	0.78	0.84 / 1.11
$\underline{i}.C_4H_9$	1.39	-0.93	1.14	1.31	1.37
$\underline{tert}. C_4H_9$	1.70	-1.54	0.94	1.39	1.37 / 1.64
$\underline{i}.C_5H_{11}$	1.85	-0.35	1.65	1.83	1.90
cyclo C_6H_{11}	2.14	-0.79	1.81	2.01	2.16
$\underline{n}.C_6H_{13}$	2.68	-0.40	2.35	2.48	2.43 / 2.69

(a) Methyl derivative taken as the parent structure and relevant π-alkyl values added.
(b) Same, but f alkyl values added.
(c) Log P values corresponding to nearest "holes"; in cases where two values are given, the former is adapted to log P(calc) [(b)] and the latter to log P(obs).

LEO et al. postulated that a steric effect of the 2-alkyl substituent might be responsible and that this substituent modifies the solubility characteristics of the nearby carbonyl group.

The following regression equation could be obtained after inclusion of E_s as an extra parameter:

$$\log P(obs) = 1.023 \log P_\pi(calc) - 0.398 E_s + 0.031 \qquad (V-7)$$
$$n = 9; \quad r = 0.993; \quad s = 0.118; \quad F = 215$$
$$\underline{t} [(\log P_\pi(calc)] = 18.1; \quad \underline{t}(E_s) = -4.33$$

Judging from eqn. V-7, the claim that E_s would effect log P might be considered proved were it not that simply by applying f instead of π the use of E_s is made entirely superfluous, as demonstrated in the following equation:

$$\log P(obs) = 1.044 \log P_f(calc) + 0.069 \qquad (V-8)$$
$$n = 9; \quad r = 0.993; \quad s = 0.104; \quad F = 550; \quad \underline{t} = 23.4$$

On comparison of the last regression equation with eqn. V-7, it can be seen that the quality of eqn. V-8 is greater, as it has a lower s, a higher F and a higher $\underline{t}$. Does this justify the conclusion that the carbonyl lipophilicity is not influenced by the nearby alkyl substituent? In answering this question, the nearest "hole" theorem put for-

Refs. pg 173

ward in the previous chapter is useful. Those log P values in Table IV,4 which agree best with log P(obs) and log P(calc) [(b)] are given in the last column of Table V,1. In six of the nine cases there is striking a-greement. In the three other cases (i.C$_3$H$_7$, tert.C$_4$H$_9$ and n.C$_6$H$_{13}$ deriv-ative), log P(obs) and log P(calc) [(b)] are, however, one key number apart; it is tempting to conclude that in these three triazinones there is a steric effect, which originates in the alkyl substituent, that in-creases the lipophilicity of the nearby C=O group by ∼ 0.27. The orig-inal claim of LEO et al. that lipophilicity in the triazinones is sub-ject to a steric influence seems correct, therefore, to the extent that such an effect does exist but cannot be described in terms of the steric parameter E$_s$, and that it is confined to three of the substituents un-der investigation, for which it has a constant value.

With the aid of the values in the last column of Table V,1, it is possible to obtain the following optimal correlation with log P(obs)[*]:

$$\log P(obs) = 0.958 \log P(calc) \left[(c)\right] + 0.075 \qquad (V-9)$$
$$n = 9; \quad r = 0.997; \quad s = 0.067; \quad F = 1,331; \quad \underline{t} = 36.5$$

It is clear that this regression procedure is not justified unless the log P measurements for these nine triazinones were accurate to 0.5 − 1%. However, the treatment given is indicative of how we can suggest a further exploration of the constant c$_M$ in practice.

De-coupling of Resonance and its Effect on the partitional Behaviour of a Structure

Introduction

The consequences of de-coupling the resonance between two moieties of a given aromatic structure can be studied very well by analysis of the conjugative absorption band in the UV spectrum (displaced benzene E-band in the nomenclature of BRAUDE [4], C-band in that of MOSER and KOHLENBERG [5], primary band in that of DOUB and VANDENBELT [6] and 'L$_a$-band in that of PLATT and KLEVENS [7], whereas MAYER and SKLAR [8] describe this band in terms of an 'A$_{1g}$ ⟶ 'B$_{1u}$ transition).

This C-band usually has a high intensity (ϵ = 10,000 or greater) in benzene derivatives, where the substituent is in conjugation with the ring, and can therefore easily be recognized in the spectral pattern.

[*] In three cases column (c) of Table V,1 gives two log P values; the highest of them were selected for introduction in eqn. V-9.

It reflects the interaction of mobile electrons of the benzene system. An essential requirement for optimal interaction (maximum resonance) in a conjugated system is a coplanar configuration of all of the bonds concerned.

When structural changes, e.g., the introduction of an ortho-substituent, reduce the coplanarity in conjugated systems (steric hindrance of resonance), a transformation of the UV absorption spectrum becomes observable. When these changes remain limited in size, the characteristic transition will be confined to vibrational states in which the appropriate bonds are sufficiently extended to allow for a reasonable degree of residual coplanarity.

The various energy levels will not change appreciably, so that the transition energy remains virtually the same and is reflected spectroscopically in unchanged λ_{max} values. The intensity of absorption is, however, decreased considerably, as the transition is restricted to an obviously smaller number of vibrational states. Typical examples occur among the acetophenones [9].

Enhanced steric hindrance, which may ultimately result in complete de-coupling of the conjugative interaction of the chromophores, increases the energy content of the resonance structures and thus the energy level of the excited state relative to that of the ground state. This effect implies a higher excitation energy and absorption at lower wavelengths. The biphenyls can be cited as an illustration [10].

The reason for describing the effect of steric hindrance on UV absorption in some detail is that the apparent continuity in the process of de-coupling of resonance requires consideration. Between the two extremes — full coplanarity and complete perpendicularity of the chromophoric moieties — there is a continuous range of possibilities.

Among other parameters that are correlated with the phenomenon of resonance there is also a continuous change in magnitude, depending on the extent of de-coupling of the resonance. A specific example is the influence of ortho-substitution in benzophenones on the stretching vibration of the C=O group [11] and on the polarographic half-wave potential of this group [12].

Refs. pg 173

Lipophilic Behaviour of some selected aromatic Structures with de-coupled Resonance Interaction between aromatic Ring and functional Group

Aromatic Structures with a CON< Group

Some compounds of the carbonamide and anilide types are listed in Table V,2. The lipophilic behaviour in these structures is rather ir-regular, but a common feature is that the partition coefficient often falls short of expectation. We believe that steric hindrance of the res-onance is mainly responsible for this discrepency, and this believe is supported by the UV spectra of the structures concerned.

The UV spectra are partially illustrated in Figs. V,1 - V,4, and further information about the conjugative absorption (C-band) is given in Table V,2.

TABLE V,2

C-BAND DATA AND log P VALUES FOR AROMATIC STRUCTURES
CONTAINING THE CO-N< GROUP

Compound	C-band[*]		log P			
	λ (nm)	ϵ	obs	ref.	clc.	res.
C_6H_5-NHCO-H	242	13800	1.15	[13]	1.06	0.09
			1.12	[14]		0.06
			1.26	[15]		0.20
C_6H_5-N(CH$_3$)CO-H	231	9900	1.09	[15]	1.30	-0.21
C_6H_5-NHCO-CH$_3$	242	14300	1.16	[16]	1.30	-0.14
			1.36	[15]		0.06
C_6H_5-N(CH$_3$)CO-CH$_3$	225	5900	1.12	[17]	1.54	-0.42
			0.97	[18]		-0.57
			1.07	[15]		-0.47
C_6H_5-NHCO-C_6H_5	265	12800	2.62	[13]	2.78	-0.16
			2.70	[15]		-0.08
C_6H_5-N(CH$_3$)CO-C_6H_5	?		2.36	[15]	3.02	-0.66
C_6H_5-CONH$_2$	268	6000	0.65	[15]	0.80	-0.15
			0.64	[16]		-0.16
lidocaine	225	3200	2.26	[19]	3.25	-0.99
			2.20	[20]		-1.05

[*] Solvent: ethanol

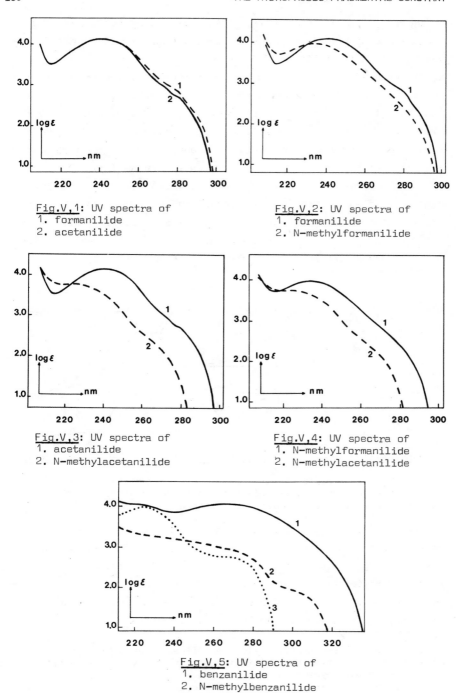

Fig.V,1: UV spectra of
1. formanilide
2. acetanilide

Fig.V,2: UV spectra of
1. formanilide
2. N-methylformanilide

Fig.V,3: UV spectra of
1. acetanilide
2. N-methylacetanilide

Fig.V,4: UV spectra of
1. N-methylformanilide
2. N-methylacetanilide

Fig.V,5: UV spectra of
1. benzanilide
2. N-methylbenzanilide
3. benzamide

Refs. pg 173

On comparison of the spectra, it is clear that substitution of a
methyl group on N in formanilide and acetanilide produces a hypso-hypo-
chromic effect[*]: see Figs. V,2 and V,3, where the N-methyl derivatives
can be compared directly with their unsubstituted analogues. It is worth
noting that whereas the spectra of formanilide and acetanilide are virtually identical in the region of the conjugative absorption band, the
spectra of the N-methyl derivatives by no means share this feature
(Fig. V,4). The presence of a C-methyl group, which by itself does not
influence the spectrum (Fig. V,1), apparently results in an extra steric
effect when an N-methyl group is simultaneously present. This effect
might be related to changes in the overall pattern of the preferential
conformations of the four structures. Fig. V,6 clearly indicates the enhanced steric effect in N-methylacetanilide. What partitional behaviour,
it may be asked, can be expected for these structures?

Fig.V,6: Steric effects in formanilide, acetanilide, N-methylformanilide and N-methylacetanilide; s.h.: steric hindrance; ⬇ denotes
rotation, i.e., de-coupling of resonance.

The log P values of formanilide and acetanilide are in reasonable
agreement with the calculated values, but the value for N-methylformanilide has a log P value that is 0.49 lower than the predicted value.
Hence lipophilicity undergoes an observable reduction that seems to parallel the magnitude of the steric effect manifested in the UV spectra
but which, curiously enough, could be quantified again.

---

[*] Spectral shifts are described in the following terms [31]:
  (a) Shifts in wavelength position: bathochromic = towards longer
      wavelengths; hypsochromic = towards shorter wavelengths.
  (b) Shifts in intensity: hypochromic = decreased intensity;
      hyperchromic = increased intensity.

On going from benzanilide to N-methylbenzanilide there is also a
very distinct decrease of 0.66 in lipophilicity (i.e., 2 key numbers).
In this instance also the UV absorption spectra are clear (Fig. V,5).
By assuming that, owing to the steric effect, the resonance in N-methylbenzanilide is fully de-coupled for the $C_6H_5-N\!\!<$ combination and has remained intact for the C_6H_5-CO- combination, log P can be calculated as
follows:

$$\log P = 2\, f(C_6H_5) + f(CH_3) + f(CON) + 3\, c_M$$
$$= 3.80 + 0.70 + (-2.89) + 0.86 = 2.47$$

which is in good agreement with the experimental value (2.36).

Complete de-coupling between C_6H_5 and the amide N, as suggested in
the above calculation, would, in our opinion, lead to such a transformation of the UV spectrum that it would approximate that of benzamide
very closely. The benzamide spectrum is shown in Fig. V,5 and it can be
seen that N-methylbenzanilide is far from spectrally identical with
benzamide, i.e.,it clearly has residual resonance possibilities remi-
niscent of those of an anilide. In conclusion, it can therefore be said
that before complete de-coupling of resonance is effected, the decrease
in lipophilicity has apparently attained its final value.

2,6-Dimethylbenzoic acid

Owing to the presence of two <u>ortho</u>-CH_3 groups, the resonance between the benzene ring and the carboxyl group in 2,6-dimethylbenzoic
acid is partially de-coupled, as illustrated in Fig. V,7, which repre-

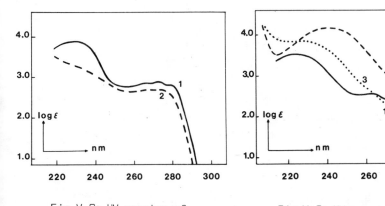

Fig.V,7: UV spectra of
1. benzoic acid
2. 2,6-dimethylbenzoic acid

Fig.V,8: UV spectra of
1. lidocaine
2. acetanilide
3. N-methylacetanilide

sents the UV spectra of benzoic acid and its 2,6-dimethyl derivative.

The spectral pattern reveals that the de-coupling is by no means complete, because the absorption in the 230 nm region is far too intense to be neglected [*].

The lipophilicity of 2,6-dimethylbenzoic acid can be calculated as 3.03, while the experimental value is 2.12 [20], which implies a reduction in the key number for the COOH group from 3 to 1.

N,N-dimethylbenzamide

In Fig. V,9, a projection of the CPK model of N,N-dimethylbenzamide is presented. The hatched intersection shows how, in a conformation where optimal interaction of C=O and the lone pair of nitrogen is maintained, one of the methyl groups attached to N prevents the ring and

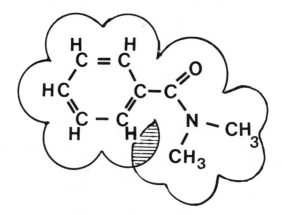

Fig.V,9: Projection of the CPK Model of N,N-dimethylbenzamide. Hatching indicates intersection of ring H and N-CH$_3$.

the carbonamide group from being fully coplanar. The illustration clearly permits the conclusion that de-coupling of resonance can be only partial. Log P, on the basis of full coplanarity, can be expected to

---

[*] Attention should also be paid to the absorption behaviour of 2,6-dimethylbenzophenone. On comparison of its UV spectrum with that of its unsubstituted derivative, it can be concluded that the conjugative absorption of the 2,6-dimethyl-substituted moiety still exists to the extent of about 30%, which, on the basis of a $\cos^2$ relationship between the angle of twist and the residual absorption, means a reduction in full coplanarity of 2,6-dimethyl-C$_6$H$_3$ and C=O by $\sim 43°$ [22].

have a value of 1.26, while partition measurements gave a value of 0.62
[21]. Again, a striking decrease in the partition value is observed and
of the three key numbers of the CON group, only one remains intact. In
these considerations, the reduction of log P has been attributed solely
to de-coupling of the resonance of C_6H_5 and C=O. In principle, it is,
of course, also conceivable that the coplanarity of C_6H_5 and C=O is
maintained and that, by rotation around the CO ♦ N bond at the expense
of part of the CO–N resonance, the steric hindrance is overcome.

Lidocaine

It appears from Fig. V,10, which shows a projection of the CPK model
of the major part of lidocaine (α-diethylamino-2,6-dimethylacetani-
lide), that the C=O is markedly hindered by the ortho-CH_3 from being
coplanar. The UV spectral data confirm this as shown in Fig. V,8, where
the UV patterns of lidocaine, acetanilide and N-methylacetanilide are
compared. The lidocaine curve has undergone a significant hypso-hypo-
chromic shift over that of N-methylacetanilide, although it certainly
cannot be concluded that there has occurred complete de-coupling of
resonance.

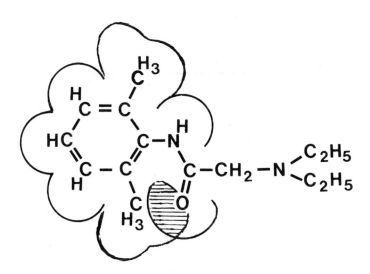

Fig.V,10: Projection of the CPK Model of Lidocaine.
Hatching indicates intersection of ring CH_3 and
Carbonamide C=O.

Refs. pg 173

The partition value of lidocaine was calculated as 3.25 but in fact the actual value is lower (2.26 [19] or 2.20 [20]), which means that the log P calculation provides a virtually correct value if the aliphatic instead of the aromatic value for *f*(CON) is used.

From the material presented here, it can be inferred that, in terms of lipophilicity, there is not a continually changing pattern but rather a quantifiable series of transitions between the extremes of complete conjugative interaction and complete de-coupling of resonance. This is, in effect, consistent with the *f*(Ar-Al) difference of a functional group, which can be described by means of a key number of 3 ± 1. An attempt to illustrate this view is given in Fig. V,11. Our line of rea-

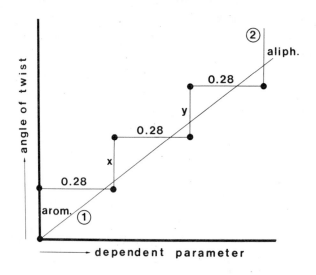

Fig.V,11: Graphical presentation of the angle of twist in a strained structure versus: (1) a dependent physico-chemical parameter changing linearly; (2) a dependent physico-chemical parameter that changes stepwise with quantifiable values (Lipophicity).

soning is based on an arbitrary structure C_6H_5–X in which the functional group can be typified by Δf(Ar-Al) = 0.84 (three key numbers). Curve 1 reflects the change in the value of, for example, the integrated absorption of the conjugative absorption band depending on increasing ortho-substitution, which leads to a larger angle of twist. The relationship between the two parameters is linear (keeping in mind that there have never been any indications that integrated UV absorption val-

ues can be quantified). Curve 2 gives an idea of how the changes in lipophilicity depend on an increasing angle of twist; in three discrete
steps, lipophilicity is lowered by the same amount (0.28), the result
being four lipophilicity species, viz., the completely aromatic one
(indicated by <u>arom</u>. in Fig. V,11), one in which de-coupling of resonance is complete (denoted by <u>aliph</u>.) and two intermediate forms, x and
y. This means that based on, for example, benzamide and by a careful
choice of the <u>ortho</u>-substituents, a series can be built up whose values
of −2.03, −2.32, −2.61 and −2.89 for f(CON) are involved in the calculation of lipophilicity. By analogy with this, it can be said that
f(COOH) = −0.09, −0.38, −0.67 and −0.95 for a series derived from benzoic acid.

In next section, a practical example is given.

Lipophilic Behaviour of a Series of alkyl-substituted para-Hydroxyacetanilides.

The above ideas on the influence of steric effects on the lipophilic behaviour of a structure, are, in a few respects, contradictory to
general principles recognized in this field. DEARDEN and O'HARA [23]
have summarized the various effects of <u>ortho</u>-substitution on lipophilicity: (a) entropy effects, (b) masking of group polarities, (c) hydrogen bonding, (d) field effects and (e) steric hindrance of resonance.
DEARDEN [24] examined the lipophilicity of a series of alkyl-substituted <u>para</u>-hydroxyanilides, and his findings are given in column 2 of
Table V,3 as π and $\Sigma\pi$ values for single and multiple substitution, respectively. DEARDEN's conclusion is that the lipophilic character of
the set of structures investigated is so complex (any substituent introduced into the parent structure acts as an <u>ortho</u>-substituent either to
OH or to NHCOCH$_3$) that the prediction of partition values for this type
of structures would be virtually impossible.

In column 3 of Table V,3 are presented the f values for various
types of substitution, calculated from the column of DEARDEN's π values
by addition of the appropriate number of f(H) values (0.175 per H atom;
see Table IV,3). The predictable f and Σf values are given in column
4, and column 5 gives the differences between the values in columns 3
and 4.

The upper part of Table V,3 shows the 3- and 3,5-alkyl-substituted
derivatives of para-hydroxyacetanilides, which are structures where the

Refs. pg 173

TABLE V,3

INFLUENCE OF STERIC HINDRANCE ON PARTITIONAL BEHAVIOUR FOR A SERIES OF ALKYL–SUBSTITUTED para–HYDROXYACETANILIDES

Substituents	π^*	f from π	f calc.	Δ	kn
3-methyl	0.482	0.66	0.70	—	0
3-ethyl	0.995	1.17	1.23	—	0
3-i.propyl	1.396	1.57	1.63	—	0
3-tert.butyl	2.046	2.22	2.25	—	0
3.5-dimethyl	0.797	1.15	1.40	0.25	1
3,5-diethyl	1.563	1.91	2.46	0.55	2
3,5-di i.propyl	2.360	2.71	3.26	0.55	2
3,5-di tert.butyl	2.869	3.22	4.48	1.26	5
2-methyl	-0.136	0.04	0.70	0.66	3
2,5-dimethyl	0.286	0.64	1.40	0.76	3
2,3-dimethyl	0.262	0.61	1.40	0.79	3
2,6-dimethyl	0.003	0.35	1.40	1.05	4
2,3,5-trimethyl	0.505	1.03	2.10	1.07	4
2,3,6-trimethyl	0.436	0.96	2.10	1.14	5
2,3,5,6-tetramethyl	0.640	1.34	2.80	1.46	6
2-methyl-5-tert.butyl	1.873	2.22	2.94	0.72	3

* These π values have been presented in the "Séminaire: Relations Structure – Activité", Paris 25-26 Mars 1974 [23] ; they were not published in the Proceedings of this Seminar.

substituent is ortho to the OH group and meta to the $NHCOCH_3$ group. The π system of HANSCH requires values of 0.50, 1.00, 1.30 and 1.60 for the four different mono-derivatives. The last value in particular, when compared with the experimentally derived value of 2.05, perturbed DEARDEN, who therefore postulated that there was a complex interplay between various interfering factors.

In our f system, there are no problems with the tert. butyl derivative, the predicted f value being in agreement with the experimental result.

By using for the constant c_M the value of 0.25, which appears to
have the greatest consistency throughout Table V,3, the decrease in the
lipophilicity values of the four 3,5-dialkyl derivatives can be established at 1, 2, 2 and 5 key numbers.

The log P value of para-hydroxyacetanilide has been measured by
DEARDEN and TOMLINSON [18] in phosphate buffer (pH 7.20). The result,
log P = 0.80, needs no correction because the pK of the OH group does
not differ much from that of phenol (σ_p of -NH-CO-CH$_3$ = 0.00 [41]),
which means that at pH 7.20 not less than 99.8% is undissociated.
DEARDEN and TOMLINSON's log P value of 0.80 is in excellent agreement
with the following calculation:

$$\text{Log P} = f(C_6H_4) + f(CH_3) + f(OH) + f(NCO) + f(H) + \left[kn(OH) + kn(NCO)\right.$$
$$\left. + kn(H)\right] c_M$$
$$= 1.688 + 0.702 + (-1.491) + (-2.894) + 0.175$$
$$+ (4 + 4 + 1)0.287 = 0.76$$

and we may conclude that the combination of OH and NCO on the aromatic
ring does not lead to extra interaction effects. Although the complete
UV data on DEARDEN's series are not available, it may be suggested that
the decrease in lipophilicity, as noted in Table V,3, is by no means
based on de-coupling of resonance alone but is also partly due to the
influence of other factors, notably hindered hydration. In this connection, attention may be drawn to the partition values of methyl-substituted phenols compared with that of phenol itself:

Phenol	Obs.:	1.46 [28]	Calc.: 1.54
		1.48 [25]	
		1.48 [26]	
2-Methylphenol		1.95 [27]	2.05
3-Methylphenol		1.96 [28]	2.05
		1.95 [27]	
		2.01 [25]	
4-Methylphenol		1.94 [28]	2.05
		1.92 [25]	
		1.95 [26]	
2,6-Dimethylphenol		2.36 [29]	2.49
		2.35 [29]	

There is no evidence of a significant ortho-effect in mono- nor di-
methyl derivatives and, by implication, any increase in the partition
values of para-hydroxyacetanilides would be due to specific structural
features present in these compounds; in other words, it would be the result of a certain degree of decoupling of resonance in the through-conjugation HO → C$_6$H$_4$ → NCO *.

___

* Indeed, there is some through-conjugation in para-hydroxyacetanilide,
  although the HAMMETT σ value for NHCOCH$_3$ has been determined to be
  virtually zero. This appears from the slight but distinct batho-hy-
  perchromic effect of the OH group in the UV spectrum of acetanilide
  [30]. We believe that this increased resonance is in itself not strong
  enough to result in an increase in log P from 0.80 to 1.08 (= 0.80 +
  1 kn).

Refs. pg 173

Alkyl substitution in the 2- and 6-positions, i.e., _ortho_ to the
NHCOCH$_3$ group, produces a much greater effect than alkyl substitution
in the 3- and 5-positions. Three key numbers operate in the 2-methyl
derivative and four in the 2,6-dimethyl derivative.

The effect of extra methyl groups in the 3- and/or 5-positions is
additive in the sense that the key numbers of the 2,5- and 2,3-dimethyl
derivatives are identical with that of the 2-methyl derivative because
3-methyl substitution would simply imply that kn = 0. The 2,3,5-trimeth-
yl derivative is additive in kn terms when this value is related to
those for the 2-methyl and 3,5-dimethyl derivatives. The extra high key
numbers of the 2,3,6-trimethyl and 2,3,5,6-tetramethyl derivatives might
be attributed to an enhanced steric effect originating in the buttressing effect of the additional methyl group.

The three key numbers found for the 2-methyl-5-_tert_. butyl derivative are in agreement with what can be expected from observations on the
2-methyl and 3-_tert_. butyl derivatives, namely the absence of any additional effect of the _tert_. butyl substituent.

Lactams and Thiolactams

In the preceding sections, some structures in which an NCO group
was directly connected via N or C to a benzene ring were discussed.
While describing their partitional behaviour, the resonance interaction
in the amide group itself was always left intact; in other words, no
room was left for taking into consideration rotation around the bond
between N and C=O. Such a de-coupling of resonance is, however, not impossible, as appears from a study of the log P values of some lactams
and thiolactams. The log P values found [45] are listed in Table V,4.

The partition values of the lactams were calculated routinely, while
those of the thiolactams were calculated on the basis of f (NCS) = -2.09,
obtained by adding 0.80 to f(NCO). This value of 0.80 was taken from a
comparison between barbiturates and thiobarbiturates (Part II - X) and
proved applicable to the structures in Table V,4 without complications
(cf., the log P differences for structures that have identical ring
sizes).

On comparison of the calculated and measured log P values, the thio-
-5-ring-lactam, curiously enough, appears to be the only structure for
which there is no discrepancy in these two values, all of the other
structures showing a lower lipophilicity. The question arises of whether

TABLE V,4

PARTITION VALUES OF LACTAMS AND THIOLACTAMS

$$(CH_2)_n \diagdown \begin{matrix} \diagup NH \\ | \\ C=O \ (S) \end{matrix}$$

Compound	log P values			
	obs.	calc. I	calc. II	res.(obs − II)
Lactams				
n = 3	n.m.	−0.84	−0.84 (0)	−
4	n.m.	−0.31	−0.61 (1)	−
5	−0.19	0.22	−0.08 (1)	−0.11
6	0.24	0.75	0.15 (2)	0.09
7	0.67	1.28	0.68 (2)	−0.01
Thiolactams				
n = 3	−0.05	−0.04	−0.04 (0)	−0.01
4	0.13	0.49	0.19 (1)	−0.06
5	0.75	1.02	0.72 (1)	0.03
6	1.00	1.55	0.95 (2)	0.05
7	1.44	2.08	1.48 (2)	−0.04

n.m. : not measured
I : from f summation
II : same, but applying the key number given in parentheses

this can be accounted for in an acceptable manner.

In saturated five-membered rings, competition between forces that tend to retain tetrahedral bond angles on the one hand, and torsional forces about the single bonds on the other, leads to a distorted conformation in which the pucker (distortion) slips around the cyclopentane ring in a low-energy vibrational motion. Appropriate substitution may cause fixation of this puckering and it seems that this can also be said to be true for 2-azacyclopentanone and 2-azacyclopentathione while maintaining coplanarity in the two moieties of the NH–CO and NH–CS units, respectively. Upwards from these five-membered rings in which, according to the above calculation of the log P data, an almost undistorted resonance interaction in the amide link seems to be present, the ring strain increases, which might explain a partial de-coupling of the amide resonance (NH↯C=O) and the resulting gain in hydrophilicity in rings that contain more than five ring atoms.

Refs. pg 173

Again there is a good indication that these enhanced lipophilicities are subject to rules of quantification, as there is consistency
for c_M = 0.30, to which are connected the key numbers 2, 2, 1, 1 and 0
for the 9-, 8-, 7-, 6- and 5-membered ring, respectively. Column 4 in
Table V,4 shows the log P values calculated on the basis of these assumptions and it can be seen from the residuals that all of the errors
are within the limits set.

LIPOPHILICITY OF POSITIVELY CHARGED N-STRUCTURES

Introduction

A literature search affords various log P values of structures of
the $\left[R_3NH\right]^+$ and $\left[R_4N\right]^+$ types in combination with a large variety of
negative counterions. Partition measurements on congeneric series are
scarce and, when available, they are often unreliable; it should not be
forgotten that N^+ structures combine extreme lipo- and hydrophilicities
in the same molecule, thus satisfying one of the requirements for mi-
celle formation.

At first sight, the lipophilic behaviour of N^+ structures seems unusual, which led LEO et al. [3] to remark that "it is somewhat unexpect-
ed to find the log P value for the $\geq N^+CH_3$ group lower than that of the
$\geq N^+H$ group". A quantitative evaluation of this phenomenon has not so
far been undertaken.

A few sets of diphenhydramine structures with a fairly large variety in the substitution pattern around the N atom were available, and
on comparing their log P values measured in 0.1 N HCl, it was found
that not only the CH_3 group but, in fact, any alkyl group showed a deviating partitional behaviour when attached to N^+.

When studying these anomalies, it is necessary to fix them either
in the N^+ centre (concept A) or in the alkyl groups substituted on the
N^+ centre (concept B), it being understood that in order to describe
the partitional behaviour in the correct way, concepts A and B have to
be superimposed. The two concepts of fixation are outlined in Fig. V,12.

The anomalous N^+ Centre (Concept A)

Four sets of structures were available for a study based on concept A. Two sets were from investigations by ELDEFRAWI and O'BRIEN [32],
(1) and (2), and the other two came from our own research.

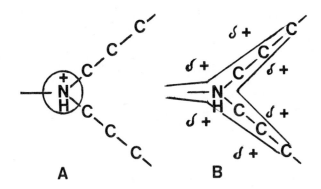

Fig.V,12: Two concepts for the study of anomalies in the
log P values of N^+ structures. <u>A</u>. Anomalies fixed in the
N^+ centre, the alkyl groups being assumed to behave normally. <u>B</u>. The N^+ centre is assumed to possess a constant
$f(N^+)$ value, the anomalies being enforced on the alkyl
groups.

(1). A set with $C_n-N^+(CH_3)_3$ as a general formula, where C_n is C_2H_5,
$\underline{n}.C_4H_9$, $\underline{n}.C_6H_{13}$, $\underline{n}.C_8H_{17}$ and $\underline{n}.C_{10}H_{21}$. The partition values measured by
ELDEFRAWI and O'BRIEN are listed in Table V,5.

By subtracting the sum of the f values of the three CH_3 groups and
the C_n group from these log P values, the $f(N^+)_{obs}$ values were obtained.
It appeared that these values correlated extremely well with the logarithm of the total number of carbon atoms in the C_n chain (eqn. V-10):

$$f(N^+) = -2.009 \log n - 5.744 \qquad (V-10)$$
$$n = 5; \quad r = 0.998; \quad s = 0.031; \quad F = 1,236; \quad \underline{t} = -36.2$$

The estimates of this regression equation are also included in
Table V,5 and show unambiguously how correctly the $f(N^+)$ values are
expressed. This means that the intermediate, unmeasured log P values
can be calculated simply by means of the equation and that the log P
value of the hitherto unmeasured tetramethylammonium chloride can be
determined from the intercept value by addition of four $f(CH_3)$ values.
This results in log P = -2.94.

As is evident from the last column of Table V,5, the anomalous effect on $f(N^+)$ has nearly disappeared with the C_{10} chain. This is also
clear from Fig. V,13, where curve 1 displays the relationship between
$f(N^+)$ and the total number of C atoms surrounding N^+.

Refs. pg 173

TABLE V,5

DEPENDENCE OF N^+ LIPOPHILICITY ON STRUCTURAL PATTERN

Compounds	log P (obs)	f_{N^+} obs.	calc.	$\Delta/.$

$$Cn-N^+-\bullet$$

				Δ
CH_3	–		-5.74	
				0.61
C_2H_5	-3.00	-6.33	-6.35	
				0.35
C_3H_7	–		-6.70	
				0.25
C_4H_9	-2.60	-6.99	-6.95	
				0.20
C_5H_{11}	–		-7.15	
				0.16
C_6H_{13}	-1.84	-7.29	-7.31	
				0.14
C_7H_{15}	–		-7.44	
				0.12
C_8H_{17}	-1.07	-7.58	-7.55	
				0.10
C_9H_{19}	–		-7.66	
				0.09
$C_{10}H_{21}$	-0.16	-7.73	-7.75	

$$\bullet-\bullet-N^+-Cn$$ (with Cn above and Cn below N)

				$\Delta/3$
CH_3	-3.00	-6.33	-6.28	
				0.52
C_2H_5	-2.82	-7.74	-7.83	
				0.30
C_3H_7	-2.19	-8.70	-8.73	
				0.21
C_4H_9	-1.30	-9.40	-9.37	
				0.17
C_5H_{11}	-0.22	-9.91	-9.87	
				0.14
C_6H_{13}	–		-10.28	

$$\begin{array}{c}\varnothing\\\varnothing\end{array}CH-O-Cn-\overset{+}{\underset{H}{N}}\begin{array}{c}\bullet\\\bullet\end{array}$$

				Δ
C_2H_4	-0.12	-5.08	-5.15	
				0.31
C_3H_6	0.03	-5.46	-5.46	
				0.22
C_4H_8	0.25	-5.75	-5.68	
				0.16
C_5H_{10}	0.58	-5.95	-5.84	
				0.14
C_6H_{12}	–		-5.98	
				0.12
C_7H_{14}	–		-6.10	
				0.10
C_8H_{16}	2.02	-6.09	-6.20	

TABLE V,5 (continued)

$$\phi\!\!\!-\!\!\!\!\overset{\phi}{\underset{\phi}{>}}\!CH-O-\bullet-\bullet-\overset{+}{N}H\quad\bigcirc C_n$$

				$\Delta/2$
C_2H_4	-0.04	-4.64	-4.64	
				0.45
C_3H_6	0.04	-5.03	-5.09	
				0.33
C_4H_8	0.25	-5.40	-5.42	
				0.25
C_5H_{10}	0.48	-5.70	-5.67	
				0.20
C_6H_{12}	0.85	-5.86	-5.87	

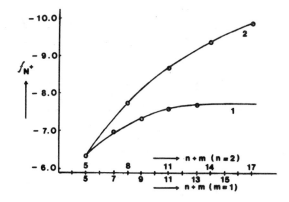

Fig.V,13: Partitional behaviour in two series of quaternary N structures investigated by ELDEFRAWI and O'BRIEN and represented in a plot of $f(N^+)$ versus the total number of carbon atoms in the substituent chains. In the general formula: $C_nN(C_m)_3^+$, m = 1 for curve (1) and n = 2 for curve (2).

(2). A set with $CH_3CH_2-N(C_n)_3^+$ as a general formula, where C_n represents CH_3, C_2H_5, $\underline{n}.C_3H_7$, $\underline{n}.C_4H_9$ or $\underline{n}.C_5H_{11}$. For log P values (measurements of ELDEFRAWI and O'BRIEN), see Table V,5. The same procedure as for set (1) was applied: from the measured log P values was subtracted the total Σf for all CH_3 and CH_2 fragments present in the four N substituents. The values so obtained again show an excellent correlation with the logarithm of the total number of carbon atoms per C_n substituent (see eqn. V-11).

$$f(N^+) = -5.148 \log n - 6.275 \qquad (V-11)$$

$$n = 5; \quad r = 0.999; \quad s = 0.066; \quad F = 1,847; \quad \underline{t} = -43.0$$

Refs. pg 173

In this instance also, the estimates (see Table V,5) are in perfect agreement with the observed $f(N^+)$ values.

There is a remarkable difference between series (1) and series (2), as is evident from the last column, where the Δ values of the successive $f(N^+)$ estimates are presented and from curve 2 in Fig. V,13: in series (2), each C_n chain is a more or less independent unit in eliciting the N^+ anomaly, one of the implications being that while for the realization of a value of e.g., -7.8 for $f(N^+)$ in series (1) a chain of 10 C atoms is required, only 3×2 C atoms are needed in series (2). This means that 13 C atoms in (1) are equivalent to 8 C atoms in (2).

(3). This series was derived from diphenhydramine by extending the C atom pair between O and N by 1, 2, 3 or 6 extra C atoms. The five structures all had the same N substitution pattern and were examined for lipophilic behaviour in 0.1 N HCl. Column 2 in Table V,5 gives the log P values found.

While drawing up column 3, the f values of all groups around N were summed. In doing so, no allowance was made for any proximity effect due to 2C separation in the first member of the series (diphenhydramine itself). It was felt that any anomaly present had to be accommodated in the $f(N^+)$ value, including those which might be associated with a proximity effect. The value of $f(H)$ was taken to be 0.00 (see below). The $f(N^+)$ values were correlated with n as follows:

$$f(N^+) \;=\; -1.734 \log n \;-\; 4.632 \tag{V-12}$$
$$n = 5; \quad r = 0.973; \quad s = 0.106; \quad F = 55.0; \quad \underline{t} = -7.41$$

Although statistically the result is less elegant than in series (1) and (2), it is satisfactory in all respects, especially when the column of the estimates is compared with the observed $f(N^+)$ values. If, after all, a proximity correction is introduced for the first member of the series, the correlation coefficient is 0.916.

(4). Like series (3), this series was derived from diphenhydramine, the two N substituents now being part of a ring system. Five derivatives were available, belonging to a non-interrupted series of a three- to a seven-membered ring. Eqn. V-13 shows the regression of the relationship between $f(N^+)$ and the logarithm of the number of C atoms in the N ring:

$$f(N^+) \;=\; -2.593 \log n \;-\; 3.856 \tag{V-13}$$
$$n = 5; \quad r = 0.999; \quad s = 0.022; \quad F = 1,937; \quad \underline{t} = -44.0$$

Again, the regression equation is of excellent quality. On comparing eqns. V-10, 11, 12 and 13, it can be seen that the measure of cooperation of the N substituents in producing the anomaly in the N^+ centre, as described above, differs from case to case. In this connection, it is worth considering the slopes of the four regression lines: a conversion of these slopes on the basis of one changing chain part affords values of 2.01, 1.72, 1.73 and 1.30, respectively, which may be regarded as the $f(N^+)$ changes per 10 C atoms per each chain part. They show a fairly wide variety, indicating that the lipophilicity of N^+ structures is determined by:

(a) the total number of C atoms included in the substitution around the N^+ centre;

(b) the mode of distribution of these C atoms over the alkyl chains (all in one chain or devided over more chains);

(c) the inclusion, if any, of the N^+ centre in a ring pattern.

It stands to reason that the presence of a hetero atom, such as the O atom in diphenhydramines, exerts its own effect on $f(N^+)$.

Anomalous Alkylchains (Concept B)

An approach that is radically different from the above is based on the assumption that the N^+ centre has its own constant f value and that the C atoms, with their corresponding H atoms, behave anomalously. In working out an arithmetic model, it is obvious to assume that its distance to the N^+ centre is decisive for any difference of a given C atom from the ordinary lipophilicity pattern. For a complete development of this concept, an extremely large variety of structures would be needed. The material available is, however, fairly limited and hence we had to make a number of concessions which, in retrospect, have not effected the final outcome too seriously, however.

Parameters were introduced according to the character of the alkyl fragment and its position relative to the N^+ centre, as illustrated by the four examples (A), (B), (C) and (D).

(A)   (B)

Greek letters indicate the positions of the fragments and the numerals denote the presence of CH_2 or CH_3. The following concessions had to be made:

(1) no specific distinction is made between primary, secondary and tertiary amines;

(2) CH_2 types can be differentiated only inadequately; thus, no difference is made between α-CH_2 in the ethyl group and α-CH_2 in the butyl group of structure (B);

(3) the α-CH_2 groups in a ring are treated as if there were no effect from cyclization; the same is true for β-CH_2 and for fragments even further away from N^+.

The four structures presented above were adapted for multiple regression analysis as indicated below:

$$\text{Log P (A)} = f(R) + f(N^+) + f(H) + f(\alpha,2) + f(\beta,2) + f(\alpha,3) + f(\gamma,3)$$

$$\text{Log P (B)} = f(R) + f(N^+) + f(H) + 2f(\alpha,2) + f(\beta,2) + f(\gamma,2) + f(\delta,3) + f(\beta,3)$$

$$\text{Log P (C)} = f(R) + f(N^+) + f(H) + 2f(\alpha,2) + f(\beta,2)$$

$$\text{Log P (D)} = f(R) + f(N^+) + 2f(H) + f(\alpha,2) + f(\beta,3)$$

In any given series of structures, e.g., diphenhydramine derivatives, R is always the same and $f(R)$ can be conceived of as a constant (concession 4); $f(R)$ and $f(N^+)$ can now be combined into one constant and the set of equations written as

$$\log P = \sum \underline{a}_n f_n + c \qquad (V-14)$$

where $c = f(R) + f(N^+)$.

In solving eqn. V-14, forcing is strictly forbidden: the intercept should possess a value $f(R) + f(N^+)$, which is unknown at this stage of the calculation.

A series of 30 structures whose lipophilicities are recorded in Table V,6 were eligible for treatment via the arithmetic pattern outlined here. Initial calculations, in which H_N+, D, α-CH_3, α-CH_2, β-CH_3, β-CH_2, γ-CH_2 and γ-CH_3 acted as parameters, showed that $f(H_N+)$, by which

TABLE V,6

PARTITION VALUES OF SOME DIPHENHYDRAMINE DERIVATIVES

$$\begin{array}{c} C_6H_5 \\ \diagdown \\ 4\text{-}R\text{-}C_6H_4 \end{array} \diagup CH\text{-}O\text{-}CH_2\text{-}CH_2\text{-}\overset{+}{N}HX_2$$

no	$-\overset{+}{N}HX_2$	R	log P obs.	calc.	res.
1	$-\overset{+}{N}H_3$	H	0.12	0.12	0.00
2		CH$_3$	0.65	0.64	0.01
3	$-\overset{+}{N}H_2CH_3$	H	-0.01	0.01	-0.02
4		CH$_3$	0.58	0.54	0.04
5	$-\overset{+}{N}H_2C_2H_5$	H	0.20	0.20	0.00
6		CH$_3$	0.72	0.73	-0.01
7	$-\overset{+}{N}H_2C_3H_7$	H	0.62	0.60	0.02
8		CH$_3$	1.09	1.13	-0.04
9	ring	H	-0.04	-0.06	0.02
10		CH$_3$	0.48	0.47	0.01
11	$-\overset{+}{N}H(CH_3)_2$	H	-0.12	-0.11	-0.01
12		CH$_3$	0.41	0.42	-0.01
13	ring	H	0.04	0.07	-0.03
14		CH$_3$	0.56	0.60	-0.04
15	$-\overset{+}{N}H(CH_3)C_2H_5$	H	0.14	0.09	0.05
16		CH$_3$	0.66	0.62	0.04
17	$-\overset{+}{N}H(CH_3)C_3H_7$	H	0.45	0.45	0.00
18		CH$_3$	0.97	0.98	-0.01
19	ring	H	0.25	0.21	0.04
20		CH$_3$	0.77	0.74	0.03
21	$-\overset{+}{N}H(C_2H_5)_2$	H	0.26	0.28	-0.02
22		CH$_3$	0.78	0.81	-0.03
23	$-\overset{+}{N}H(C_2H_5)C_3H_7$	H	0.67	0.64	0.03
24		CH$_3$	1.19	1.17	0.02
25	ring	H	0.48	0.52	-0.04
26		CH$_3$	1.01	1.06	-0.05
27	$-\overset{+}{N}H(C_3H_7)_2$	H	1.05	1.01	0.04
28		CH$_3$	1.53	1.54	-0.01
29	ring	H	0.85	0.83	0.02
30		CH$_3$	1.39	1.36	0.03

Refs. pg 173

TABLE V,6 (continued)

no	$-\overset{+}{N}HX_2$	R	log P		
			obs.	calc.	res.
31	$-\overset{+}{N}H_2-$ cyclo C_5H_9	H	1.03	1.04	−0.01
32		CH_3	1.59	1.53	0.06
33	$-\overset{+}{N}H_2-$ cyclo C_6H_{11}	H	1.47	1.44	0.03
34		CH_3	1.94	1.97	−0.03
35	$-\overset{+}{N}H_2C_4H_9$	H	1.12		

TABLE V,7

f VALUES OF THE ALKYL OR ALKYLENE FRAGMENTS OF N^+ IN
A SERIES OF DIPHENHYDRAMINES WITH VARIATIONS IN THE
STRUCTURAL PATTERN OF THE N ATOM

Fragment	f	s.d.	$\underline{t}$
$(4-CH_3)$	0.530	0.011	47.4
$\alpha-CH_3$	−0.115	0.010	11.6
$\alpha-CH_2$	−0.093	0.013	7.4
$\beta-CH_3$	0.172	0.013	13.2
$\beta-CH_2$	0.135	0.014	9.3
$\gamma-CH_3$	0.400	0.012	32.1
$\gamma-CH_2$	0.314	0.014	21.7

$n = 30$	$r = 0.998$	$F = 912$
	$s' = 0.030$	c.t. $= 0.122$

$\alpha-CH$	0.02	
$\delta-CH_2$	0.40	} secondary f values
$\alpha-H$	0.00*	

* Fixed value

is meant H connected directly to N^+, was virtually zero. For this reason, we decided to exclude this H from our computations, in other words,
to fix this fragment at $f = 0.00$. By neglecting this most frequently occurring fragment, it is possible to achieve more congruency in the series as a whole and, in consequence, to improve the computation results.
The constant c in eqn. V-14 will now agree with the log P of structure
No. 1 in Table V,6. As in this instance the correct value of the constant term is known, forcing can be applied when necessary. This can be

effected by adding a few extra cards to the data package featuring c, 0,
0, 0 . . . 0. D is a dummy parameter to permit the inclusion of several
4-methyl-substituted derivatives in the regression. This parameter D is
taken to be zero in the absence and unity in the presence of a 4-methyl
group. In the output, the regressor value connected with D should have a
value equal to $f(CH_3) - f(H)$.

All data concerning the regression equation are given in Tables V,6
and 7. A few words of comment may be called for on the procedure:

(1). Forcing was not necessary. The constant term amounts to 0.122
   and this is exactly the log P value measured for structure
   No. 1.

(2). The regressor value of D appears to be 0.530, which agrees with
   the predicted value: 0.702 - 0.175 = 0.527.

(3). The values obtained for r, s, F and $\underline{t}$ are excellent.

(4). All f values of CH_2 and CH_3 are anomalous. The nearer they are
   to the N^+ centre, the more they deviate; α-CH_3 and α-CH_2 are
   even significantly negative and the γ-situated fragments just
   fail to reach their normal levels of 0.70 and 0.53. Structures
   Nos. 31, 32, 33 and 34 possess a CH group and lend themselves
   to the calculation of a secondary α-CH value, while for δ-CH_2
   also a value becomes available.

The two methods of calculation that were applied to express the deviations in the partitional behaviour of N^+ as a function of the substitution pattern gave, in fact, a fully comparable final result. There
appears to be a gradual development in extra hydrophilicity, the extent
of which is determined by the length and number of the substituting
chains. The following reasoning can also be given. Owing to its charge,
the N^+ centre has a considerable hydrophilicity, which, owing to the
small diameter of this centre, cannot manifest itself fully in the limited periphery of the N atom; hence we distinguish between "active" and
"non-active" hydrophilic potency. By introducing substituting alkyl
chains on to N, it is possible to achieve re-distribution of the N^+
charge along these chains; this may be associated with an activation of
an amount of initially "non-active" hydrophilicity and, as soon as these
chains become sufficiently extended, the situation will tend towards
complete N^+ hydrophilicity. Fig. V,12(B) attempts to express this principle by distributing the original positive charge over the chains in
the form of a series of $\delta+$ charges. Provided with such partial charges,

Refs. pg 173

these chains, in which the numerical value of $\delta+$ decreases as the distance to the periphery of the N atom increases, might be conceived of
as hydrophilic "landing stages", which are wide at the beginning and
then taper off.

As stated above, the complete description of the lipophilic behaviour of N^+ structures will have to be based on a combination of the two
concepts A and B. This implies that the increase in hydrophilicity that
can be brought about by an N^+ structure through the introduction of one
or more chains and/or the extension of chains that are already present
have to be discounted partially in a lowered $f(N^+)$ value and for the
remaining part in a decrease in the *f* values of CH_3, CH_2, CH and C. We
believe that the establishment of the proper population (a/b ratio) in
eqn. V-15:

$$\Delta \log P \;=\; a\,\Delta f(N^+) \;+\; b\,\Sigma\Delta f(CH_n) \qquad\qquad (V\text{-}15)$$

does not differ essentially from the determination of the exact charge
distribution in and around N in quaternary and protonated nitrogen
structures with increasing complexity of the substituents. In fact, it
is the same as the problem that at the moment is of great interest in
quantum pharmacology [33,34,35,36].

"Magic" Constant and N^+ Lipophilicity

One means of treating the problem of N^+ lipophilicity remains, which
is to express anomalies in partitional behaviour of this centre, if possible, in terms of the magic constant, c_M, and a series of key numbers.
For this purpose, it is preferable to start from an arrangement as presented in eqn. V-16:

$$\log P = a\,n(CH_3) \;+\; b\,n(CH_2) \;+\; c\,n(CH) \;+\; d\,n(C) \;+\; e \qquad (V\text{-}16)$$

The first four members in the right-hand term of eqn. V-16 contain
the numbers of CH_3, CH_2, CH and C fragments in the variable substituents of N^+ and e is a constant equal to the residual lipophilicity of
the structure after removal of the variable chain parts. In explanation,
the parameters for structures (A), (B), (C) and (D) are presented below:

$$\log P(A) = 2\,a + 2\,b + e$$
$$\log P(B) = 2\,a + 4\,b + e$$
$$\log P(C) = 3\,b + e$$
$$\log P(D) = a + b + e$$

The constant term e is equal to the log P value of $R-NH_3^+$ (absent substituents can be supplemented with Hs as one thinks fit).

By resolving eqn. V-16 via MRA, the regressors a, b, c and d will deviate more from $f(CH_3)$, $f(CH_2)$, $f(CH)$ and $f(C)$ and the constant term e will deviate more from $\log P(R-NH_3^+)$ as the anomalies become more prominent, in which case the problem narrows down to finding an acceptable method of correction that is based on the constant c_M.

Table V,8 records a collection of structures that lend themselves to the procedure outlined here. The result of the regression analysis is as follows:

$$\log P = 0.123\ n(CH_3) + 0.167\ n(CH_2) + 0.686\ n(CH) - 0.185 \qquad (V-17)$$

$$n = 23; \quad r = 0.885; \quad s = 0.24; \quad F = 22.9$$

$$\underline{t}\left[n(CH_3)\right] = 1.82$$
$$\underline{t}\left[n(CH_2)\right] = 4.81$$
$$\underline{t}\left[n(CH)\right] = 4.23$$

TABLE V,8

MAGIC CONSTANT AND LIPOPHILICITY OF DIPHENHYDRAMINES

$-NHX_2^+$	log P (obs)	kn	log P (calc.)			
			I	res.	II	res.
$-NH_3^+$	0.12	0	-0.18	0.30	0.16	-0.04
$-NH_2CH_3^+$	-0.01	3	-0.06	0.05	0.01	-0.02
$-NH_2C_2H_5^+$	0.21	4	0.11	0.10	0.24	-0.03
$-NH^+$ (cyclopropyl)	-0.04	4	0.15	-0.19	0.08	-0.12
$-NH_2C_3H_7^+$	0.54	5	0.27	0.27	0.47	0.07
$-NH_2^+$ (cyclopentyl)	1.03	5	1.17	-0.14	1.00	0.03
same, 4-CH$_3$-deriv.	1.59	5	1.68	-0.09	1.51	0.08
$-NH_2^+$ (cyclohexyl)	1.47	5	1.34	0.13	1.51	-0.04
same, 4-CH$_3$-deriv.	1.94	5	1.84	0.10	2.01	-0.07
$-NH_2C_4H_9^+$	1.12	5	0.44	0.68	0.98	0.14
$-NH(CH_3)_2^+$	-0.12	6	0.06	-0.18	-0.15	0.03
same, 4-CH$_3$-deriv.	0.43	6	0.57	-0.14	0.36	0.07

Refs. pg 173

TABLE V,8 (continued)

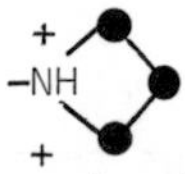 (ring) +$-NH$	0.04	6	0.32	−0.28	0.03	0.01
+$-NH(CH_3)(C_2H_5)$	0.11	7	0.23	−0.12	0.08	0.03
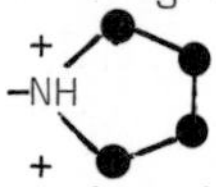 (ring) +$-NH$	0.23	7	0.49	−0.26	0.27	−0.04
+$-NH(C_2H_5)_2$	0.24	8	0.40	−0.16	0.31	−0.07
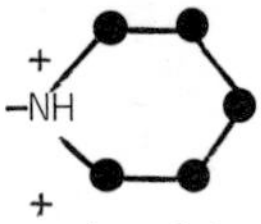 (ring) +$-NH$	0.49	8	0.65	−0.16	0.50	−0.01
+$-NH(CH_3)(C_3H_7)$	0.47	8	0.40	0.07	0.31	0.16
+$-NH(C_2H_5)(C_3H_7)$	0.47[a]	9	0.56	−0.09	0.55	−0.08
	0.47[a]	9		−0.09		−0.08
	0.41[a]	9		−0.15		−0.14
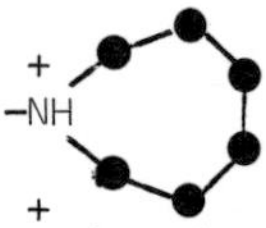 (ring) +$-NH$	0.85	9	0.82	0.03	0.73	0.12
+$-NH(C_3H_7)_2$	1.05	9	0.73	0.32	1.06	−0.01

[a]   Results from three independent experiments.

TABLE V,9

CH_n FRAGMENT VALUES OF THE N^+ SUBSTITUENTS
IN DIPHENHYDRAMINES

CH_n fragment	*f* values		
	I	II	III
CH_3	0.702 (0.021)	0.123 (0.068)	0.675 (0.054)
CH_2	0.527 (0.006)	0.167 (0.035)	0.512 (0.032)
CH	0.236 (0.022)	0.686 (0.162)	0.178 (0.073)

I   Fragment analysis from Table III,2
II Eqn. V-17
III Eqn. V-18
In parentheses: standard deviations

Among the 23 data points introduced, there are three from 4-methyl-
-substituted derivatives. For these three, $f(CH_3)$ was taken to be 0.53

(which means that we carried out a simultaneous correction for the conversion $C_6H_4 \rightarrow C_6H_5$) so that these 4-methyl derivatives could be included in the pattern of eqn. V-16 without further modification.

Eqn. V-17 has a much too low correlation coefficient, about 22% of
the variance being unexplained by the regression analysis. The constant
deviates widely from the predicted value ($\Delta = 0.31$) and the regressor
values also appear to be out of the range, the term with $n(CH_3)$ not
even being significant because of its extremely low t value. For more
details, see Table V,8, in which column 4 shows the estimates, and
Table V,9 for further information on the standard errors in the fragment values.

The column of residuals has only one real outlier, but five of
them are distinctly above the desirable level of 0.20. Starting from
these six structures, a trial and error procedure was applied to force
the whole series to an optimal fit on the basis of eqn. V-16 with the
difference that the constant c_M was introduced as an extra parameter.
The results are given in eqn. V-18:

$$\log P = 0.675\ n(CH_3) + 0.512\ n(CH_2) + 0.178\ n(CH) - 0.278\ kn$$
$$+ 0.165 \qquad (V\text{-}18)$$

$$n = 23; \quad r = 0.986; \quad s = 0.087; \quad F = 169$$

$$t\left[n(CH_3)\right] = 12.6$$
$$t\left[n(CH_2)\right] = 15.8$$
$$t\left[n(CH)\right] = 2.44$$
$$t(kn) = -11.5$$

Eqn. V-18 shows an appreciable improvement over eqn. V-17, as illustrated below:

(1). The correlation coefficient increased from 0.885 to 0.986, so
that the regression equation leaves only 5% of the variances
unexplained.

(2). All four regressors operate significantly. This is particularly
true for the freshly introduced parameter c_M, which appears to
have the highly acceptable value of 0.278. The three other regressors also lie within the limits of accuracy found previously for CH_3, CH_2 and CH; see Table V,9.

(3). The values of s and F have undergone considerable improvement.

It will be clear that eqn. V-18 is not acceptable unless there is a
substantial amount of logic in assigning the key numbers in the transformation of eqn. V-17 into eqn. V-18: the sequence of these key numbers

Refs. pg 173

needs to reflect closely the changes in the structural pattern around the N atom. Fig. V,14 presents a survey of the key numbers assigned to each structure.

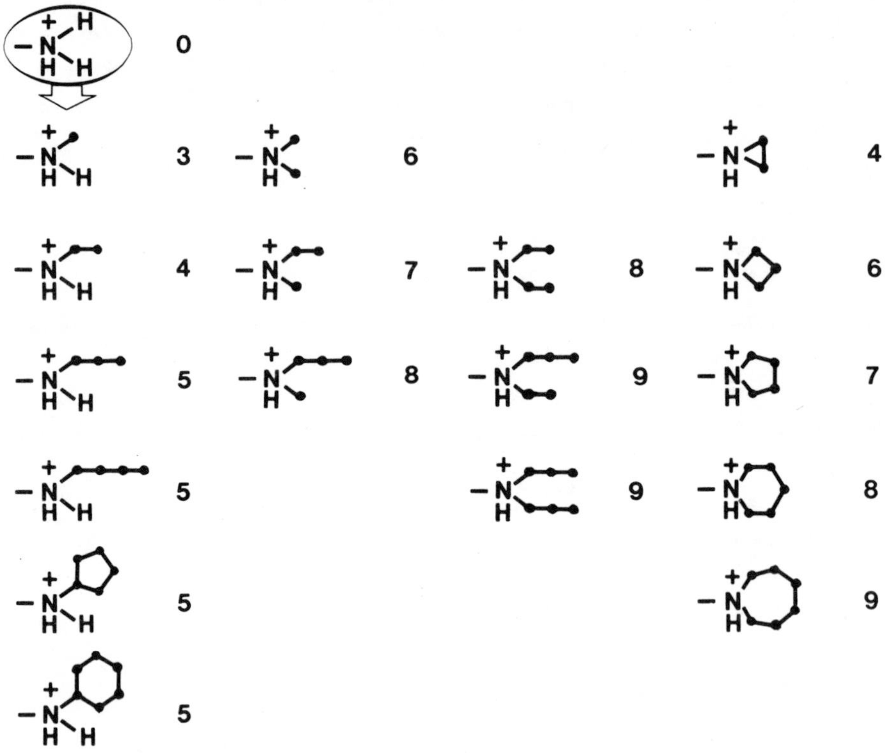

$$-\overset{+}{N}\begin{smallmatrix}H\\H\\H\end{smallmatrix} \quad 0$$

3	6		4
4	7	8	6
5	8	9	7
5		9	8
5			9
5			

Fig.V,14: Key number pattern for some related N^{+} structures of the diphenhydramine type.

The series starts in the top left-hand corner with the structure $R-NH_3^+$ (kn = 0). For the introduction of one CH_3 group, key number 3 is required. An additional CH_2 group, extending CH_3 to C_2H_5, needs one extra key number and then the pattern is continued with a regular additivity:

$$-NH^+(CH_3)_2 \quad : \quad 2 \times 3$$
$$-NH^+(CH_3)(C_2H_5) \quad : \quad 3 + 4$$
$$-NH_2^+(C_3H_7) \quad : \quad 4 + 1$$
$$-NH^+(CH_3)(C_3H_7) \quad : \quad 3 + 5, \text{ etc.}$$

As appears from our previous calculations (concepts A and B), dis-

tance exerts a distinct effect. This is also true in the present calculations; cf., C_2H_5 and CH_3. The fact that C_3H_7 receives one key number
more than C_2H_5 is felt to be the consequence of the argument in Chapter
IV concerning the nearest hole theorem; a key number of 4 being just
not sufficient for C_3H_7 implies that C_4H_9 will not require the inclusion of an extra key number. Hence, $kn(C_3H_7) = kn(C_4H_9)$ in the underlying series of structures, and the parallelism with the observation
$\log P(\text{helium}) = \log P(\text{neon})$ is obvious.

On the proposed line of reasoning, the key number of $R-NH^+(C_3H_7)_2$
is likewise of interest; it appears to be 9 while 2 x 5 is to be expected. We observe, however, that for $R-NH_2^+(C_3H_7)$ it is only just possible
to reach 5 key numbers and, therefore, 4 + 5 = 9 key numbers for the
structure $R-NH^+(C_3H_7)_2$ is not surprising.

The cyclic N structures in our series of diphenhydramines also appear to fit admirably in the concept of quantification, on the one hand
by comparing them with the corresponding open structures and on the
other by comparing them mutually. The three-membered-ring is two lower
in key number than the open analogue $R-NH^+(CH_3)_2$; the four-, five- and
six-membered-rings are one key number lower than the open structural
analogues $R-NH^+(CH_3)(C_2H_5)$, $R-NH^+(C_2H_5)_2$ and $R-NH^+(C_2H_5)(C_3H_7)$, respectively, while the key number of the seven-membered-ring is the same as
that of $R-NH^+(C_3H_7)_2$.

The major features of the magic constant emerge most clearly when
the structures are considered a whole; as substitution is extended,
hydrophilicity develops, operating as a multiple of c_M and conforming
to what the nearest hole theorem calls for.

Considering Chapters IV and V as a whole, we certainly admit that
more than one speculative element remains, but there is no other way to
account in a comprehensible way for a number of curious facts that occur
in lipophilic behaviour of several compounds.

APPLICATION OF THE f SYSTEM IN A FEW SAR STUDIES

Denaturizing Activity of Aliphatic Alcohols

HERSKOVITS et al. tested the denaturizing activity of alcohols on
(a) myoglobin from whale sperm and (b) α-chymotrypsinogen [37], and the
activities obtained were correlated by HANSCH and DUNN [38] with the
octanol − water partition values of the alcohols involved. Their re-

Refs. pg 173

TABLE V,10

DENATURIZING ACTIVITY OF ALIPHATIC ALCOHOLS

Compound	log 1/C obs	log P I	log P II	log 1/C (est) I	log 1/C (est) res.	log 1/C (est) II	log 1/C (est) res.
Myoglobin (from sperm of whale)							
CH_3–OH	−1.09	−0.66[a]	−0.74	−0.75	−0.34	−0.92	−0.17
C_2H_5–OH	−0.72	−0.16[b]	−0.21	−0.53	−0.19	−0.61	−0.11
n. C_3H_7–OH	−0.30	0.34[a]	0.32	−0.31	0.01	−0.31	0.01
i. C_3H_7–OH	−0.53	0.14[b]	0.20	−0.39	−0.14	−0.38	−0.15
n. C_4H_9–OH	0.10	0.88[a]	0.84	−0.07	0.17	−0.01	0.11
i. C_4H_9–OH	0.10	0.65[a]	0.73	−0.17	0.27	−0.07	0.17
t. C_4H_9–OH	−0.38	0.37[a]	0.37	−0.29	−0.09	−0.28	−0.10
$HOCH_2CH_2OH$	−1.15	−1.93[a]	−1.37*	−1.31	0.16	−1.29	0.14
$CH_3CH(OH)CH_2OH$	−0.93	−1.43[b]	−0.95*	−1.09	0.16	−1.04	0.11
α-Chymotrypsinogen							
CH_3–OH	−0.86	−0.66[a]	−0.74	−0.65	−0.21	−0.81	−0.05
C_2H_5–OH	−0.58	−0.16[b]	−0.21	−0.43	−0.15	−0.52	−0.06
n. C_3H_7–OH	−0.20	0.34[a]	0.32	−0.21	0.01	−0.23	0.03
i. C_3H_7–OH	−0.42	0.14[b]	0.20	−0.30	−0.12	−0.30	−0.12
n. C_4H_9–OH	0.15	0.88[a]	0.84	0.02	0.13	0.05	0.10
i. C_4H_9–OH	0.10	0.65[a]	0.73	−0.08	0.18	−0.01	0.11
t. C_4H_9–OH	−0.30	0.37[a]	0.37	−0.20	−0.10	−0.21	−0.09
sec. C_4H_9–OH	−0.04	0.61[a]	0.73	−0.10	0.06	−0.01	−0.03
$HOCH_2CH_2OH$	−1.04	−1.93[a]	−1.37*	−1.20	0.16	−1.15	0.11
$CH_3CH(OH)CH_2OH$	−0.93	−1.43[b]	−0.95*	−0.98	0.05	−0.92	−0.01

* Proximity effect taken into account: p.e.(2) = 2 × 0.28
a Measured log P values
b Calculated log P values (π method)

sults permit the following regression equations:

(a) $\log 1/C$ = 0.441 log P − 0.456   (V−19)

n = 9;   r = 0.901;   s = 0.217;   F = 30.2;   t = 5.50

(b) $\log 1/C$ = 0.434 log P − 0.360   (V−20)

n = 10;   r = 0.947;   s = 0.146;   F = 69.6;   t = 8.34

In eqns. V−19 and V−20, C is the molar concentration of alcohol in which the protein (a) or enzyme (b) was tested.

A number of partition values needed for regression were determined

directly, and the others were calculated. Those for straight-chain alcohols were obtained by using the experimental log P for the nearest
homologue below it as a base, applying $\pi(CH_2) = 0.50$. For branched
structures, e.g., $\underline{i}.C_3H_7-OH$, the necessary branching corrections were
taken into account. The results of the regression analysis vary from
acceptable to good; outliers are absent in both series. It is notable,
however, that in series (a), two log 1/C values have rather high residuals,
namely those of CH_3-OH and $\underline{i}.C_4H_9-OH$. Surprisingly, these alcohols also
have the highest residuals in series (b), despite its much better regression.

For two reasons, this pair of regression equations was chosen as a
test object for our f constants:

(1) to establish in a simple series of structures whether the exclusive use of f values can compete with a mixed set consisting of experimental log P values and data obtained by π-addition; and

(2) on introduction of the proximity effect (Chapter III), ethylene
glycol emerged as an outlier. It is easy to argue that a measuring
fault must have entered into the log P determination, but an effect as
yet not understood may also have given rise to the outlier. In the two
series, the log P of propylene glycol is derived from that of ethylene
glycol by mere π-addition and these two log P values fit easily into
the regression; and apparently there seems to be no reason to mistrust
the log P value of glycol.

It is clear that a regression analysis based solely on f summations,
due to the presence of ethylene glycol and propylene glycol, will present a very interesting test case for our f system.

The results of our calculations are expressed in eqns. V-21 and
V-22:

(a) $\log 1/C = 0.612 \sum f - 0.502$ (V-21)

        $n = 9; \quad r = 0.966; \quad s = 0.128; \quad F = 99.6; \quad \underline{t} = 9.98$

(b) $\log 1/C = 0.569 \sum f - 0.418$ (V-22)

        $n = 10; \quad r = 0.983; \quad s = 0.083; \quad F = 232; \quad \underline{t} = 15.2$

The following conclusions can be drawn:

(1) The regression equations with $\sum f$ as parameter have substantially improved quality over those with log P.

(2) The alcohols that deviate most clearly in eqns. V-19 and V-20
(CH_3-OH and $\underline{i}. C_4H_9OH$) now behave normally.

Refs. pg 173

(3) The structures of ethylene glycol and propylene glycol can be correctly fitted into the series by virtue of their Σf values.

The regression lines belonging to the (a) series are depicted graphically in Fig.V,15. Curve 1 shows clearly that the reasonable fit of the two glycols (see the residuals in Table V,10) is attained mainly at the

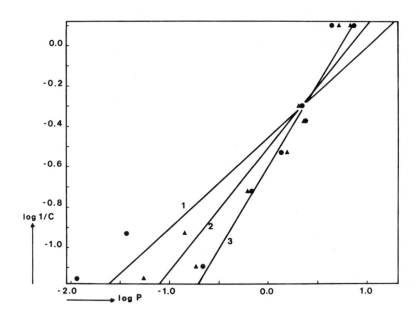

<u>Fig.V,15</u>: Graphical presentation of eqn. V-19 (curve 1), eqn. V-21 (curve 2) and eqn. V-23 (curve 3)

expense of CH_3–OH and, to a lesser extent, <u>i</u>. C_4H_9–OH. This explains why, by leaving out the data points of the two glycols, eqn. V-19 is improved considerably by eqn. V-23:

$$\log 1/C \;=\; 0.821 \log P \;-\; 0.586 \tag{V-23}$$
$$n = 7; \quad r = 0.982; \quad s = 0.088; \quad F = 137; \quad \underline{t} = 11.7$$

Eqn. V-23 (cf., curve 3 in Fig.V,14) is even of a significantly better quality than eqn. V-21; by excluding the data points of the two glycols from eqn. V-21, eqn. V-24 is obtained:

$$\log 1/C \;=\; 0.775 \, \Sigma f \;-\; 0.569 \tag{V-24}$$
$$n = 7; \quad r = 0.983; \quad s = 0.084; \quad F = 149; \quad \underline{t} = 12.2$$

Eqns. V-23 and V-24 do not differ significantly in describing activ-

ity <u>versus</u> lipophilicity, although the intercepts and slopes do not co-
incide completely.

It can be assumed that in the experiments reported, Σf acts as an
excellent replacement for the log P value used originally, and, in all
likelihood, a mistake was made in the measurement of log P for ethylene
glycol.

An improvement analogous to that described above for myoglobin denaturation can also be attained for α-chymotrypsinogen denaturation. The
result is not as significant because the regression eqn. V-20 is already
so much better than eqn. V-19.

Lipophilic Behaviour of Carbamates

HOUSTON <u>et al</u>. [39] examined the utility of partition coefficients
in the gastrointestinal absorption of a homologous series of carbamates.
Within the context of this section, the partitional behaviour rather
than the pharmacokinetic aspects of the structures studied by HOUSTON
<u>et al</u>. (reference is made to their original publication as well as to an
article by LIEN [40], where the same subject comes up for discussion) is
worth considering.

The lipophilicity values determined by HOUSTON <u>et al</u>. are presented
in column 2 in Table V,11. A system of octanol — 0.1 M phosphate buffer,
pH 7.4, was used for these determinations. As an ester of carbamic acid
has a pK value lower than 0 , the apparent partition coefficients determined in these experiments refer exclusively to the free carbamate
species and the second column in Table V,11 can, therefore, be compared
directly with the log P values given in column 3 and obtained by f addition. The correlation between experimental and calculated log P values
is not unreasonable (eqn. V-25), but there are four distinctly deviating
structures, viz., Nos 4, 7, 9 and 13. The first three of these struc-

$$\log P(obs) = 0.925 \log P(clc) + 0.036 \qquad (V-25)$$
$$n = 13; \quad r = 0.981; \quad s = 0.192; \quad F = 294; \quad \underline{t} = 17.1$$

tures have experimental values that are much lower than the calculated
values, but for No. 13 the experimental value is higher. Although
HOUSTON <u>et al</u>. assign two different names to structures 7 and 9[*], there
is every reason to assume that the correct formulation of these two
compounds should be as indicated in Table V,11. LIEN [40] seems to

[*]) The authors speak of <u>tert</u>. butyl (2,2-dimethyl ethyl), <u>tert</u>. pentyl
(2,2-dimethylpropyl) and <u>tert</u>. hexyl (3,3-dimethylbutyl), respectively.

share this opinion, considering his selection of steric TAFT parameter
values for the structures.

TABLE V,11

LIPOPHILIC BEHAVIOUR OF CARBAMATES

no.	Ester group	log P values				
		obs	clc.I	clc.II	est.II	res.
1	CH_3-	-0.66	-0.78	-0.78	-0.64	-0.02
2	CH_3-CH_2-	-0.15	-0.25	-0.25	-0.14	-0.01
3	$CH_3-CH_2-CH_2-$	0.36	0.28	0.28	0.36	0.00
4	$CH_3-\overset{\overset{CH_3}{\vert}}{\underset{\underset{CH_3}{\vert}}{C}}-$	0.48	0.78	0.32	0.40	0.08
5	$CH_3-\underset{\underset{CH_3}{\vert}}{CH}-CH_2-$	0.65	0.69	0.69	0.75	-0.10
6	$CH_3-(CH_2)_3-$	0.85	0.81	0.81	0.86	-0.01
7	$CH_3-CH_2-\overset{\overset{CH_3}{\vert}}{\underset{\underset{CH_3}{\vert}}{C}}-$	0.94	1.31	0.85	0.90	0.04
8	$CH_3-(CH_2)_4-$	1.35	1.34	1.34	1.36	-0.01
9	$CH_3-(CH_2)_2-\overset{\overset{CH_3}{\vert}}{\underset{\underset{CH_3}{\vert}}{C}}-$	1.45	1.84	1.38	1.40	0.05
10	$CH_3-(CH_2)_5-$	1.85	1.87	1.87	1.86	-0.01
11	$CH_3-(CH_2)_6-$	2.36	2.40	2.40	2.36	0.00
12	$CH_3-(CH_2)_7-$	2.85	2.93	2.93	2.86	-0.01
13	$C_6H_5-CH_2-$	1.23	0.94	0.94	(0.98)	0.25

Value in brackets not included in eqn. V-27

As stated in Chapter III, structures with a quaternary C atom bound
to a negative functional group have a partition value that is 0.46
$(= 2 \times c_M$?) lower than the normal value.

When log P(obs) is correlated with the thus corrected series of
log P(clc) values (column 4), a considerably improved regression equation (eqn. V-26) appears in which No. 13 (the benzyl ester) is clearly
an outlier: residual = 0.23, compared with a standard error of 0.084.

$$\log P(obs) = 0.943 \log P(clc) + 0.116 \qquad (V-26)$$
$$n = 13; \quad r = 0.996; \quad s = 0.084; \quad F = 1,583; \quad \underline{t} = 39.8$$

Removal of the outlier modifies this equation to

$$\log P(\text{obs}) = 0.944 \log P(\text{clc}) + 0.096 \qquad (\text{V-27})$$
$$n = 12; \quad r = 0.999; \quad s = 0.046; \quad F = 5,198; \quad \underline{t} = 72.1$$

The estimates calculated on the basis of eqn. V-27 are given in column 6 and the residuals in column 7.

This example shows clearly that the hydrophobic fragmental constant not only permits a check on the experimental partition values but also can replace them adequately.

Antihistamine Activity in a Series of Diphenhydramine Derivatives

NAUTA et al. [42] investigated the antihistamine activities of several diphenhydramine derivatives, whose structures are shown in Table V,12 together with all data relevant for a regression analysis. For information about their lipophilicities, see Tables V,7 and V,8.

The antihistamine values (anti-H_1-antagonism tested on the guinea pig ileum) can be rendered most correctly with the aid of eqn. V-27:

$$pA_2 = 1.135 \, D + 2.825 \, \ell - 0.430 \, \ell^2 - 0.577 \text{-} f(N^+)$$
$$- 0.835 \log P + 0.402 \qquad (\text{V-27})$$
$$n = 27; \quad r = 0.906; \quad s = 0.402; \quad F = 26.4; \quad \underline{t} \text{ values of the}$$
$$\text{five regressors: } 5.20, \ 3.95, \ -4.37, \ -2.25 \text{ and } -2.65, \text{ respectively.}$$

In this equation, D is a dummy parameter by which two series of structures are coupled together, viz., one with a para-methyl substituent (D = 1) and one without this substituent (D = 0), log P is the total lipophilicity of the structure, as measured in 0.1 N HCl, ℓ is the dimension of the structure in the periphery of the N atom and is measured at the greatest width perpendicular to the main axis of the structure; for a flexible chain pattern, the energetically most favourable conformations were chosen (in general, these are the forms with optimal staggering). The values of the $f(N^+)$ parameter in this equation were taken from Table V,6.

As pointed out previously, $f(N^+)$ is, in fact, a lipophilicity parameter that is closely connected with the charge distribution pattern, and it could therefore be qualified as a disguised parameter of the σ^* type. A large negative $f(N^+)$ value is associated with the diversion of the original N^+ charge over the substituents, which by itself promotes activity. Introduction of these substituents, however, leads to a relative log P increase, thus decreasing the activity. It follows that

Refs. pg 173

TABLE V,12

ANTIHISTAMINE ACTIVITIES AND RELEVANT PARAMETERS OF DIPHENHYDRAMINES

No	Alkylene-imines	Alkyl-amines	log P		ℓ in A	$f(N^+)$	pA$_2$	
	DIPHENHYDRAMINES		a	b			a	b
1		$-N^+(H)(H)(H)$	0.64*	0.12	1.68	−3.45	6.8	5.6
2			0.58*	0.06	2.67	−4.20	7.7	7.2
3			0.72*	0.20	2.69	−4.59	8.3	6.8
4			1.09*	0.57*	3.42	−4.75	8.0	7.1
5			0.48*	−0.04	2.20	−4.66	8.0(c)	7.4
6			0.41	−0.12	3.65	−5.08	8.7	8.0
7			0.56*	0.04	3.37	−5.11	10.0	8.7
8			0.66*	0.14*	3.66	−5.35	8.5	8.2
9			0.97	0.45	4.41	−5.57	7.8	7.7
10			0.77	0.25	3.55	−5.43	9.5	8.6
11			0.78	0.26*	3.67	−5.76	8.7	7.4
12			1.19*	0.67*	4.42	−5.88	6.5	5.7
13			1.01	0.48	3.65	−5.73	8.5	8.0
14			1.53	1.01*	5.17	−6.07	6.8	d

TABLE V,12 (continued)

15		1.39	0.85	3.96	-5.89	8.1 7.6

a Para-methyl-substituted derivatives
b Unsubstituted derivatives
c Irreversible short-acting antagonists
d Not determined because of too high pD_2' values
* Interpolated or derived log P values

there exists a critical interplay between charge and lipophilicity in
the cationic tail of the diphenhydramine molecule, the result being an
optimal level of activity. By reason of certain dimensional relations,
there are, however, some restrictions on the interplay, as illustrated
in the parameters ℓ and ℓ^2. The optimal value for ℓ is ℓ_0 = 3.30, which
dimensions the structure with an N-four-membered-ring. For further information on the antihistamine activities of diphenhydrames and the results of modifications in the substitution pattern of the aromatic
rings, reference is made to the publications by HARMS et al. [43],
REKKER et al. [44] and NAUTA and REKKER [46].

REFERENCES

1 R.W. Taft, in M.S. Newman (Editor), Steric Effects in Organic
   Chemistry, John Wiley & Sons, Inc., New York, Chapman & Hall,
   Ltd., New York , 1956.

2 W. Draber, K. Büchel, K. Dickore, A. Trebst and E. Pistorius,
   Progr. Photosyn. Res., 3(1969)1789.

3 A. Leo, C. Hansch and D. Elkins, Chem. Rev., 71(1971)525.

4 E.A. Braude, Ann. Reps. on Progr. Chem. (Chem. Soc. London),
   42(1945)105.

5 C.M. Moser and A.I. Kohlenberg, J. Chem. Soc., (1951)854.

6 L. Doub and J.M. Vandenbelt, J. Amer. Chem. Soc., 69(1947)2714.

7 J.R. Platt and H.B. Klevens, J. Chem. Phys., 17(1949)470.

8 M.G. Mayer and A.L. Sklar, J. Chem. Phys., 6(1938)645.

9 E.A. Braude and F. Sondheimer, J. Chem. Soc., (1955)3754.

10 E.A. Braude and W.F. Forbes, J. Chem. Soc., (1955)3776.

11 A.M. de Roos, Rec. trav. chim. Pays Bas, 87(1968)1359.

12 P. Zuman, O. Exner, R.F. Rekker and W.Th. Nauta, Coll. Czech.
    Chem. Commun. 33(1968)3213.

13 C. Church, unpublished analysis (see ref. [45]).

14 J. Fastrez and A.R. Fersht, Biochem., 12(1973)1067.

15 H.M. de Kort, W.F.C. Vlaardingerbroek and R.F. Rekker, unpublished analysis.

16 T. Fujita, J. Iwasa and C. Hansch, J. Med. Chem., 8(1965)150.

17 W.J. Dunn III, private communication to Hansch (see ref. [45]).

18 J.C. Dearden and E. Tomlinson, J. Pharm. Pharmac., 73S(1971)73.

19 W.L. McKenzie and W.O. Foye, J. Med. Chem., 15(1972)291.

20 H.M. de Kort and R.F. Rekker, unpublished analysis.

21 K. Kim and C. Hansch, unpublished analysis (see ref. [45]).

22 R.F. Rekker and W.Th. Nauta, Rec. trav. Chim., Pays Bas, 80(1961)747.

23 J.C. Dearden and J.H. O'Hara, Relations Structure – Activité, Séminaire Paris – 25/26 Mars 1974; Edité par la Société de Chimie Thérapeutique.

24 J.C. Dearden and J.H. Tubby, J. Pharm. Pharmac., 26S(1974)73P.

25 M. Tichy and K. Bocek, communicated to Hansch (see ref. [45]).

26 S. Roth and P. Seeman, Biochem. Biophys. Acta,

27 W.J. Dunn III and C. Hansch, unpublished analysis (see ref. [45]).

28 T. Fujita, J. Iwasa and C. Hansch, J. Amer. Chem. Soc., 86(1964)5175.

29 E. Kutter, unpublished results (see ref. [45]).

30 J.N. Smith and R.T. Williams, Biochem. J., 42(1948)538.

31 R.A. Friedel and M. Orchin, Ultraviolet Spectra of Aromatic Compounds, John Wiley & Sons Inc., New York, Chapman & Hall, Ltd., London, 1951.

32 M.E. Eldefrawi and R.D. O'Brien, J. Experim. Biol., 46(1967)1.

33 L.B. Kier, J. Med. Chem., 11(1968)441.

34 B. Pullman and P. Courrière, Théoret. Chim. Acta, 31(1973)19.

35 A. Pullman and G.N.J. Port, Théoret. Chim. Acta, 32(1973)77.

36 A. Pullman and B. Pullman, Quart. Rev. Biophysics, 7(1975)505.

37 T.T. Herskovits, B. Gadegbeku and H. Jaillet, J. Biol. Chem., 245(1970)2588.

38 C. Hansch and W.J. Dunn, J. Pharm. Sci. 61(1972)1.

39 J.H. Houston, D.G. Upshall and J.W. Bridges, J. Pharmacol. Exp. Therap. 189(1974)244.

40 E.J. Lien, in Medicinal Chemistry IV, Proceedings of the 4th International Symposium on Medicinal Chemistry (Editor J. Maas), Noordwijkerhout, The Netherlands, September 9–13, 1974.

41 D.H. McDaniel and H.C. Brown, J. Org. Chem., 23(1958)420.

42 W.Th. Nauta, T. Bultsma, R.F. Rekker and H. Timmerman, in Medicinal Chemistry – Special Contributions presented at the 3rd International Symposium on Medicinal Chemistry – Milan 1972 (Editor P. Pratesi), London, Butterworths 1973.

43 A.F. Harms, W. Hespe, W.Th. Nauta, R.F. Rekker, H. Timmerman and
    J. de Vries, in Medicinal Chemistry, a Series of Monographs (Editor
    G. deStevens), VOLUME 11: Drug Design (Editor E.J. Ariëns),Vol. VI,
    Academic Press, New York, San Francisco, London, 1975.

44 R.F. Rekker, W.Th. Nauta, T. Bultsma and C.G. Waringa,
    Eur. J. Med. Chem., 10(1975)557.

45 Lipophilicity data issued by Pomona College Medicinal
    Chemistry Project.

46 W.Th. Nauta and R.F. Rekker, Structure – Activity Relationships
    of H_1-Receptor Antagonists, in Handbook of Experimental Pharma-
    cology, Vol. XVIII/2 (Editor: M. Rocha e Silva), Springer Verlag,
    Heidelberg (in press).

PART II

MUTUAL DIFFERENCES BETWEEN PARTITIONING SOLVENT SYSTEMS
AND THE CHARACTERIZATION OF MEMBRANE SYSTEMS

CHAPTER VI

SYSTEMS OTHER THAN OCTANOL — WATER

INTRODUCTION

In SAR studies, preference is given to a partition coefficient that
has either been determined or calculated in the octanol — water system,
where lipophilicity is considered to play a role in finding satisfactory interpretations. HANSCH et al. in particular have frequently decided in favour of this system. The octanol — water system has, no
doubt, its merits.

An important point seems to be that the larger part of the octanol
and the water molecules which it dissolves (a relatively high concentration of water in the octanol phase, amounting to 1.72 M, should be
borne in mind) are tied up in a tetrahedral hydrogen-bonded complex,
which, because of the presence of the four 8-C non-polar chains enclos-
ing the polar core of the complex, has retained much of its hydrophobicity. With its fairly high polarity, the system as a whole precludes
self-association of partitioned solutes, such as carboxylic acids, while
in addition the stability of the tetrahedral complex is sufficient to
prevent a solute molecule from being exchanged for an alcohol molecule
of the complex [1].

To summarize, it can be said that in the complexed organic phase of
the octanol — water system, hydrophilicity and lipophilicity are well
balanced. However, this fact does not necessarily imply that the octanol — water system would be ideally suited for describing general drug
transport, including passage through membranes. Biological systems
building up membranes are infinitely complicated in comparison with

Refs. pg 198

TABLE VI,1

PARTITIONING SYSTEMS

lipoid partner	number of structures	number of measurements
octanol	2384	2898
chloroform	857	1142
cyclohexane	723	806
diethylether	597	833
various oils	330	454
norm. heptane	276	309
benzene	232	317
oleyl alcohol	188	213
iso butanol	158	176
carbon tetrachloride	154	179
others (61)	1567	2695
	7466	10022

Note: This table presents a condensation of 605 references,
240 of which were published before 1960 and the others
in the period between 1960 and 1974.

such a simple system as octanol — water. Such complications are encountered especially in the polar portion of these biological systems,
which, by virtue of their ionic structure, can retain large amounts of
water and, as a consequence, possess a hydrophilicity — lipophilicity
ratio that is bound to be of a more pronounced character than in the
octanol — water system.

The importance of the octanol — water system under practical conditions can be seen from the survey of partitioning systems in Table VI,1.

Until 1973, more than 30% of partition coefficients had been measured in the octanol — water system, and it even appears to have been
used in 40% of all partition measurements if only the 10 most common
solvent systems are considered. It would not be entirely justifiable,
however, to conclude from these figures that nowadays medicinal chemists
unanimously prefer the octanol — water system to any other.

Since 1960, at least 53 publications have appeared where other systems were opted for, namely chloroform — water (18), cyclohexane — water (18), diethylether — water (8), benzene — water (6) and oil — water
(3). It seems that there are investigators who, presumably on no more

than intuitive grounds, consider, for example, chloroform — water or cyclohexane — water as the system of choice rather than octanol — water.

The aims in Part II are:

(1) to deal with the various solvent systems by analogy with our discussion of the octanol — water system in Part I;

(2) to establish the extent to which the information to be collected about hydrophobic fragmental constants could possibly assist us in characterizing membrane lipids.

We shall first consider some attempts made previously to correlate partitioning data from two different solvent systems.

THE COLLANDER EQUATION

Although SMITH [2] has pointed out that, in principle, it should be possible to establish a link between partition coefficients measured in different solvent systems, it being assumed that the solute sets are not too dissimilar, COLLANDER [3,4] was the first to elaborate this type of relationship.

While investigating the partitional behaviour of about 50 different structures (relatively weak acids and bases as well as neutral compounds) in the system _iso_ butanol — water, _iso_ pentanol — water, octanol — water and oleyl alcohol — water, he inferred that with any given solute the partition coefficient measured in system a (P_a) should be capable of being related to the value measured in system b (P_b) using the equation

$$\log P_a = \rho \log P_b + q \qquad \qquad (VI-1)$$

where ρ and q are constants that are characteristic of the solvent systems employed. An example of such a relationship is given in eqn. VI-2, which connects the partition values in the octanol — water system with those in the _iso_ butanol — water system:

$$\log P(\text{octanol}) = 1.24 \log P(\underline{iso} \text{ butanol}) - 0.42 \qquad (VI-2)$$

COLLANDER found the fit to be poorest with those structures which contained two or more hydrophilic groups, and he suggested that this effect could be due to differences in hydrogen bonding by pointing out that in butanol comparatively more hydrogen bonds will be formed between solute and solvent molecules than in octanol.

Refs. pg 198

SOLVENT REGRESSION EQUATIONS OF LEO et al.

HANSCH and co-workers [5] extended COLLANDER's work, and stressed
that a similarity between the solvent systems is not strictly required,
so that a reasonable correlation may exist between partition coefficients in the chloroform — water system on the one hand and carbon
tetrachloride — water, xylene — water, benzene — water and iso amyl
acetate — water on the other. The quality of the correlation appears to
be best when in the COLLANDER equation two solvents of equal polarity
are involved. The simultaneous presence of a polar and a non-polar solvent may cause the correlation to deteriorate severely. When viewed in
the light of hydrogen bonding differences between polar and non-polar
solvents, this observation agrees well with what COLLANDER noted with
respect to the less correct fit of structures possessing several hydrophilic groups in eqn. VI-2.

LEO et al. [6,7] compared 20 different partitioning systems with
octanol — water and derived correlations of the COLLANDER type from the
results obtained. In doing so, they began to appreciate that in order
to secure workable statistics, they had to introduce some differentiation among the structures to be applied in a given regression equation.
The necessity to differentiate or not depends on the question of whether the COLLANDER equation refers to a combination of a polar and a
non-polar solvent or to a combination of two polar solvents.

The 103 structures in the diethylether — water versus octanol — water regression afford the following equation:

$$\log P(\text{ether}) = 1.186 \log P(\text{octanol}) - 0.472 \qquad (\text{VI-3})$$
$$n = 103; \quad r = 0.929; \quad s = 0.477; \quad F = 640; \quad \underline{t} = 25.3$$

By dividing these 103 structures into so-called donor and acceptor
compounds, two different equations are obtained:

$$\log P(\text{ether}) = 1.133 \log P(\text{octanol}) - 0.168 \qquad (\text{VI-4})$$
$$n = 71 \text{ donors}; \quad r = 0.988; \quad s = 0.185; \quad F = 2,829; \quad \underline{t} = 53.2$$

$$\log P(\text{ether}) = 1.141 \log P(\text{octanol}) - 1.068 \qquad (\text{VI-5})$$
$$n = 32 \text{ acceptors}; \quad r = 0.956; \quad s = 0.328; \quad F = 320; \quad \underline{t} = 17.9$$

It will be noted that the statistics of eqns. VI-4 and VI-5 have
been improved considerably in every respect.

In some instances, the two solvent systems cannot be correlated

correctly unless a further differentiation is made, e.g., in the system
chloroform – water _versus_ octanol – water. The regression equation for
all 72 compounds taken into consideration can be written as

$$\log P(\text{chloroform}) = 1.012 \log P(\text{octanol}) - 0.513 \qquad \text{(VI-6)}$$
$$n = 72; \quad r = 0.811; \quad s = 0.733; \quad F = 135; \quad \underline{t} = 11.6$$

Eqns. VI-7, VI-8 and VI-9 are obtained by differentiation of the 72
structures into 28 donor, 21 acceptor and 23 neutral compounds:

$$\log P(\text{chloroform}) = 1.127 \log P(\text{octanol}) - 1.343 \qquad \text{(VI-7)}$$
$$n = 28 \text{ \underline{donors}}; \quad r = 0.967; \quad s = 0.307; \quad F = 375; \quad \underline{t} = 19.4$$

$$\log P(\text{chloroform}) = 1.276 \log P(\text{octanol}) + 0.171 \qquad \text{(VI-8)}$$
$$n = 21 \text{ \underline{acceptors}}; \quad r = 0.975; \quad s = 0.251; \quad F = 377; \quad \underline{t} = 19.4$$

$$\log P(\text{chloroform}) = 1.105 \log P(\text{octanol}) - 0.649 \qquad \text{(VI-9)}$$
$$n = 23 \text{ \underline{neutrals}}; \quad r = 0.971; \quad s = 0.292; \quad F = 354; \quad \underline{t} = 18.8$$

Again, the statistical improvement is appreciable. It is worth
noting that at first only 66% but now about 94% of the variances can be
accounted for. The poorest COLLANDER regression emerges from the cyclohexane – water system: for 56 structures, a correlation coefficient as
low as 0.649 can be obtained. The differentiation into donor and acceptor compounds merely improves the fit of the acceptor structures (see
eqn. VI-10). Having a regression equation where r = 0.761, the donor
structures appear to undergo no material improvement.

$$\log P(\text{cyclohexane}) = 1.061 \log P(\text{octanol}) - 0.728 \qquad \text{(VI-10)}$$
$$n = 30 \text{ \underline{acceptors}}; \quad r = 0.957; \quad s = 0.358; \quad F = 307; \quad \underline{t} = 17.5$$

Similarly poor results are obtained with the heptane – water system.
Again, the donor structures fail to give an acceptable regression,
whereas the acceptor solutes can correctly be described by eqn. VI-11:

$$\log P(\text{heptane}) = 1.848 \log P(\text{octanol}) - 2.223 \qquad \text{(VI-11)}$$
$$n = 11 \text{ \underline{acceptors}}; \quad r = 0.954; \quad s = 0.534; \quad F = 91.6; \quad \underline{t} = 9.57$$

On practical grounds, and prompted by findings of HIGUCHI _et al._
[8] and TAFT _et al._ [9], LEO _et al._ divided the structures into donor
and acceptor compounds. In principle, this division provides a differentiation between "minus deviants" and "plus deviants", implying that
hydrogen-bond donors belong to the former and hydrogen-bond acceptors
to the latter series, an exceptional case being the diethylether – water system where the deviations have the opposite signs.

Refs. pg 198

A complete review of donors and acceptors, as drawn up by LEO et al., is presented in Table VI,2.

TABLE VI,2

GENERAL SOLUTE SYSTEMS

H donors	
	1. Acids
	2. Phenols
	3. Barbiturates
	4. Alcohols
	5. Amides (negatively substituted, but not di-N-substituted)
	6. Sulfonamides
	7. Nitriles
	8. Imides
	9. Amides
H acceptors	10. Aromatic amines (not di-N-substituted)
	11. Miscellaneous acceptors
	12. Aromatic hydrocarbons
	13. Intramolecular H bonds
	14. Ethers
	15. Esters
	16. Ketones
	17. Aliphatic amines and imines
	18. Tertiary amines (including ring N compounds)

When applying Table VI,2, it should be borne in mind that

(1) Nos. 3–9 are "neutral" in the chloroform – water and carbon tetra-
     chloride – water systems;

(2) Nos. 9 and 10 need to be reversed for the diethylether – water and
     oil – water systems;

(3) if both a donor and an acceptor group occur in the same solute,
     the effect of either as a rule predominates. Which of the two
     groups will eventually determine the final classification depends
     on, amongst others, the solvent system. In the chloroform – water
     and carbon tetrachloride – water systems, both of which have a
     "neutral" equation, such a mixed structure will perhaps fit best
     in the "neutral" equation.

THE APPLICATION OF SOLVENT REGRESSION EQUATIONS

The solvent regression equations were used by HANSCH et al. in the calculation of log P(octanol) values that could not be gathered from direct measurements [10]. A few words of comment would seem to be called for on the procedure employed in this conversion. The solvent regression equations, in their computed form, give log P(octanol) as an implicit term, whereas in the calculation of log P(octanol) the term purports to be explicit. Such changes of im- and explicit terms are allowed in algebraic functions but not in a regression equation. As the correlation coefficients become smaller, the deviations will grow larger, and cannot be avoided unless the roles of dependent and independent parameters are interchanged prior to the computation of the solvent regression equation. This interchange of parameters was carried out for eqns. VI-12 and VI-13 and they will enable us to find "correct" log P(octanol) values. This is not true, however, for eqns. VI-14 and VI-15, which are obtained simply by changing implicit for explicit and vice versa in eqns. VI-3 and VI-6, respectively.

$$\log P(\text{octanol}) = 0.728 \log P(\text{ether}) + 0.407 \qquad (\text{VI-12})$$
$$n = 103; \quad r = 0.929; \quad s = 0.374; \quad F = 640; \quad \underline{t} = 25.3$$

$$\log P(\text{octanol}) = 0.650 \log P(\text{chloroform}) + 0.731 \qquad (\text{VI-13})$$
$$n = 72; \quad r = 0.811; \quad s = 0.588; \quad F = 135; \quad \underline{t} = 11.$$

$$\log P(\text{octanol}) = 0.843 \log P(\text{ether}) + 0.398 \qquad (\text{VI-14})$$

$$\log P(\text{octanol}) = 0.988 \log P(\text{chloroform}) + 0.507 \qquad (\text{VI-15})$$

It will be noted from the examples given that especially the slopes of the regressions can deviate considerably. Where necessary, we shall make use exclusively of equations in which log P(calc) functions explicitly.

Table VI-3 shows the results of some calculations on structures with a carbonamide or related group. The reason for choosing this group was that in Table VI,2 it represents a borderline case.

When we realize that in an aliphatic amine the lone pair of electrons is freely available (for hydrogen bonding etc.) and that in an aniline this lone pair is involved in the resonance of the benzene ring, there is a rational basis for the fact that an aromatic amine occurs as a markedly weakened acceptor relative to an aliphatic amine.

Refs. pg 198

$$H_2\ddot{N} - CH_3$$

acceptor (no. 17)

$$H_2\ddot{N} \text{—} \text{(benzene ring)}$$

acceptor (no. 10)

$$H_2\ddot{N} - \underset{\underset{O}{\|}}{C} - CH_3$$

donor (no. 9)

$$H_2\ddot{N} - \underset{\underset{O}{\|}}{C} - \ddot{O} - R$$

acceptor (?)

This will apply *mutatis mutandi* to a carbonamide. It is, therefore, not
surprising that compared with an aniline, a carbonamide might act as a
donor and that under certain conditions even a reversal of the positions of the two structures in the donor–acceptor scheme may occur.
With this in mind, urethane will be an interesting structure to consider; through the competitive interaction of the lone pair on the –O–
atom with C=O, the interaction of the N-lone pair with C=O is impaired
and there is a great chance that a urethane with its NH_2 group acts as
an acceptor again, and in combination with the COO group (also denoted
as an acceptor in Table VI,2), the overall character of a urethane
should then be clearly accepting. It can be concluded from Table VI,3
that this reasoning, sound as it may seem, hardly works in practice;
obviously, the partitional behaviour of a urethane, as far as it is
concerned with donor–acceptor properties, is unpredictable.

It is possible to point to other difficult cases. Ketones, for example, operate as acceptors in Table VI,2, and this is in entire agreement with the practice of organic chemistry. No chemist will doubt the
correctness of this designation; even the qualification "stronger acceptor than an O–C=O" is evident for the C=O group. Yet a comparison of
the partition values reveals that the C=O group as a rule acts as a
donor rather than as an acceptor.

The intercept value in a solvent regression equation is obviously
linked with the lipophilicity of the organic phase of the solvent system. By definition, the intercept is the log P value for any solute
that is distributed equally between octanol and water, i.e., log P(oct)
= 0, and intercept values may be either negative or positive. A negative
intercept value implies that the solvent is more lipophilic than octanol, whereas a positive value denotes a relatively more hydrophilic behaviour.

LEO et al. stress that the intercept values of the solvent regres-

TABLE VI,3

EXAMPLES OF CONVERTING log P(solv) VALUES INTO log P(oct)

VALUES USING D, A, N OR S-EQUATIONS

Solvent system */water	log P	ref.	log P(oct) D	A	N	S	calc
Acetamide							
octanol	nm						-1.25
dieth.ether	-2.60	[11]	-2.15	-1.34			
dieth.ether	-2.60	[12]	-2.15	-1.34			
chloroform	-2.00	[12]	-0.58	-1.70	-1.26		
oil	-3.08	[13]	-1.61	-2.46			
oil	-3.08	[14]	-1.61	-2.46			
Thioacetamide							
octanol	nm						-0.45
dieth.ether	-0.55	[12]	-0.34	0.46			
chloroform	-1.14	[12]	0.18	-1.03	-0.48		
Barbituric acid							
octanol	-1.41	[15]					
octanol	-1.47	[16]					
dieth.ether	-1.63	[17]	-1.29	-0.49			
chloroform	-2.10	[18]	-0.67	-1.78	-1.35		
n. butanol	-1.16	[19]				-2.21	
Methyl urethane							
octanol	nm						-0.73
dieth.ether	-0.85	[11]	-0.63	0.09			
oil	-1.60	[13]	-0.26	-0.88			
oil	-1.40	[20]	-0.04	-0.72			
oil	-1.40	[21]	-0.04	-0.72			
Ethyl urethane							
octanol	-0.15	[16]					-0.20
dieth.ether	-0.19	[11]	-0.04	0.68			
oil	-1.12	[13]	0.22	-0.49			
oil	-0.92	[22]	0.38	-0.32			
oil	-1.52	[23]	-0.15	-0.82			
oil	-0.85	[21]	0.44	-0.26			
oil	-1.00	[24]	0.30	-0.39			

Refs. pg 198

Notes to Table VI,3:

    D-eqn. = donor equation
    A-eqn. = acceptor equation
    N-eqn. = neutral equation
    S-eqn. = sole equation (no differentiation among the structures
             is necessary)
    nm = not measured
    Underlined values are consistent with experimental log P(oct)
    values (barbituric acid and ethyl urethane) within 0.20 unit
    or approximate predictable log P(oct) values (acetamide, thio-
    acetamide and methyl urethane)

sion equations are clearly related to the extent to which water is dissolved in the organic phase of the solvent system. Regression equation
VI-16 points this way:

$$\log (H_2O) = 1.076 \, [\text{intercept}] + 0.249 \qquad\qquad (VI\text{-}16)$$
$$n = 17; \quad r = 0.978; \quad s = 0.216; \quad F = 340; \quad \underline{t} = 18.4$$

where (H_2O) is the water concentration at saturation and $[\text{intercept}]$
represents the intercept values of seven "donor" and ten "sole" equations.

As regards the slope of a solvent equation, LEO et al. are of the
opinion that if hydrogen bonding were accounted for separately, the
slopes of all the equations connecting any lipophilic solvent system
with the octanol — water system would be unity. An important consequence
of this statement will be dealt with in more detail in the following
section, which discusses the increments to hydrogen bonding, as reported by SEILER.

In conclusion to the discussion of solvent regression equations, we
wish to draw attention to the subject "outliers"; some discussion seems
useful and we will take the solvent combinations diethylether — octanol
and chloroform — octanol as specific instances.

Tables VI,4 and VI,5 give the details of how we manipulated the material described by LEO et al. The outlier criterion proposed in Part I
for deriving hydrophobic fragmental constants was applied unchanged; if
the double s-value is exceeded by the residual value (= observed minus
estimate), the data point in question should be qualified as an outlier. The outliers are removed one at a time, and each time the regression equation is recalculated and the table of residuals checked as
indicated.

H-donor solutes in diethylether — octanol. Apart from the two outliers
removed by LEO et al., we had to exclude nine others from the regres-

TABLE VI,4

OUTLIERS IN DIETHYLETHER — OCTANOL SOLVENT REGRESSION

No	Solute	log P (octanol)	log P(ether) obs.	calc. I	calc. II
	H-donor solutes				
1	alanine*	-2.94 [16]	-5.85 [11]	-3.50 (-2.35)	-3.52 (-2.33)
37	chloramphenicol*	1.14 [16]	0.62 [25]	1.12 (-0.50)	1.10 (-0.48)
49	azelaic acid	1.57 [4]	1.20 [26]	1.61 (-0.41)	1.58 (-0.38)
39	bromoacetic acid	0.41 [4]	0.69 [2, 17]	0.30 (0.39)	0.27 (0.47)
48	α-bromopropionic acid	0.92 [4]	1.24 [2, 11, 27]	0.87 (0.37)	0.85 (0.39)
16	formic acid	-0.54 [4]	-0.44 [2, 17, 28 29, 30]	-0.78 0.34	-0.80 0.36
71	o-nitrophenol	1.76 [31, 32]	2.18 [11]	1.83 (0.35)	1.80 (0.38)
24	acetonitrile	-0.34 [33]	-0.22 [11]	-0.55 (0.33)	-0.58 (0.36)
62	m-nitroaniline	1.37 [31, 32]	1.71 [12]	1.38 (0.33)	1.36 (0.35)
44	p-aminobenzoic acid	0.68 [16]	0.88 [11]	0.60 (0.26)	0.58 (0.30)
7	2,3-butylene glycol	-0.92 [4]	-1.54 [11]	-1.21 (-0.33)	-1.23 (0.31)
	H-acceptor solutes				
31	strychnine	1.93 [16]	0.34 [11]	1.13 (-0.79)	1.96 (-1.62)
32	methyl acetate*	0.18 [4]	0.43 [11]	-0.86 (1.29)	-0.88 (1.31)
22	acetone*	-0.24 [4]	-0.21 [11]	-1.34 (1.13)	-1.40 (1.19)
23	urethane*	-0.15 [16]	-0.19 [11]	-1.24 (1.05)	-1.26 (1.07)
36	ethyl acetate*	0.70 [4, 33]	0.93 [11]	-0.27 (1.20)	0.06 (0.87)
35	atropine	1.79 [16]	0.61 [11]	0.97 (-0.36)	1.75 (-1.14)
26	benzimidazole	1.34 [16]	-0.02 [12]	0.46 (-0.48)	1.05 (-1.07)
38	ethylether*	0.80 [4, 16]	1.00 [11]	-0.16 (1.16)	0.21 (0.79)

Refs. pg 198

TABLE VI,4 (continued)

No	Solute	log P (octanol)	log P (ether) obs.	calc. I	calc. II
16	morphine	0.70 [16]	0.93 [33]	−0.27 (−0.41)	0.06 (−0.74)
37	dipropylamine	1.73 [16]	0.95 [11]	0.91 (0.04)	1.75 (−0.67)
7	thiourea	−1.14 [16]	−2.14 [11, 12, 34]	−2.37 (0.23)	−2.84 (0.70)
11	dimethylamine	−0.50[a]	−1.22 [11]	−1.64 (0.42)	−1.84 (0.62)
1	piperazine*	−1.17 [4]	−3.28 [11]	−2.40 (−0.88)	−2.85 (−0.43)
39	acetophenone*	1.58 [31]	1.75 [35]	0.73 (1.02)	1.42 (0.33)
40	methyliodide*	1.69 [4]	1.92 [11]	0.86 (1.06)	1.59 (0.33)
41	ethyliodide*	2.00 [33]	2.45 [11]	1.21 (1.24)	2.07 (0.38)
42	anisole*	2.08 [31, 46]	2.46 [36]	1.31 (1.15)	2.20 (0.26)

Numbering corresponds to Tables XIIIa and XIIIb from the microfilm version of ref.[6].
*) The asterisked items were regarded as outliers by LEO et al.;
"Calc. I" values were obtained from eqn. VI-4 (donors) and eqn. VI-5 (acceptors).
"Calc. II" values were obtained from eqn. VI-17 (donors) and eqn. VI-18 (acceptors)
a) Not measured; log P = log P(diethylamine) − 2 x 0.50 [4, 47].
References are given in square brackets; values in parentheses are residuals.

TABLE VI,5

OUTLIERS IN CHLOROFORM — OCTANOL SOLVENT REGRESSION

No	Solute	log P (octanol)	log P (chloroform) obs.	calc. I	calc.II
H-donor solutes					
7	azelaic acid*	1.57 [4]	−0.58 [37]	0.43 (−1.01)	0.53 (−1.11)
17	salicylic acid	2.21 [16]	0.40 [38, 39, 40]	1.15 (−0.75)	1.29 (−0.89)
10	p-nitrophenol	1.96 [31]	0.18 [41, 42]	0.87 (−0.69)	0.99 (−0.81)
14	m-nitrophenol	2.00 [31, 32]	0.41 [41]	0.91 (−0.50)	1.04 (−0.63)

TABLE VI,5 (continued)

27	p-toluic acid	2.27 [31]	1.80 [38]	1.22 (0.58)	1.36 (0.44)
22	o-iodobenzoic acid	2.40 [16]	1.09 [2]	1.36 (-0.27)	1.51 (-0.42)
16	m-nitrobenzoic acid	1.83 [31]	0.48 [38]	0.72 (-0.24)	0.84 (-0.36)
	H-acceptor solutes				
2	benzimidazole*	1.34 [16]	-0.10 [12]	1.88 (-1.98)	1.88 (-1.98)
3	benzotriazole*	1.34 [33]	-0.05 [12]	1.88 (-1.93)	1.88 (-1.93)
17	nitropropane*	0.65 [33]	1.91 [43]	1.00 (0.91)	1.00 (0.91)
	H-bond neutral solutes				
7	chloramphenicol*	1.14 [16]	-0.65 [25]	0.61 (-1.26)	0.55 (-1.20)
1	thiourea*	-1.14 [16]	-3.10 [12]	-1.91 (-1.19)	-1.93 (-1.17)
6	phenylurea*	0.83 [16]	-0.72 [44]	0.27 (-0.99)	0.21 (-0.93)
22	m-nitroaniline	1.37 [31]	1.61 [12]	0.86 (0.75)	0.80 (0.81)
19	phthalimide	1.15 [16]	1.08 [44]	0.62 (0.46)	0.56 (0.52)

Numbering corresponds to Tables IXa, IXb and IXc from the microfilm
version of ref. [6].
*) The asterisked items were regarded as outliers by LEO et al.;
"Calc I" values were obtained from eqn. VI-7 (donors), VI-8 (acceptors)
and VI-9 (neutrals).
"Calc. II" values were obtained from eqn. VI-20 (donors),VI-8 (acceptors) and VI-21 (neutrals).
References are given in square brackets; values in parentheses are
residuals.

sion equation (see Table VI-4). The regression equation for the 62 remaining structures is

$$\log P(ether) = 1.130 \log P(octanol) - 0.193 \qquad (VI-17)$$

$$n = 62 \text{ donors}; \quad r = 0.993; \quad s = 0.142; \quad F = 4,272; \quad \underline{t} = 65.4$$

The quality of this regression equation shows a good improvement over
that of eqn. VI-4, although the intercept and slope are virtually unchanged.

H-acceptor solutes in diethylether - octanol. On application of the procedure described above, five of the ten structures removed by LEO et al.

Refs. pg 198

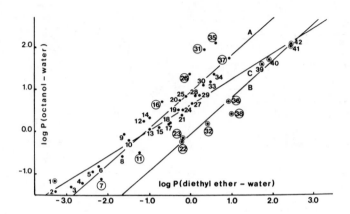

Fig.VI,1: Diethylether – octanol solvent regression of H-acceptor solutes. __A__. Omission of structures denoted by circled dots
(see eqn. VI-5); __B__. Regression of all data points denoted by
circled dots, except No. 1 (eqn. VI-19); __C__. Omission of structures denoted by circled figures (eqn. VI-18).

1. piperazine	15. i-valeramide	29. piperidine
2. diethanolamine	16. morphine	30. benzylamine
3. ethanolamine	17. propylamine	31. strychnine
4. sulfaguanidine	18. quinone	32. methyl acetate
5. i-propanolamine	19. dipterex	33. acetanilide
6. di-i-propanolamine	20. trimethylamine	34. benzotriazole
7. thiourea	21. benzamide	35. atropine
8. methylamine	22. acetone	36. ethyl acetate
9. caffeine	23. urethane	37. dipropylamine
10. n-butyramide	24. diethylamine	38. ethyl ether
11. dimethylamine	25. phenylurea	39. acetophenone
12. antipyrine	26. benzimidazole	40. methyl iodide
13. sulfathiazole	27. pyridine	41. ethyl iodide
14. nicotinic acid	28. n-butylamine	42. anisole

as well as eight data points that originally fitted correctly in eqn.
VI-5 disappeared from the data set, and there was no reason to remove
the five compounds given at the bottom of Table VI,4. The final regression equation can be written as

$$\log P(\text{ether}) \;=\; 1.552 \log P(\text{octanol}) \;-\; 1.031 \qquad (\text{VI-18})$$
$$n = 30; \quad r = 0.983; \quad s = 0.279; \quad F = 847; \quad \underline{t} = 29.1$$

Again, there is a considerable improvement in quality compared with eqn.
VI-5. The slope shows a striking increase, but really noteworthy is the
sequence of removal of the 13 outliers in comparison with the original
block of 11. This made us decide to plot eqns. VI-5 and VI-18 graphi-

cally for closer inspection (see Fig. VI,1). From this graph, it becomes
apparent that the elimination procedure of LEO et al. for outliers implies that from the total gamma a subset is split off which can be
fitted in eqn. VI-19; this equation has a slope that is not significantly different from that of the remaining data points (eqn. VI-5):

$$\log P(\text{ether}) = 1.134 \log P(\text{octanol}) + 0.082 \qquad (\text{VI-19})$$
$$n = 9; \quad r = 0.996; \quad s = 0.097; \quad F = 899; \quad \underline{t} = 30.0$$

H-donor solutes in chloroform - octanol. The H-donor solutes equation
appears with seven instead of one outlier:

$$\log P(\text{chloroform}) = 1.186 \log P(\text{octanol}) - 1.332 \qquad (\text{VI-20})$$
$$n = 22; \quad r = 0.993; \quad s = 0.151; \quad F = 1{,}481; \quad \underline{t} = 38.5$$

H-acceptor solutes in chloroform - octanol. The H-acceptor solutes equation is not changed by applying our outlier criterion.
H-bond neutral solutes in chloroform - octanol. Our H-bond neutral solutes equation rejects two outliers more than the three that were removed by LEO et al.:

$$\log P(\text{chloroform}) = 1.090 \log P(\text{octanol}) - 0.692 \qquad (\text{VI-21})$$
$$n = 21; \quad r = 0.984; \quad s = 0.202; \quad F = 580; \quad \underline{t} = 24.1$$

Eqns. VI-20 and VI-21 show a good deal of improvement over eqns.
VI-7 and VI-9. The other solvent regression equations can be improved
appreciably in an analogous manner. The total picture remains virtually
unchanged, however, i.e., the slopes do not definitely approach unity
more closely, even though this might have been expected on considering
the marked improvements in the quality of the solvent regression equations; in addition, eqn. VI-16 undergoes no improvement. We are inclined
to conclude that although the differentiation referred to is based on
significant grounds, several questions have remained unanswered.

In this connection, it is not so much the question of why the behaviour of certain structures (for example, the ketones referred to previously in this chapter) is obscure, but rather the remarkable difference between the cyclohexane - water and heptane - water systems on the
one hand and octanol - water on the other that is worthy of note.

According to eqns. VI-10 and VI-11, nearly all H-acceptor solutes
should possess considerably different partition values, depending on
whether they are measured in the cyclohexane - water or in the heptane -
- water system, which does not seem plausible.

Refs. pg 198

Examples:

I: log P(octanol) values chosen arbitrarily	−1.0 0.0 +1.0
II: log P(cyclohexane)	−1.80 −0.73 +0.33
III: log P(heptane)	−4.07 −2.22 −0.38

$$\Delta(\text{II} - \text{III}): \quad 2.27 \quad 1.49 \quad 0.71$$

SEILER's INCREMENTS TO HYDROGEN BONDING

SEILER [45] carried out an interesting investigation into the relationships between the log P values in different saturated hydrocarbon —
— water systems on the one hand and the cyclohexane — water system on
the other. It appears that these relationships can be fitted in the
usual COLLANDER pattern:

$$\log P(\text{cyclohexane}) = a \log P(\text{hydrocarbon}) + b \qquad (\text{VI--22})$$

a and b being remarkably constant:

hexane	a = 1.05 ± 0.05	b = 0.02 ± 0.05
heptane	0.99 ± 0.02	0.13 ± 0.02
octane	1.00 ± 0.05	0.20 ± 0.04
hexadecane	1.03 ± 0.02	0.19 ± 0.02

The four regressions were statistically correct; thus, r varied from
0.98 to 1.00 and s from 0.04 to 0.24.

SEILER's next step was to correlate as many log P values measured
in saturated hydrocarbon — water systems as possible with their octanol — water values. To this end, all available hexane, heptane, octane
and hexadecane partition values were first converted into log P(c.hexane)
values by means of eqn. VI-22 using the appropriate a and b values. In
order to derive optimal information from the largest possible data set,
the dodecane — water system was also taken into consideration by relating it to hexadecane — water via eqn. VI-23:

$$\log P(\text{hexadecane}) = (0.95 \pm 0.04) \log P(\text{dodecane}) -$$
$$(0.12 \pm 0.03) \qquad (\text{VI--23})$$

$$n = 5[*]; \quad r = 0.99; \quad s = 0.07; \quad F = 970$$

As a consequence, a total of 230 structures[**] became available for
a comparison on the basis of eqn. VI-24:

---

[*] Structures used for this correlation included iodine, pentanol,
pentafluorophenol, hexanol and heptanol.
[**] When more than one log P value was available for any given compound,
they were averaged.

$$\log P(\text{cyclohexane}) \; = \; f(\log P[\text{octanol}]) \qquad (\text{VI-24})$$

SEILER resolved this function with reference to the postulate of LEO et
al. that if hydrogen bonding were accounted for separately, the slopes
of all solvent regression equations would be unity. This transforms eqn.
VI-24 into

$$\log P(\text{cyclohexane}) \; + \; \Sigma I_H \; = \; \log P(\text{octanol}) \; + \; b \qquad (\text{VI-25})$$

where I_H = the increment to hydrogen bonding of a given part of the
structure. Instead of eqn. VI-25, it is possible to write

$$\Delta \log P \; = \; \log P(\text{octanol}) - \log P(\text{cyclohexane}) = \Sigma I_H - b \qquad (\text{VI-26})$$

TABLE VI,6

SEILER's INCREMENTS TO HYDROGEN BONDING FOR log P(CYCLOHEXANE)
          - log P(OCTANOL) CONVERSION

No.	molecular segment	I_H
1	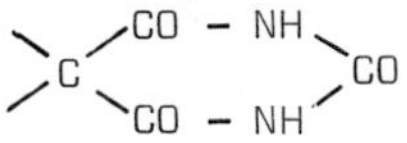 $>C<^{CO-NH}_{CO-NH}>CO$	3.06
2	COOH(al)	2.88
3	COOH(ar)	2.87
4	OH(ar)	2.60
5	CONH	2.56
6	SO_2NH	1.93
7	OH(al)	1.82
8	NH_2(al)	1.33
9	NH_2(ar)	1.18
10	N	1.01
11	NHR (R $\neq$ H)	0.61
12	$CO-CH_2-CO$	0.59
13	NR_1R_2 (R $\neq$ H)	0.55
14	NO_2	0.45
15	C=O	0.31
16	CN	0.23
17	O	0.11

Although SEILER's calculations were concerned primarily with the
conversion of the cyclohexane — octanol system, it will be clear that
in principle any solvent system lends itself for this mode of manipulation.

Refs. pg 198

The final resolution of the equation set VI-26 was performed by using the FREE — WILSON method [48], with omission of what is called the restrictive equation. The most important increment values found by SEILER are presented in Table VI,6.

After removal of 35 outliers, 195 structures remained. The correlation coefficient was r = 0.967, which meant that the model accounted for over 93% of the total of variances. The standard error of estimate was 0.333, and the result of the F-test pointed to significance at the 0.1% level. In addition to the increments proposed for pertinent structure fragments, it appeared necessary to introduce a number of corrective terms for ortho-substituents; phenols and anilines also needed their own corrections, depending on their pK values. The intercept (b) was −0.16.

What renders this investigation so important is that it demonstrates unambiguously that in designing a solvent regression equation there is no urgent need for a differentiation between donor and acceptor solutes provided that correct use is made of an increment value that reflects correctly the hydrogen bonding capacity of the various (functional) groups. There are good reasons to suggest that this manipulation is not restricted to cyclohexane — octanol conversions but will have wider applicability.

Some objections may be raised to SEILER's proposals:

(1) Adoption of the postulate of LEO et al. that in equations of the type VI-25 the slope is unity as soon as adequate corrections for hydrogen bonding have been applied means that for systems which lack these hydrogen bonding groups the following equation will be valid:

$$\Delta \log P = \log P(\text{octanol}) - \log P(\text{cyclohexane}) = -b = 0.16$$

(VI-27)

In other words, the log P value of all aliphatic, aromatic and mixed hydrocarbons as well as the f values of alkyl and aryl groups will not differ from each other in cyclohexane — water and octanol — — water by more than the small amount of 0.16.

(2) Several groups were scarcely represented in SEILER's correlations. Thus, the aliphatic COOH group occurred only once, the aliphatic NH_2 group twice and the SO_2NH group three times. This low multiplicity will certainly have contributed to the lack of overall significance of the increment values observed.

ACCOMMODATION OF MEMBRANE LIPIDS IN THE COLLECTION OF SOLVENT SYSTEMS

Although, in this context, it is not possible yet to deal quantitatively with the question of whether the octanol — water system is really the model of choice in qualifying solvent characteristics of membrane lipids, it looks inviting to speculate on some qualitative aspects.

Data on partition experiments in membrane lipid — water (buffer) systems appear to be extremely scanty and in fact only what SEEMAN et al.[49] have reported on experiments on lipids from human erythrocyte membrane (HEML) is suitable for our purpose.

HANSCH and DUNN [50] included HEML in their considerations in a review in 1972, with the following equation in which the partition values measured by SEEMAN et al. are incorporated (these values refer to five aliphatic alcohols, namely pentanol, heptanol, octanol, nonanol and decanol):

$$\log P(HEML) = 0.981 \log P(octanol) - 0.883 \quad ^* \qquad (VI-28)$$
$$n = 5; \quad r = 0.997; \quad s = 0.082; \quad F = 586; \quad \underline{t} = 24.2$$

The slope of eqn. VI-28 indicates a close parallelism between the HEML — buffer system on the one hand and the octanol — water system on the other, while the overtly negative intercept value qualifies the HEML system clearly to be comparable with a solvent system the organic phase of which has a much higher water solubility than octanol.

Extension of eqn. VI-28 to material reported later by SEEMAN [51] affords the equation

$$\log P(HEML) = 0.871 \log P(octanol) - 0.438 \qquad (VI-29)$$

This equation is taken from the "Pomona file". Statistical details are not reported and the number of data points is uncertain.

Starting from eqn. VI-28, we added stepwise all relevant data reported by SEEMAN and arrived at eqns. VI-30, VI-31 and VI-32:

(a) Addition of the structures benzyl alcohol, phenol, hexanol, 4-bromophenol and 3-methyl-4-chlorophenol[**] gives

$$\log P(HEML) = 0.906 \log P(octanol) - 0.445 \qquad (VI-30)$$

---

*) This equation is slightly different from that originally reported by HANSCH and DUNN; we used f summations instead of experimentally determined log P(oct) values
**) log P values from f summations.

Refs. pg 198

$$n = 10; \quad r = 0.988; \quad s = 0.143; \quad F = 330; \quad \underline{t} = 18.2$$

This equation does not differ appreciably from eqn. VI-29, which could suggest that eqn. VI-29 was calculated from the same data set as our eqn. VI-30. For hexanol we used log P(HEML) = 1.11 instead of the value of 0.81 employed for eqn. VI-29, the latter value apparently being incorrect.

(b) Addition of the structures urethane[*], diphenylhydantoin[**], barbital[**], phenobarbital[**] and pentobarbital[**] to eqn. VI-30 results in

$$\log P(\text{HEML}) = 0.872 \log P(\text{octanol}) - 0.375 \qquad (\text{VI-31})$$
$$n = 15; \quad r = 0.967; \quad s = 0.256; \quad F = 191; \quad \underline{t} = 13.8$$

(c) Addition of the structures chlorpromazine[**], morphine[**] and methadone[**] to eqn. VI-31 gives rise to

$$\log P(\text{HEML}) = 0.666 \log P(\text{octanol}) - 0.040 \qquad (\text{VI-32})$$
$$n = 18; \quad r = 0.909; \quad s = 0.430; \quad F = 76.5; \quad \underline{t} = 8.74$$

No outliers occur in eqns. VI-28, VI-30 and VI-31. In eqn. VI-32, methadone with a residual value of -1.07 should be regarded as a true outlier. The log P value we used for introduction in eqn. VI-32 was 3.93, which is the value for the basic species of methadone. By removing methadone from the data set, eqn. VI-33 is obtained:

$$\log P(\text{HEML}) = 0.735 \log P(\text{octanol}) - 0.121 \qquad (\text{VI-33})$$
$$n = 17; \quad r = 0.951; \quad s = 0.328; \quad F = 144; \quad \underline{t} = 12.0$$

Considering (1) the large drop in the quality of the regression when the number of data points belonging to one class of structures is extended by compounds belonging to different classes, (2) the continuously increasing difference of the slope from unity and (3) the gradual decrease in intercept value, it is pertinent to ask if the properties of HEML can really be related so directly to the octanol — water system, as eqn. VI-28 was likely to suggest.

DISCRIMINATING PROPERTIES OF SOLVENT SYSTEMS

In judging solvent systems as described in the preceding section, the need for a qualification for relating the partitional behaviour of

---

[*] log P values from *f* summations
[**] log P values from experiments (where more than one value was available, they were averaged).

one of those systems to another made itself felt. In this connection,
we propose the term "discriminating power", which denotes the spread any
given solvent system imparts to the partition values of a set of structures presented to that system for partitioning. This discriminating
power is designated by the distances p, q and r in Fig.VI,2.

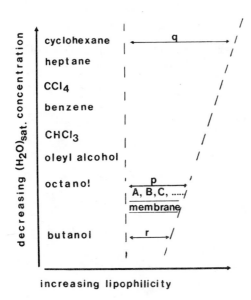

increasing lipophilicity

Fig.VI,2: Graphical representation of the discriminating properties of a number of solvent systems
(only indicated by their organic phases) as a function of the water-saturation concentration of the
organic phases; p, q and r denote the discriminating
powers of the octanol — water, cyclohexane — water
and butanol — water systems, respectively, towards
a set of partitioned structures A, B, C .... The
approximate location of the human erythrocyte membrane
lipid is indicated by: membrane.

On the whole, the lipophilicities of A, B, C ...   will be closer
together in the butanol — water system than in the octanol — water system, while in a system such as cyclohexane — water they show a relatively greater spread. Fig.IV,2 represents this by a distance r$<$p and a
distance q$>$p, respectively. The discriminating powers of a set of solvent systems may be supposedly arranged along a wedge-shaped pattern as
a function of some physico-chemical parameter that is closely connected
with the partitioning behaviour of the solvent system. It remains to be
seen, however, whether the parameter used for this purpose in Fig.VI,2

Refs. pg 198

(the $\left[H_2O\right]_{sat.}$ concentration of the organic phase) is indeed most suitable. The illustration also shows the approximate location of HEML as, in our opinion, this needs to be qualified in connection with the preceding section: it is thought to be slightly hypodiscriminative towards the octanol - water system and hyperdiscriminative towards the butanol - water system, and not isodiscriminative towards the octanol - water system as is frequently supposed.

The problem concerned with HEML are closely connected with the following questions:

(a) how to scale correctly the discriminative behaviour of a solvent system towards another;

(b) which factors do actually govern discriminative behaviour; is the $\left[H_2O\right]_{sat.}$ concentration in the organic phase an exclusively determining factor?

REFERENCES

1 R.N. Smith, C. Hansch and M.A. Ames, J. Pharm. Sci., 64(1975)599.

2 H.W. Smith, J. Phys. Chem., 25(1921)204,605.

3 R. Collander, Acta Chem. Scand., 4(1950)1085.

4 R. Collander, Acta Chem. Scand., 5(1951)774.

5 C. Hansch, Il Farmaco, Ed. Sci., 23(1968)294.

6 A. Leo and C. Hansch, J. Org. Chem., 36(1971)1539.

7 A. Leo, C. Hansch and D. Elkins, Chem. Rev., 71(1971)525.

8 T. Higuchi, J.C. Richards, S.S. Davis, A. Kamada, J. Hou, M. Nakanc, N. Nakano and I.H. Pitman, J. Pharm. Sci., 58(1969)661.

9 R.W. Taft, D. Gurka, L. Joris, R. von Schleyer and J.W. Rakshys, J. Amer. Chem. Soc., 91(1969)4794,4801.

10 C. Hansch and A. Leo, Pomona College Medicinal Chemistry Project, Issue 6, January 1975.

11 R. Collander, Acta Chem. Scand, 3(1949)717.

12 K.B. Sandell, Naturwissensch., 53(1966)330.

13 R. Collander, Phys. Plant., 7(1954)420.

14 R. Hober and J. Hober, J. Cell and Physiol., 10(1937)401.

15 E. Kutter, unpublished analysis (see ref. [10]).

16 S. Anderson, unpublished analysis (see ref. [10]).

17 O.C. Dermer, W.G. Markham and H.M. Trimble, J. Amer. Chem. Soc., 63(1941)3524.

18 C.A.M. Hogben, D.J. Tocco, B.B. Brodie and L.S. Schanker, J. Pharmacol. & Exptl. Ther., 125(1959)275.

19 J. Tinker and G.B. Brown, J. Biol. Chem., 173(1948)585.

20 K.H. Meyer and H. Gottlieb-Billroth, Z. Physiol. Chem., 112(1921)55.

21 F. Baum, Arch. Exp. Path. Pharmakol., 42(1889)119.

22 R. Macy, J. Ind. Hyg. Toxicol., 30(1948)140.

23 K.H. Meyer and H. Hemmi, Biochem. Z., 277(1935)39.

24 E. Knaffl-Lenz, Arch. Exptl. Path. Pharmacol., 84(1919)66.

25 A. Brunzell, J. Pharm. Pharmacol., 8(1956)329.

26 E.E. Chandler, J. Amer. Chem. Soc., 30(1908)696.

27 O.C. Dermer and V.H. Dermer, J. Amer. Chem. Soc., 65(1943)1653.

28 N. Schilow and L. Lepin, Z. Phys. Chem., 101(1922)353.

29 R.C. Archibald, J. Amer. Chem. Soc., 54(1932)3178.

30 F. Auerbach and H. Zeglin, Z. Phys. Chem., 103(1922)200.

31 T. Fujita, J. Iwasa and C. Hansch, J. Amer. Chem. Soc., 86(1964)5175.

32 M. Tichy and K. Bocek, communication to Hansch (see ref. [10]).

33 C. Hansch and S.M. Anderson, J. Org. Chem., 32(1967)2583.

34 E.J. Ross, J. Physiol., 112(1951)229.

35 K.B. Sandell, Monatsh. Chem., 89(1958)36.

36 J.J. Lindberg, Soc. Sci. Fennica, Commentations, Phys. Math., 21(1958)1.

37 C.S. Marvel and J.C. Richards, Anal. Chem., 21(1949)1480.

38 H.W. Smith and T.A. White, J. Phys. Chem., 33(1929)1953.

39 B. Hök, Svensk Kem. Tidskr., 65(1953)182.

40 W.S. Hendrixson, Z. Anorg. Chem., 13(1897)73.

41 W. Kemula and H. Buchowski, J. Phys. Chem., 63(1959)155.

42 W. Kemula, H. Buchowski and J. Teperek, Bull. Acad. Polon. Sci., 12(1964)347.

43 W. Kemula, H. Buchowski and J. Teperek, Bull. Acad. Polon. Sci., 12(1964)491.

44 K.B. Sandell, Monatsh. Chem., 92(1961)1066.

45 P. Seiler, Eur. J. Med. Chem., 9(1974)473.

46 K.S. Rogers and A. Cammarata, J. Med. Chem., 12(1969)692.

47 K.B. Sandell, Naturwissensch., 49(1962)12.

48 S.M. Free and J.W. Wilson, J. Med. Chem., 7(1961)395.

49 P. Seeman, S. Roth and H. Schneider, Biochim. Biophys. Acta, 225(1971)17.

50 C. Hansch and W.J. Dunn III, J. Pharm. Sci., 61(1972)1.

51 P. Seeman, Pharmacol. Rev., 24(1972)583.

52   R.F. Rekker and G.G. Nys, Proceedings of the 7th Jerusalem
     Symposium on Molecular and Quantum Pharmacology (Editors:
     E. Bergmann and B. Pullman), 1974, D. Reidel Publishing Comp.,
     Dordrecht — Holland.

CHAPTER VII

DERIVATION OF APPROPRIATE SETS OF HYDROPHOBIC FRAGMENTAL CONSTANTS FOR SEVERAL SOLVENT SYSTEMS

INTRODUCTION

As was shown in the preceding chapter, an equation of the COLLANDER type can be employed to transform information on the partition value of a solute for any given solvent system into data pertinent to another system:

$$\log P_a = \rho \log P_b + q \qquad \text{(VII-1)}$$

This correlation deteriorates as the polarities of the two organic phases involved become further apart. As soon as the difference between these polarities can be neglected, it can be taken as certain that the regression equation is of excellent quality.

It has also been demonstrated that both by differentiation in the solutes (the introduction of a donor besides acceptor groups and, if necessary, the introduction of neutral groups, as performed by LEO et al. [159]) and the application of a group increment to hydrogen bonding (as done by SEILER [160]), the correlation level can be much improved. Many problems still remain to be solved, however, especially when one wishes to class human erythrocyte membrane lipid (HEML) in the collection of solvent systems.

Our approach to these problems is as follows. The hydrophobic fragmental constant enables us to transform eqn. VII-1 into eqn. VII-2:

$$\Sigma f_a = \rho \Sigma f_b + q \qquad \text{(VII-2)}$$

provided that in each solvent system a set of hydrophobic fragmental constants can be defined. Σf_a then represents the summation of a series of fragmental constants valid for solvent system a and Σf_b the summation of the same fragmental constants in solvent system b.

The advantage of eqn. VII-2 over eqn. VII-1 is that it can be used for any desired differentiation in Σ terms. An example is the equation

$$\Sigma f_{1(a)} + \Sigma f_{2(a)} = \rho \left[\Sigma f_{1(b)} + \Sigma f_{2(b)} \right] + q \qquad \text{(VII-3)}$$

Refs. pg 262

where the subscripts 1 and 2 represent fragment qualifications. Thus, subscript 1 may denote a non-polar fragment and 2 a polar fragment.

Eqn. VII-3 permits a further differentiation that is more subtle than the procedure of LEO et al.; for instance:

$$\Sigma f_{1(a)} \;+\; \Sigma f_{2(a)} \;=\; \rho' \,\Sigma f_{1(b)} \;+\; \rho'' \,\Sigma f_{2(b)} \;+\; q' \qquad\qquad (VII-4)$$

When for the sake of brevity, the method of LEO et al. is expressed as "differentiation in the solute group", ours may be qualified as "differentiation in the solute structure". In the latter instance, it remains possible that the a/b transformations of non-polar and polar fragments are connected with different slopes, which explains the use of ρ' and ρ'' in eqn. VII-4. On application of this procedure, we even wish to allow for a possible intercept change (hence q' instead of q).

The following sections deal with the derivation of sets of hydrophobic fragmental constants for all pertinent systems, provided that enough material with sufficient accuracy is available. Chapter VIII will show how the data obtained can be arranged on the basis of eqn. VII-4.

DIETHYLETHER – WATER SYSTEM

Our first calculations were carried out by analogy with the octanol – water system considered in Chapter III, the aliphatic and aromatic structures being dealt with separately. The aliphatic series afforded an *f*-set the statistical quality of which proved remarkably good whereas, mainly as a result of a lack of workable data points, the *f*-set of the aromatic series was statistically of much poorer quality. On application of the *f* values so found to structures where our observations with the octanol – water system suggests that proximity effects occur (this type of structure had been deliberately excluded from the first diethylether – water calculations), both a p.e.(1) and a p.e.(2) appeared. As far as verification was possible, the difference between aromatic and aliphatic functional groups was found to approximate to a multiple of the same factor that apparently also operates in the proximity effects; in other words, there was every reason to perform an advanced calculation with the constant c_M as a regressor and the key number as a parameter. This calculation referred to 273 log P values belonging to 172 different structures; 108 structures (189 measurements) were of the aliphatic type and the remainder were phenyl derivatives. A complete review of all structures taken into consideration is given in Table VII,1,

and Table VII,2 (2nd column) shows which fragments were included in the
analysis.

The same criterion as described in Chapter III for establishing outliers was applied, with the result that 58 structures covered by 87
measurements and divided over 28 aliphatic structures and 30 phenyl derivatives had to be excluded from the final analysis.

The 4th column in Table VII,2 shows the key number representing the
factor in terms of the constant c_M by which the f(al) value of a functional group should be increased in order to secure its f(ar) value.

On the whole, the results are extremely good (for details, see Table
VII,2). Some points should be noted here:

(1) The congruity of the data set sometimes leaves much to be desired,
    certain structure types being insufficiently represented.

(2) As a consequence of point (1), the picture presented by the regres-
    sion analysis is awry on certain points. This is especially true for
    structures with a C≡N group, which occurs six times in the data
    set: twice in $CNCH_2CN$ (No. 15), once in $CNCH_2COOH$ (No. 16) (which
    features with three measurements in the series) and once in CH_3CN
    (outlier, No. 3). In the first two structures, a proximity effect
    on the magnitude of three key numbers has to be allowed for, while
    the third structure can be included in the computation without fur-
    ther processing. The first two structures with four data points and
    five C≡N fragments will determine the f(CN) value (f = -0.673),
    forcing CH_3CN into the position of an outlier (residual = -0.47). It
    will be shown in the following chapter that the correctness of the
    value -0.673 is open to doubt and that the real value will be in the
    region of -1.2. The implication is that the value of -0.22 found for
    the partition value of CH_3CN is reasonably correct and that the par-
    tition values for $CNCH_2CN$ and $CNCH_2COOH$ are open to criticism. It is
    not so much an error in measuring the partition values as an extra
    proximity effect[*] that could be responsible for this phenomenon,
    especially as it is unlikely that four different investigators
    would have made mistakes of exactly the same magnitude in measuring
    the partition data.

(3) The value for f(NO_2)$_{arom.}$, namely 0.952, is also interesting. It
    results from the "statistical interplay" of a number of nitro deriv-

[*)] It must be borne in mind that the CN group introduces an exceedingly
high negative inductive effect. This finds reflection in its σ^* (Taft)
constant which, with a value of ∼ 3.6, can be counted among the highest known σ^* values.

Refs. pg 262

TABLE VII,1

STRUCTURES INTRODUCED INTO STATISTICAL TREATMENT
(ETHER — WATER SYSTEM)

I,D: Log P(oct) values calculated with LEO et al.'s "H-donor solute
equation: log P(oct) = 0.862 log P(ether) + 0.158
I,A: Log P(oct) values calculated with LEO et al.'s "H-acceptor sol-
ute equation: log P(oct) = 0.801 log P(ether) + 0.880
II: Log P(oct) values obtained from *f*(oct) summation
III: Log P(oct) values derived from an equation described in Chapter
VIII (pg 274)

no	Compound	log P(ether) obs. ref.	log P(ether) clc. res.	log P(octanol) I	log P(octanol) II	log P(octanol) III
1	CH_3-I	1.92 [1]	1.73 0.19	1.81 D	1.29	1.59
2	CH_3-OH	-0.85 [1]	-0.96 0.11	-0.57 D	-0.79	-0.71
3	CH_3-NH_2	-1.64 [1]	-1.64 0.00	-0.43 A	-0.73	-0.71
4	$Cl-CH_2-COOH$	0.41 [3]	0.36 0.05	0.51 D	0.50	0.34
		0.42 [4]	0.36 0.06	0.52 D		0.35
		0.39 [5]	0.36 0.03	0.49 D		0.32
		0.37 [7]	0.36 0.01	0.48 D		0.31
5	$I-CH_2-COOH$	0.86 [1]	1.00 -0.14	0.90 D	1.02	0.71
		0.83 [8]	1.00 -0.17	0.87 D		0.69
6	CH_3-COOH	-0.36 [8]	-0.32 -0.04	-0.15 D	-0.25	-0.30
		-0.35 [9]	-0.32 -0.03	-0.14 D		-0.29
		-0.30 [1]	-0.32 0.02	-0.10 D		-0.25
		-0.33 [10]	-0.32 -0.01	-0.13 D		-0.27
		-0.30 [5]	-0.32 0.02	-0.10 D		-0.25
		-0.34 [6]	-0.32 -0.02	-0.14 D		-0.28
		-0.34 [11]	-0.32 -0.02	-0.14 D		-0.28

TABLE VII,1 (continued)

		-0.34 [12]	-0.32	-0.14 D		-0.28
7	$HO-CH_2-COOH$	-1.55 [3]	-1.69 0.14	-1.18 D	-1.05	-1.29
8	CH_3-CH_2-I	2.45 [1]	2.26 0.19	2.27 D	1.82	2.03
9	CH_3-CONH_2	-2.60 [1]	-2.48 -0.12	-1.20 A	-1.27	-1.29
		-2.60 [8]	-2.48 -0.12	-1.20 A		-1.29
10	CH_3-CH_2-OH	-0.58 [1]	-0.43 -0.15	-0.34 D	-0.26	-0.48
		-0.57 [2]	-0.43 -0.14	-0.33 D		-0.47
11	$HO-CH_2-CH_2-OH$	-2.27 [1]	-2.10 -0.17	-1.80 D	-1.35	-1.88
12	$CH_3-NH-CH_3$	-1.22 [1]	-1.12 -0.10	-0.10 A	-0.42	-0.36
13	$CH_3-CH_2-NH_2$	-1.18 [1]	-1.11 -0.07	-0.07 A	-0.20	-0.33
14	$HO-CH_2-CH_2-NH_2$	-2.89 [1]	-2.78 -0.11	-1.43 A	-1.29	-1.74
15	$NC-CH_2-CN$	0.04 [14]	0.08 -0.04	0.19 D	-0.74	-0.84
16	$NC-CH_2-COOH$	-0.52 [15]	-0.49 -0.03	-0.29 D	-0.63	-0.87
		-0.43 [8]	-0.49 0.06	-0.21 D		-0.79
		-0.44 [11]	-0.49 0.05	-0.22 D		-0.80
17	$HOOC-CH_2-COOH$	-0.99 [16]	-1.05 0.06	-0.70 D	-0.52	-0.82
		-1.08 [15]	-1.05 -0.03	-0.77 D		-0.90
		-0.91 [17]	-1.05 0.14	-0.63 D		-0.76
		-0.89 [18]	-1.05 0.16	-0.61 D		-0.74
18	$CH_3-CHCl-COOH$	0.95 [15]	1.02 -0.07	0.98 D	0.91	0.79
19	$Cl-CH_2-CH_2-COOH$	0.62 [15]	0.59 0.03	0.69 D	0.74	0.51
20	$I-CH_2-CH_2-COOH$	1.15 [5]	1.23 -0.08	1.15 D	1.27	0.95

Refs. pg 262

TABLE VII,1 (continued)

no	Compound	log P(ether) obs. ref.	log P(ether) clc. res.	log P(octanol) I	log P(octanol) II	log P(octanol) III
21	$CH_3-CO-CH_3$	-0.21 [1]	-0.10 -0.11	-0.02 D	-0.30	-0.17
22	$CH_3-COO-CH_3$	0.43 [1]	0.28 0.15	0.53 D	0.11	0.36
23	CH_3-CH_2-COOH	0.13 [9]	0.21 -0.08	0.27 D	0.28	0.11
		0.20 [15]	0.21 -0.01	0.33 D		0.17
		0.23 [8]	0.21 0.02	0.36 D		0.19
		0.18 [5]	0.21 -0.03	0.31 D		0.15
		0.23 [20]	0.21 0.02	0.36 D		0.19
		0.27 [12]	0.21 0.06	0.39 D		0.22
24	$CH_3-CH(OH)-COOH$	-1.09 [19]	-1.03 -0.06	-0.78 D	-0.65	-0.90
		-0.99 [21]	-1.03 0.04	-0.70 D		-0.82
		-0.96 [8]	-1.03 0.07	-0.67 D		-0.80
		-1.09 [20]	-1.03 -0.06	-0.78 D		-0.90
		-1.07 [22]	-1.03 -0.04	-0.76 D		-0.89
25	CH_3-O-CH_2-COOH	-0.76 [15]	-0.76 0.00	-0.50 D	-0.44	-0.63
		-0.62 [8]	-0.26 0.14	-0.38 D		-0.51
26	$HO-CH_2-CH(OH)-CH_2-Cl$	-1.10 [23]	-1.05 -0.05	-0.79 D	-0.48	-0.91
		-1.10 [1]	-1.05 -0.05	-0.79 D		-0.91
27	$CH_3-CH_2-CONH_2$	-1.89 [1]	-1.95 0.06	-0.63 A	-0.74	-0.70
28	$CH_3-(CH_2)_2-OH$	0.28 [1]	0.10 0.18	0.40 D	0.27	0.23
		-0.03 [2]	0.10 -0.13	0.13 D		-0.02

TABLE VII,1 (continued)

29	$CH_3-O-CH_2-CH_2-OH$	-1.21 [1]	-1.17 -0.04	-0.88 D	-0.74	-1.00
30	$HO-(CH_2)_3-OH$	-2.00 [1]	-2.16 0.16	-1.57 D	-1.39	-1.66
31	$HO-CH_2-CH(OH)-CH_2-OH$	-3.18 [1]	-3.11 -0.07	-2.58 D	-2.03	-2.64
		-2.96 [13]	-3.11 0.15	-2.39 D		-2.46
32	$CH_3-CH_2-CH_2-NH_2$	-0.54 [1]	-0.58 -0.04	0.45 A	0.33	0.21
33	$N(CH_3)_3$	-0.34 [1]	-0.34 0.00	0.61 A	-0.05	0.15
		-0.26 [24]	-0.34 0.08	0.67 A		0.22
		-0.39 [7]	-0.34 -0.05	0.57 A		0.11
34	$H_2N-CH_2-CH(NH_2)-CH_3$	-2.94 [1]	-2.80 -0.14	-1.47 A	-0.82	-1.13
35	$HOOC-(CH_2)_2-COOH$	-0.87 [16]	-0.82 -0.05	-0.59 D	-0.27	-0.72
		-0.82 [1]	-0.82 0.00	-0.55 D		-0.68
		-0.82 [1]	-0.82 0.00	-0.55 D		-0.68
		-0.89 [3]	-0.82 -0.07	-0.61 D		-0.74
		-0.90 [17]	-0.82 -0.08	-0.62 D		-0.75
		-0.65 [5]	-0.82 0.17	-0.40 D		-0.54
		-0.83 [25]	-0.82 -0.01	-0.56 D		-0.69
		-0.84 [22]	-0.82 -0.02	-0.57 D		-0.70
		-0.86 [12]	-0.82 -0.04	-0.58 D		-0.71
36	$HOOC-CH(OH)-CH_2-COOH$	-1.88 [15]	-2.06 0.18	-1.46 D	-1.20	-1.56
37	$HOOC-CH_2-O-CH_2-COOH$	-1.52 [1]	-1.50 -0.02	-1.15 D	-0.71	-1.26
		-1.54 [15]	-1.50 -0.04	-1.17 D		-1.28
38	$CH_3-COO-CH_2-CH_3$	0.93 [1]	0.82 0.11	0.96 D	0.64	0.77

TABLE VII,1 (continued)

no	Compound	log P(ether) obs. ref.	log P(ether) clc. res.	log P(octanol) I	log P(octanol) II	log P(octanol) III
39	$CH_3-(CH_2)_2-COOH$	0.66 [9]	0.74 -0.08	0.73 D	0.81	0.55
		0.68 [15]	0.74 -0.06	0.74 D		0.56
		0.66 [5]	0.74 -0.08	0.73 D		0.55
		0.81 [20]	0.74 0.07	0.86 D		0.67
40	$HO-CH(CH_3)-COO-CH_3$	-0.43 [1]	-0.42 -0.01	-0.21 D	-0.28	-0.36
41	$CH_3-CH_2-CH(OH)-COOH$	-0.48 [26]	-0.50 0.02	-0.26 D	-0.12	-0.40
42	$CH_3-CH_2-O-CH_2-COOH$	-0.34 [15]	-0.23 -0.11	-0.14 D	0.09	-0.28
43	$CH_3-(CH_2)_2-CONH_2$	-1.24 [1]	-1.42 0.18	-0.11 A	-0.21	-0.16
44	$CH_3-(CH_2)_3-OH$	0.57 [2]	0.63 -0.06	0.65 D	0.80	0.47
45	$(CH_3)_2CH-CH_2-OH$	0.84 [1]	0.77 0.07	0.88 D	0.68	0.70
46	$CH_3-CH_2-CH(OH)-CH_3$	0.65 [1]	0.77 -0.12	0.72 D	0.68	0.54
47	$CH_3-CH_2-O-CH_2-CH_3$	1.00 [1]	1.03 -0.03	1.02 D	0.88	0.83
48	$CH_3-CH(OH)-CH_2-CH_2-OH$	-1.38 [23]	-1.50 0.12	-1.03 D	-0.98	-1.14
49	$HO-(CH_2)_4-OH$	-1.72 [23]	-1.63 -0.09	-1.32 D	-0.86	-1.43
50	$CH_3-CH_2-O-(CH_2)_2-OH$	-0.70 [23]	-0.64 -0.06	-0.44 D	-0.21	-0.58
51	$HO-(CH_2)_2-O-(CH_2)_2-OH$	-2.36 [1]	-2.31 -0.05	-1.88 D	-1.30	-1.96
52	$CH_3-(CH_2)_3-NH_2$	0.11 [24]	-0.05 0.16	0.97 A	0.86	0.75
53	$CH_3-CH_2-NH-CH_2-CH_3$	-0.07 [24]	-0.05 -0.02	0.82 A	0.64	0.60
54	$HO-(CH_2)_2-NH-(CH_2)_2-OH$	-3.27 [1]	-3.40 0.13	-1.73 A	-1.54	-2.06
55	$H_2N-(CH_2)_4-NH_2$	-2.89 [1]	-3.00 0.11	-1.43 A	-0.74	-1.09

TABLE VII,1 (continued)

No.	Compound						
56	$CH_3-CO-CH_2-CH_2-COOH$	-0.58 [1]	-0.60	0.02	-0.34 D	-0.32	-0.48
		-0.64 [15]	-0.60	-0.04	-0.39 D		-0.53
		-0.64 [8]	-0.60	-0.04	-0.39 D		-0.53
		-0.49 [5]	-0.60	0.11	-0.26 D		-0.41
57	$CH_3-(CH_2)_3-COOH$	1.24 [9]	1.27	-0.03	1.23 D	1.34	1.03
		1.17 [5]	1.27	-0.10	1.17 D		0.97
		1.36 [20]	1.27	0.09	1.33 D		1.13
58	$(CH_3)_2CH-CH_2-CONH_2$	-0.77 [1]	-0.76	-0.01	0.26 A	0.20	0.23
59	$(CH_3)_2CH-(CH_2)_2-OH$	1.28 [1]	1.30	-0.02	1.26 D	1.21	1.06
60	$HO-(CH_2)_5-OH$	-1.26 [23]	-1.10	-0.16	-0.93 D	-0.33	-1.05
61	$CH_3-O-(CH_2)_2-O-(CH_2)_2-OH$	-1.43 [1]	-1.38	-0.05	-1.07 D	-0.68	-1.19
62	$CH_3-CH_2-O-CH_2-CH(OH)-CH_2$, OH	-1.58 [1]	-1.65	0.07	-1.20 D	-0.89	-1.31
63	$H_2N-(CH_2)_5-NH_2$	-2.56 [1]	-2.47	-0.09	-1.17 A	-0.21	-0.82
64	$4-NO_2-C_6H_4-OH$	2.04 [1]	2.13	-0.09	1.92 D	1.27	1.04
		2.01 [8]	2.13	-0.12	1.89 D		1.01
65	$2-NO_2-C_6H_4-OH$	2.18 [1]	2.13	0.05	2.04 D	1.27	1.16
66	$3-NO_2-C_6H_4-OH$	2.20 [1]	2.13	0.07	2.05 D	1.27	1.17
		2.18 [8]	2.13	0.05	2.04 D		1.16
67	$4-NO_2-C_6H_4-NH_2$	1.48 [8]	1.44	0.04	1.43 D	0.76	0.57
68	C_6H_5-OH	1.64 [1]	1.58	0.06	1.57 D	1.54	1.58
		1.58 [31]	1.58	0.00	1.52 D		1.53
69	$1,3-diOH-C_6H_4$	0.62 [1]	0.49	0.13	0.69 D	1.00	0.95

TABLE VII,1 (continued)

no	Compound	log P(ether)		log P(octanol)		
		obs. ref.	clc. res.	I	II	III
		0.67 [32]	0.49 0.18	0.74 D	1.00	0.99
70	1,4-diOH-C_6H_4	0.46 [1]	0.49 -0.03	0.55 D	1.00	0.82
		0.36 [34]	0.49 -0.13	0.47 D		0.73
		0.37 [35]	0.49 -0.12	0.48 D		0.74
		0.38 [32]	0.49 -0.11	0.48 D		0.75
71	1,3,5-triOH-C_6H_3	-0.35 [1]	-0.39 0.04	-0.14 D	0.40	0.36
		-0.35 [32]	-0.39 0.04	-0.14 D		0.36
72	C_6H_5-NH_2	0.85 [36]	0.89 -0.04	0.89 D	1.03	0.92
73	HOOC-CH_2-CH(COOH)-CH_2-COOH	-1.22 [1]	-1.19 -0.03	-0.89 D	-0.42	-1.01
		-1.30 [15]	-1.19 -0.11	-0.96 D		-1.08
74	CH_3-CO-$(CH_2)_2$-CO-CH_3	-0.35 [23]	-0.38 0.03	-0.14 D	-0.37	-0.29
75	HOOC-$(CH_2)_4$-COOH	-0.29 [3]	-0.35 0.06	-0.09 D	0.21	-0.24
		-0.29 [27]	-0.35 0.06	-0.09 D		-0.24
		-0.24 [17]	-0.35 -0.11	-0.05 D		-0.20
76	CH_3-COO-$(CH_2)_2$-OOC-CH_3	0.30 [1]	0.39 -0.09	0.42 D	0.45	0.25
77	CH_3-$(CH_2)_4$-COOH	1.88 [9]	1.80 0.08	1.78 D	1.87	1.56
		1.97 [1]	1.80 0.17	1.86 D		1.64
		1.95 [20]	1.80 0.15	1.84 D		1.62
78	HO-$(CH_2)_4$-CH(OH)-CH_2-OH	-2.51 [23]	-2.52 0.01	-2.00 D	-1.01	-2.08
79	CH_3-$(CH_2)_2$-NH-$(CH_2)_2$-CH_3	0.95 [1]	1.01 -0.06	1.64 A	1.70	1.44

TABLE VII,1 (continued)

No.	Compound					
80	$HO-(CH_2)_2-N(CH_2-CH_3)_2$	-0.46 [1]	-0.42 -0.04	0.51 A	0.45	0.05
81	$NH[CH_2-CH(OH)-CH_3]_2$	-2.23 [1]	-2.07 -0.16	-0.91 A	-0.73	-1.20
82	C_6H_5-COOH	1.89 [1]	1.92 -0.03	1.79 D	1.79	1.79
		1.78 [5]	1.92 -0.14	1.69 D		1.70
		1.85 [12]	1.92 -0.07	1.75 D		1.75
83	$4-OH-C_6H_4-COOH$	1.00 [5]	0.83 0.17	1.02 D	1.25	1.27
84	$3-NH_2-C_6H_4-COOH$	0.18 [1]	0.15 0.03	0.31 D	0.74	0.59
85	$2-NH_2-C_6H_4-COOH$	0.05 [5]	0.15 -0.10	0.20 D	0.74	0.48
86	$C_6H_5-O-CH_3$	2.46 [31]	2.51 -0.05	2.28 D	2.16	2.26
87	$3,4-diOH-C_6H_3-CH_3$	1.23 [33]	1.22 0.01	1.22 D	1.45	1.46
88	$2-OH-C_6H_4-CH_2-OH$	0.00 [37]	-0.17 0.17	0.16 D	0.38	0.22
89	$2-OH-C_6H_4-O-CH_3$	1.36 [31]	1.42 -0.06	1.33 D	1.61	1.56
90	$4-OH-C_6H_4-O-CH_3$	1.36 [31]	1.42 -0.06	1.33 D	1.61	1.56
91	$C_6H_5-CH_2-NH_2$	0.28 [1]	0.23 0.05	1.10 A	0.99	0.89
		0.32 [5]	0.23 0.09	1.14 A		0.92
		0.36 [28]	0.23 0.13	1.17 A		0.95
92	$HOOC-(CH_2)_5-COOH$	0.17 [16]	0.18 -0.01	0.30 D	0.74	0.14
		0.18 [16]	0.18 0.00	0.31 D		0.15
		0.14 [15]	0.18 -0.04	0.28 D		0.12
		0.04 [17]	0.18 -0.14	0.19 D		0.03
93	$CH(OH)[CH_2-OOC-CH_3]_2$	-0.66 [1]	-0.62 -0.04	-0.41 D	-0.23	-0.54
94	$CH(OH)[CH_2-O-CH_2-CH_3]_2$	-0.07 [1]	-0.18 0.11	0.10 D	0.25	-0.06

Refs. pg 262

TABLE VII,1 (continued)

no	Compound	log P(ether) obs. ref.	log P(ether) clc. res.	log P(octanol) I	log P(octanol) II	log P(octanol) III
95	$C_6H_5-CH_2-COOH$	1.57 [1]	1.56 0.01	1.51 D	1.46	1.30
		1.44 [5]	1.56 -0.12	1.40 D		1.20
96	$3,4-diOH-C_6H_3-CH_2-CH_3$	1.65 [33]	1.75 -0.10	1.58 D	1.98	1.81
97	$3-CH_3O-4-OH-C_6H_3-CH_2-OH$	-0.20 [31]	-0.12 -0.08	-0.01 D	0.40	0.27
98	$C_6H_5-CH_2-NH-CH_3$	0.78 [5]	0.76 0.02	1.50 A	1.29	1.30
		0.85 [28]	0.76 0.09	1.56 A		1.36
99	$HOOC-(CH_2)_6-COOH$	0.67 [16]	0.71 -0.04	0.74 D	1.27	0.56
100	$HO-(CH_2)_2-O-(CH_2)_2-O$ $CH_3-(CH_2)_3$	0.04 [23]	0.21 -0.17	0.19 D	0.03	0.91
101	$HO-(CH_2)_2-O-(CH_2)_2-O$ $HO-(CH_2)_2-O-(CH_2)_2$	-2.62 [1]	-2.73 0.09	-2.10 D	-1.19	-2.17
102	$CH_3-CH(CH_3)-CH_2-NH$ $(CH_3)_2CH-CH_2$	2.52 [1]	2.34 0.18	2.90 A	2.51	2.75
103	$C_6H_5-CH_2-NH-CH_2-CH_3$	1.21 [28]	1.29 -0.08	1.85 A	1.82	1.66
104	$3,4-diOH-C_6H_3-(CH_2)_2-CH_3$	2.37 [33]	2.28 0.09	2.20 D	2.51	2.40
105	$HOOC-(CH_2)_7-COOH$	1.20 [16]	1.24 -0.04	1.19 D	1.80	1.00
106	$HOOC-(CH_2)_8-COOH$	1.76 [16]	1.77 -0.01	1.68 D	2.33	1.46
107	$C_6H_5-NHCOO-(CH_2)_2-N(C_2H_5)_2$	0.50 [38]	0.41 0.09	1.28 A	3.03	2.81
108	same, $3-CH_3$-derivative	0.78 [38]	0.93 -0.15	1.50 A	3.53	3.04
109	same, $4-CH_3$-derivative	0.75 [38]	0.93 -0.18	1.48 A	3.53	3.02
110	same, $3-CH_3O$-derivative	0.36 [38]	0.26 0.10	1.17 A	3.10	2.91
111	same, $4-CH_3O$-derivative	0.35 [38]	0.26 0.09	1.16 A	3.10	2.91
112	same, $3-C_2H_5O$-derivative	0.78 [38]	0.79 -0.01	1.50 A	3.63	3.26

TABLE VII,1 (continued)

113	same, 2-C_2H_5O-derivative	0.85 [38]	0.79 0.06	1.56 A	3.63	3.32
114	same, 4-C_2H_5O-derivative	0.78 [38]	0.79 -0.01	1.50 A	3.63	3.26

Outliers:

1	CH_3-OH	-1.29 [2]	-0.96 -0.33	-0.95 D	-0.79	-1.07
2	Cl-CH_2-COOH	0.02 [6]	0.36 -0.34	0.18 D	0.02	0.50
3	CH_3-CN	-0.22 [1]	0.25 -0.47	-0.03 D	-0.36	-0.62
4	HOOC-CH_2-NH_2	-2.08 [3]	-2.38 0.30	-0.79 A	-0.99	-1.07
5	CH_3-CH_2-OH	0.28 [13]	-0.43 0.71	0.40 D	-0.26	0.23
6	HOOC-CH_2-COOH	-0.34 [5]	-1.05 0.71	-0.14 D	-0.52	-0.28
7	CH_3-CH(OH)-COOH	-0.64 [5]	-1.03 0.39	-0.39 D	-0.65	-0.53
8	HO-CH_2-CH(OH)-COOH	-2.05 [1]	-2.70 0.65	-1.61 D	-1.74	-1.70
9	CH_3-OOC-CH_2-NH_2	-1.14 [1]	-1.78 0.64	-0.03 A	-0.63	-0.29
10	H_2N-CH(CH_3)-COOH	-5.85 [1]	-1.71 -4.14	-3.81 A	-0.58	-4.20
11	$(CH_3)_2$CH-OH	-0.19 [1]	0.24 -0.43	-0.01 D	0.15	-0.16
		-0.33 [2]	0.24 -0.57	-0.13 D		-0.27
12	HO-CH(CH_3)-CH_2-OH	-1.74 [1]	-1.46 -0.28	-1.34 D	-0.94	-1.44
13	H_2N-CH_2-CH(CH_3)-OH	-2.37 [1]	-2.14 -0.23	-1.02 A	-0.88	-1.31
14	H_2N-CH_2-CH(OH)-CH_2-NH_2	-3.70 [1]	-4.48 0.78	-2.08 A	-1.90	-1.76
15	HOOC-CH(OH)-CH_2-COOH	-1.85 [22]	-2.06 0.21	-1.44 D	-1.20	-1.54
16	HOOC-CH(OH)-CH(OH)-COOH	-2.43 [3]	-3.30 0.87	-1.94 D	-2.12	-2.02
		-1.01 [5]	-3.30 2.29	-0.71 D		-0.84
		-2.42 [22]	-3.30 0.88	-1.93 D		-2.01

Refs. pg 262

TABLE VII,1 (continued)

no	Compound	log P(ether)		log P(octanol)		
		obs. ref.	clc. res.	I	II	III
		-2.34 [12]	-3.30 0.96	-1.86 D	-2.12	-1.94
17	$HOOC-CH_2-CH(CH_3)-OH$	-0.40 [5]	-0.80 0.40	-0.19 D	-0.40	-0.33
18	$CH_3-CH_2-CH(OH)-COOH$	-0.08 [5]	-0.50 0.42	0.09 D	-0.12	-0.07
19	$CH_3-CH_2-CH(NH_2)-COOH$	-5.58 [1]	-1.18 -4.40	-3.59 A	-0.05	-3.98
20	$CH_3-(CH_2)_3-OH$	0.89 [1]	0.63 0.26	0.93 D	0.80	0.74
21	$(CH_3)_2CH-CH_2-OH$	0.53 [2]	0.77 -0.24	0.61 D	0.68	0.44
22	$CH_3-CH_2-CH(CH_3)-OH$	0.28 [2]	0.77 -0.49	0.40 D	0.68	0.23
23	$HO-CH(CH_3)-CH(CH_3)-OH$	-1.54 [1]	-0.77 -0.77	-1.17 D	-0.53	-1.28
24	$CH_3-O-CH_2-CH(OH)-CH_2-OH$	-1.72 [1]	-2.18 0.46	-1.32 D	-1.42	-1.43
25	$CH_3-CH_2-NH-CH_2-CH_3$	-0.28 [1]	-0.05 -0.23	0.66 A	0.64	0.42
26	$HOOC-(CH_2)_3-COOH$	-0.55 [16]	-0.88 0.33	-0.32 D	-0.32	-0.46
		-0.57 [15]	-0.88 0.31	-0.33 D		-0.47
		-0.60 [17]	-0.88 0.28	-0.36 D		-0.50
		-0.47 [5]	-0.88 0.41	-0.25 D		-0.39
27	$CH_3-COO-CH_2-CH(OH)-CH_2OH$	-1.39 [1]	-1.86 0.47	-1.04 D	-1.13	-1.15
28	$(CH_3)_2CH-CH_2-CH_2-NH_2$	0.30 [1]	0.61 -0.31	1.12 A	1.27	0.90
29	$3-NH_2-C_6H_4-NO_2$	1.71 [8]	1.44 0.27	1.63 D	0.76	0.77
30	$2-NH_2-C_6H_4-NO_2$	1.95 [8]	1.44 0.51	1.84 D	0.76	0.96
31	$1,2-diOH-C_6H_4$	1.04 [1]	0.48 0.56	1.05 D	1.00	1.30
		0.86 [33]	0.48 0.38	0.90 D		1.15

TABLE VII,1 (continued)

#	Compound					
		0.89 [31]	0.48 0.41	0.93 D	1.00	1.17
32	1,2,3-triOH-C_6H_3	0.23 [1]	-0.39 0.62	0.36 D	0.40	0.84
		0.09 [32]	-0.39 0.48	0.24 D		0.73
33	3-OH-C_6H_4-NH_2	0.11 [32]	-0.20 0.31	0.25 D	0.49	0.53
34	1,2-diNH_2-C_6H_4	-0.06 [32]	-0.88 0.82	0.83 A	-0.02	0.39
35	$(CH_3)_2$CH-CH_2-CH(NH_2)-COOH	-4.92 [1]	0.01 -4.93	-3.06 A	0.88	-3.43
36	HO-$(CH_2)_6$-OH	-0.92 [1]	0.71 -1.63	-0.64 D	0.20	-0.76
37	CH_3-CH_2-O-$(CH_2)_2$-O-$(CH_2)_2$ OH	-1.19 [23]	-0.85 -0.34	-0.87 D	-0.15	-0.99
38	HO-$(CH_2)_3$-O-$(CH_2)_3$-OH	-1.45 [23]	-2.44 0.99	-1.09 D	-1.38	-1.20
39	N(CH_2-CH_2-OH)$_3$	-2.96 [1]	-3.77 0.81	-1.49 A	-1.73	-2.02
40	2,4-diNO_2-C_6H_3-COOH	1.18 [5]	3.23 -1.05	1.18 D	1.18	-0.55
41	3-NO_2-C_6H_4-COOH	1.97 [5]	2.47 -0.50	1.86 D	1.52	0.98
42	2-NO_2-C_6H_4-COOH	1.59 [5]	2.47 -0.88	1.53 D	1.52	0.67
43	3-OH-C_6H_4-COOH	1.32 [1]	0.83 0.49	1.30 D	1.25	1.53
44	2-OH-C_6H_4-COOH	2.37 [1]	0.83 1.54	2.20 D	1.25	2.40
		2.53 [5]	0.83 1.70	2.34 D		2.53
45	4-OH-C_6H_4-COOH	1.42 [1]	0.83 0.69	1.38 D	1.25	1.61
46	2,5-diOH-C_6H_3-COOH	1.35 [5]	-0.05 1.40	1.32 D	0.65	1.77
47	3,5-diOH-C_6H_3-COOH	1.45 [5]	-0.05 1.50	1.41 D	0.65	1.86
48	2-NH_2-C_6H_4-COOH	1.43 [1]	0.14 1.29	1.39 D	0.74	1.62
		1.48 [8]	0.14 1.34	1.43 D		1.66
49	4-NH_2-C_6H_4-COOH	0.88 [1]	0.14 0.74	0.92 D	0.74	1.17

Refs. pg 262

TABLE VII,1 (continued)

no	Compound	log P(ether) obs. ref.	log P(ether) clc. res.	log P(octanol) I	II	III
50	$3-CH_3-C_6H_4-OH$	1.80 [36]	2.09 -0.29	1.71 D	2.05	1.71
51	$CH_3-(CH_2)_6-NH_2$	1.30 [5]	1.54 -0.24	1.92 A	2.45	1.73
52	$1,2-diCOOH-C_6H_4$	0.20 [16]	1.17 -0.97	0.33 D	1.50	0.60
		0.10 [15]	1.17 -1.07	0.24 D		0.52
		0.28 [5]	1.17 -0.89	0.40 D		0.67
53	$1,3-diCOOH-C_6H_4$	1.46 [16]	1.17 0.29	1.42 D	1.50	1.65
54	$C_6H_5-CH_2-COOH$	0.47 [17]	1.56 -1.09	0.56 D	1.46	0.39
55	$2-CH_3O-C_6H_4-COOH$	0.78 [8]	1.76 -0.98	0.83 D	1.86	1.08
56	$2,6-diCH_3-C_6H_3-OH$	2.53 [31]	2.82 -0.29	2.34 D	2.49	2.32
57	$2,6-diCH_3O-C_6H_3-OH$	0.74 [31]	1.47 -0.73	0.80 D	1.63	1.27
58	$HOOC-(CH_2)_6-COOH$	0.47 [17]	0.71 -0.24	0.56 D	1.27	0.39
59	$CH(OH)-COO-CH_2-CH_3$ | $CH(OH)-COO-CH_2-CH_3$	-0.19 [1]	-1.03 0.84	-0.01 D	-0.34	-0.16
		-0.35 [29]	-1.03 0.68	-0.14 D		-0.29
60	$1,3,5-triCOOH-C_6H_3$	1.04 [1]	0.64 0.40	1.05 D	1.15	1.52
61	$C_6H_5-CH_2-NH-CH_2-CH_3$	1.08 [5]	1.29 -0.21	1.75 A	1.82	1.55
62	$HOOC-(CH_2)_7-COOH$	0.97 [17]	1.24 -0.27	0.99 D	1.80	0.81
63	$H_2N-CH(CH_3)-(CH_2)_3$ | $N(CH_2-CH_3)_2$	-0.24 [1]	0.03 -0.27	0.69 A	1.58	1.11
64	$C_6H_5-CH_2-CH(COOH)_2$	1.18 [1]	1.49 -0.31	1.18 D	1.60	0.98
65	$C_6H_5-CH_2-NH-(CH_2)_2-CH_3$	1.49 [28]	1.82 -0.33	2.07 A	2.35	1.89
66	$C_6H_5-(CH_2)_3-NH-CH_3$	1.46 [28]	1.82 -0.36	2.05 A	2.35	1.87

TABLE VII,1 (continued)

67	$C_6H_5-CH_2-CH(CH_3)-NH-CH_3$	1.46 [28]	1.96 −0.50	2.05 A	2.23	1.87
68	$HO-(CH_2)_{10}-OH$	1.32 [1]	1.55 −0.23	1.30 D	2.32	1.10
69	$C_6H_5-(CH_2)_3-NH-CH_2-CH_3$	1.80 [28]	2.36 −0.56	2.32 A	2.88	2.15
70	$C_6H_5-(CH_2)_3-NH-(CH_2)_2-CH_3$	2.11 [28]	2.89 −0.78	2.57 A	3.41	2.41
71	$C_6H_5-(CH_2)_3-NH-(CH_2)_3-CH_3$	2.43 [28]	3.42 −0.99	2.83 A	3.94	2.67
72	$C_6H_5-CH_2-NH-CH_2-C_6H_5$	1.16 [30]	2.64 −1.48	1.81 A	3.01	1.62

atives that also possess an OH or NH_2 group; in other words, structures that were qualified in Chapter III as outliers in the derivation of aromatic f values in the octanol – water system. The reason
that they do not manifest themselves as outliers in the ether – water regression is that in this system they lack a number of "normally" behaving correcting partners.

The data in Table VII,2 make it clear that alkyl and phenyl fragments in the ether – water system show a higher lipophilicity than in
the octanol – water system. The remaining (negative) fragments in the
ether – water system are far more hydrophilic than in the octanol- water system.

The three last columns in Table VII,1 show the results of two conversions of log P(ether) into log P(octanol) values in comparison with
the log P values obtained from an f (oct) summation. For those values
denoted by "I" and "II" the solvent regressions of LEO et al. are used,
with the difference that dependent and independent parameters are interchanged correctly, following the instructions given in the preceding
chapter (see eqns. VII-5 and VII-6), while column "III" is based on an
equation that will be discussed in the following chapter.

$$\log P(octanol) = 0.862 \log P(ether) + 0.158 \qquad (VII-5)$$
$$n = 71 \underline{donors}; \quad r = 0.988; \quad s = 0.161; \quad F = 2,829; \quad \underline{t} = 53.2$$

$$\log P(octanol) = 0.801 \log P(ether) + 0.880 \qquad (VII-6)$$
$$n = 32 \underline{acceptors}; \quad r = 0.956; \quad s = 0.275; \quad F = 320; \quad \underline{t} = 17.9$$

Table VII,3 shows that in the ether – water system also, a hydrogen
atom attached to an electron-withdrawing centre has markedly increased

TABLE VII,2

f VALUES IN THE ETHER – WATER SYSTEM

No	fragment	*f*	kn.	st.dev.	$\underline{t}$	*f*(oct)
1	C_6H_5	2.266	0	0.056	40.2	1.886
2	C_6H_4	1.866	0	0.055	34.0	1.688
3	C_6H_3	1.677	0	0.094	17.9	1.431
4	CH_3	0.919	0	0.051	18.0	0.702
5	CH_2	0.531	0	0.007	71.3	0.530
6	CH	0.277	0	0.056	4.95	0.235
7	(al)COOH	−1.238	3	0.052	−23.8	−0.954
8	(al)COO	−1.553		0.039	−39.4	−1.292
9	(al)CO	−1.937		0.042	−46.6	−1.703
10	(al)$CONH_2$	−3.403		0.068	−49.9	−1.970
11	(al)CN	(−0.673)		0.064	−10.6	−1.066
12	(ar)NO_2	(0.952)		0.071	13.3	−0.078
13	(ar)NHCOO *	−3.309		0.075	−43.9	−0.795
14	(al)O	−1.867	4	0.034	−55.1	−1.581
15	(al)OH	−1.879	4	0.053	−35.5	−1.491
16	(al)NH_2	−2.563	4	0.055	−46.4	−1.428
17	(al)NH	−2.954		0.045	−65.0	−1.825
18	(al)N	−3.099		0.082	−37.6	−2.160
19	(al)Cl	0.176		0.062	2.82	0.061
20	(al)I	0.810		0.067	12.0	0.587
21	c_M	0.297		0.010	30.9	0.287

*) Aromatic fragment connected to N
Values in parentheses are suspect (see text)

n = 186 (+1);   s = 0.097;
r = 0.997;   F = 1,464;   const. term = 0.000.

lipophilicity. The *f*(H) value can be taken to be 0.59 in this instance.

The set of primary *f* values from Table VII,2 could be extended by some secondary *f* values. The structures suitable for this purpose are given in Table VII,4 and the secondary *f* values in Table VII,5. Again, it is clear that the *f* values that were already positive or negative in the octanol – water system have become increasingly so.

TABLE VII,3

APPLICATION OF $f(H) = 0.59$ IN STRUCTURES WITH HYDROGEN ATOMS ATTACHED TO ELECTRON-WITHDRAWING CENTRES (ETHER – WATER SYSTEM)

no	compound	obs.	ref.	calc.	res.
			log P values		
1	NH_3	-1.96	[2]	-1.97	0.01
2	H-CO-H	-0.96	[7]	-0.76	-0.20
3	H-COOH	-0.52	[9]	-0.65	0.13
		-0.40	[39]		0.25*
		-0.45	[3]		0.20
		-0.40	[5]		0.25*
		-0.43	[12]		0.22*
4	CH_3-CO-H	-0.48	[2]	-0.43	-0.05
5	CH_3-CH_2-CO-H	0.30	[1]	0.10	0.20
6	H-$CONH_2$	-2.85	[1]	-2.81	-0.04
7	$-SO_2N$	-2.94	T.VII,4		
8	$-SO_2NH-$ a	-2.50	T.VII,4	-2.35	-0.15
9	$-SO_2NH_2$ b	-1.97	T.VII,4	-1.76	-0.21*

*) Outliers. a) $f(SO_2NH) = f(SO_2N) + f(H)$. b) $f(SO_2NH_2) = f(SO_2N) + f(H)$.

TABLE VII,4

DERIVATION OF SECONDARY f VALUES (ETHER – WATER SYSTEM)

fragment	compound	obs.	ref.	calc.	res.
			log P values		
$(ar)SO_2NH_2$	$C_6H_5-SO_2NH_2$	0.30	[4]	(0.30)	
$(ar)SO_2NH-$	$C_6H_5-SO_2NH-CH_3$	0.80	[4]	0.69	0.11
	$C_6H_5-SO_2NH-CH_2-CH_3$	1.11	[4]	1.22	-0.11
$(ar)SO_2N$	$C_6H_5-SO_2N(CH_3)_2$	1.16	[4]	1.17	-0.01
	$C_6H_5-SO_2N(CH_2-CH_3)_2$	2.22	[4]	2.23	-0.01
$(ar)CO$	C_6H_5-COH	1.74	[32]	1.92	-0.18
	$3-CH_3-O-4-OH-C_6H_3-COH$	0.97	[1]	0.88	0.09
		0.91	[31]		0.03
		0.93	[184]		0.05
		0.96	[32]		0.08
	$3-C_2H_5-O-4-OH-C_6H_3-COH$	1.34	[32]	1.41	-0.07
$(ar)I$	$2-I-C_6H_4-COOH$	3.11	[5]	3.34	-0.23*
	$4-I-C_6H_4-SO_2NH-CH-COOH$ HOOC-CH_2	1.18	[185]	1.00	0.18
	$4-I-C_6H_4-SO_2NH-CH-COOH$ $CH_3-CH(OH)$	0.90	[185]	1.03	-0.13

Refs. pg 262

TABLE VII,4 (continued)

	4-I-C_6H_4-SO_2NH-CH-COOH HOOC-CH_2-$\overset{\shortmid}{C}H_2$	1.23	[41]	0.94	0.29*
	4-I-C_6H_4-OH	2.87	[182]	2.99	-0.12
(al)Br	Br-CH_2-COOH	0.64	[3]	0.61	0.03
	Br-CH_2-CHBr-COOH	1.79	[5]	1.85	-0.06
	CH_3-CHBr-COOH	1.18	[1]	1.28	-0.10
		1.04	[15]		-0.24*
		1.50	[5]		0.22*
	HOOC-CHBr-CHBr-COOH	1.73	[16]	1.52	0.21*
	HOOC-CHBr-CH_2-COOH	0.46	[16]	0.57	-0.11
		0.84	[5]	0.76	0.08

*) Outliers
Parentheses: only one data-point available

TABLE VII,5

SECONDARY *f* VALUES (ETHER — WATER SYSTEM)

no	fragment	(ether)	(oct)
1	H	0.59	0.47
2	(ar)SO_2NH_2	-1.97	-1.53
3	(ar)SO_2NH-	-2.50	-1.99
4	(ar)SO_2N$<$	-2.94	-2.45
5	(ar)CO	-0.94	-0.84
6	(ar)I	1.82	1.44
7	(al)Br	0.43	0.27

CHLOROFORM — WATER SYSTEM

In conformity with the ether — water system, 138 log P values belonging to 89 different structures were incorporated in calculations for the chloroform — water system. Thirty-five structures (57 measurements) were purely aliphatic in nature and the others were of the mixed aliphatic-aromatic type.

For a complete review of all structures considered, see Table VII,6; the second column in Table VII,7 shows the fragments to which the analysis refers.

TABLE VII,6

STRUCTURES INTRODUCED INTO STATISTICAL TREATMENT
(CHLOROFORM — WATER SYSTEM)

no	Compound	log P(obs)	ref.	log P(clc)	res.
1	CH_3-OH	−1.36	[2]	−1.40	0.04
2	CH_3-NH_2	−0.90	[41]	−0.97	0.07
		−1.09	[42]		−0.12
3	$Br-CH_2-COOH$	−1.14	[5]	−1.04	−0.10
4	CH_3-COOH	−1.54	[43]	−1.52	−0.02
		−1.52	[8]		0.00
		−1.60	[5]		−0.08
		−1.70	[6]		−0.18
		−1.58	[44]		−0.06
5	CH_3-CH_2-OH	−0.85	[2]	−0.77	−0.08
6	$Br-CH_2-CH(Br)-COOH$	−0.42	[5]	−0.37	−0.05
7	$CH_3-CH(Br)-COOH$	−0.44	[45]	−0.54	0.10
8	$Br-CH_2-CH_2-COOH$	−0.61	[45]	−0.73	0.12
9	$CH_3-CO-CH_3$	0.72	[46]	0.79	−0.07
10	CH_3-CH_2-COOH	−0.78	[19]	−0.89	0.11
		−0.79	[47]		0.10
		−0.79	[48]		0.10
		−0.85	[8]		0.04
		−0.80	[45]		0.09
11	$CH_3-(CH_2)_2-OH$	−0.21	[2]	−0.15	−0.06
12	$(CH_3)_2CH-OH$	−0.35	[2]	−0.27	−0.08
13	$CH_3-CH_2-CH(Br)-COOH$	0.08	[45]	0.09	−0.01
14	$CH_3-(CH_2)_2-COOH$	−0.27	[45]	−0.26	−0.01
		−0.27	[5]	−0.26	−0.01
15	$(CH_3)_2CH-COOH$	−0.25	[19]	−0.39	0.14
		−0.28	[45]	−0.39	0.11
16	$CH_3-(CH_2)_3-OH$	0.45	[2]	0.48	−0.03
17	$(CH_3)_2CH-CH_2-OH$	0.34	[2]	0.35	−0.01
18	$CH_3-CH_2-CH(CH_3)-OH$	0.30	[2]	0.35	−0.05
19	$CH_3-(CH_2)_3-NH_2$	0.99	[24]	0.91	0.08
20	$CH_3-CH_2-NH-CH_2-CH_3$	0.81	[24]	0.80	0.01
		0.89	[5]		0.09
21	$2-OH-C_5H_4N$	−1.21	[32]	−1.15	−0.06

Refs. pg 262

TABLE VII,6 (continued)

no	Compound	log P(obs)	ref.	log P(clc)	res.
22	$CH_3-CO-CH_2-CO-CH_3$	1.40	[51]	1.24	0.16
23	$CH_3-(CH_2)_3-COOH$	0.34	[45]	0.36	-0.02
		0.32	[5]		-0.04
24	$(CH_3)_2CH-CH_2-COOH$	0.21	[48]	0.24	-0.03
		0.17	[45]		-0.07
25	$3-NO_2-C_6H_4-OH$	0.41	[52]	0.22	0.19
26	$4-NO_2-C_6H_4-OH$	0.08	[52]	0.22	-0.14
		0.27	[53]		0.05
27	$4-NO_2-C_6H_4-NH_2$	1.23	[8]	1.30	-0.07
		1.30	[54]		0.00
28	$3-NO_2-C_6H_4-SO_2NH_2$	-0.36	[55]	-0.29	-0.07
29	C_6H_5-OH	0.34	[56]	0.30	0.04
		0.37	[57]		0.07
		0.36	[54]		0.06
		0.38	[58]		0.08
		0.41	[59]		0.11
30	$C_6H_5-NH_2$	1.42	[54]	1.37	0.05
		1.23	[60]		-0.16
		1.32	[61]		-0.05
31	$2-CH_3-C_5H_4N$	1.79	[62]	1.86	-0.07
32	$3-CH_3-C_5H_4N$	1.89	[62]	1.86	0.03
33	$4-CH_3-C_5H_4N$	1.88	[62]	1.86	0.02
34	$C_6H_5-SO_2NH_2$	-0.24	[4]	-0.22	-0.02
		-0.24	[55]		-0.02
35	$4-NH_2-C_6H_4-SO_2NH_2$	-1.39	[63]	-1.55	0.16
		-1.40	[64]		0.15
		-1.63	[4]		-0.08
		-1.52	[65]		0.03
		-1.69	[55]		-0.14
		-1.57	[66]		-0.02
36	$CH_3-CH_2-CO-CH_2-CO-CH_3$	1.75	[50]	1.87	-0.12
37	$CH_3-(CH_2)_4-COOH$	1.05	[45]	0.99	0.06
		0.85	[5]		-0.14
38	$(CH_3)_2CH-(CH_2)_2-COOH$	0.90	[45]	0.86	0.04
39	$3-NO_2-C_6H_4-COOH$	0.48	[45]	0.42	0.06
		0.41	[54]		-0.01

TABLE VII,6 (continued)

40	C_6H_5-COOH	0.46	[54]	0.50	-0.04
		0.50	[58]		0.00
		0.54	[68]		0.04
41	4-OH-C_6H_4-COOH	-2.00	[54]	-1.90	-0.10
42	4-CH_3-C_6H_4-NH_2	1.99	[54]	1.98	0.01
43	3-CH_3-C_6H_4-SO_2NH_2	0.32	[55]	0.39	-0.07
44	2-CH_3-C_6H_4-SO_2NH_2	0.46	[55]	0.39	0.07
45	4-CH_3-C_6H_4-SO_2NH_2	0.33	[55]	0.39	-0.06
46	4-OH-C_6H_4-CO-CH_3	0.08	[75]	0.09	-0.01
47	C_6H_5-CH_2-COOH	0.45	[45]	0.49	-0.04
		0.42	[58]		-0.07
48	4-CH_3-CO-C_6H_4-SO_2NH_2	-0.36	[55]	-0.42	0.06
49	CH_3-CH(CH_3)-CH_2-CO-CH_2-CO-CH_3	2.97	[50]	3.00	-0.03
50	C_6H_5-CH_2-CH_2-COOH	1.10	[5]	1.12	-0.02
51	4-OH-C_6H_4-CO-CH_2-CH_3	0.71	[54]	0.72	-0.01
52	C_6H_5-CH_2-CH(CH_3)-NH_2	2.20	[70]	2.17	0.03
		2.17	[72]		0.00
53	4-CH_3-$(CH_2)_3$-C_5H_4N	3.74	[73]	3.75	-0.01
54	HOOC-$(CH_2)_7$-COOH	-0.58	[17]	-0.58	0.00
55	C_6H_5-CH_2-CH(CH_3)-NH-CH_3	2.75	[72]	2.68	0.07
56	4-CH_3-$(CH_2)_4$-C_5H_4N	4.38	[73]	4.37	0.01
57	HOOC-$(CH_2)_8$-COOH	0.04	[17]	0.05	-0.01
58	C_6H_5-CH_2-CH(CH_3)-NH-CH_2-CH_3	3.25	[72]	3.31	-0.06
59	C_6H_5-CH_2-CH_2-NH-CH(CH_3)$_2$	3.23	[74]	3.31	-0.08
60	4-CH_3-$(CH_2)_5$-C_5H_4N	5.00	[73]	5.00	0.00
61	4-CH_3-$(CH_2)_6$-C_5H_4N	5.65	[73]	5.63	0.02
62	C_6H_5-CH_2-CH(CH_3)-NH-$(CH_2)_2$-CH_3	3.91	[72]	3.94	-0.03
63	4-CH_3-$(CH_2)_7$-C_5H_4N	6.33	[73]	6.26	0.07

Outliers:

1	CH_3-NH_2	-0.56	[40]	-0.97	0.41
2	CH_3-COOH	-1.19	[19]	-1.52	0.33
3	CH_3-CH(OH)-COOH	-2.23	[8]	-2.77	0.54
		-1.81	[5]		0.96
4	HOOC-$(CH_2)_2$-COOH	-1.92	[5]	-3.08	1.16
5	3-OH-C_5H_4N	-1.40	[32]	-1.15	-0.35
6	4-NH_2-C_5H_4N	-0.70	[49]	-0.08	-0.62
7	CH_3-CO-CH_2-CO-CH_3	0.77	[50]	1.24	-0.47

Refs. pg 262

TABLE VII,6 (continued)

no	Compound	log P(obs)	ref.	log P(clc)	res.
8	$HOOC-CH_2-CH_2-CO-CH_3$	-1.19	[8]	-0.76	-0.43
		-1.32	[5]		-0.58
9	$HOOC-(CH_2)_3-COOH$	-1.81	[5]	-3.09	1.28
10	$2-COOH-C_5H_4N$	-1.64	[8]	-0.95	-0.69
11	$3-COOH-C_5H_4N$	-2.05	[8]	-0.95	-1.10
12	$2-NO_2-C_6H_4-OH$	2.54	[52]	0.22	2.32
13	$3-NO_2-C_6H_4-NH_2$	1.61	[8]	1.29	0.32
		1.59	[54]		0.30
14	$2-NO_2-C_6H_4-NH_2$	2.13	[8]	1.29	0.84
15	$2-NO_2-C_6H_4-SO_2NH_2$	0.14	[55]	-0.29	0.43
16	$4-NO_2-C_6H_4-SO_2NH_2$	-0.60	[55]	-0.29	-0.31
17	C_6H_5-OH	0.00	[46]	0.30	-0.30
18	$4-NH_2-C_6H_4-SO_2NH_2$	-1.85	[65]	-1.55	-0.30
19	$2-NO_2-C_6H_4-COOH$	0.03	[45]	0.42	-0.39
		-0.19	[5]		-0.61
20	$4-NO_2-C_6H_4-COOH$	0.86	[45]	0.42	0.44
21	C_6H_5-COOH	0.30	[67]	0.50	-0.20
		0.71	[45]		0.21
22	$2-OH-C_6H_4-COOH$	0.48	[69]	-1.90	2.38
		0.50	[45]		2.40
		0.34	[67]		2.24
		0.46	[54]		2.36
		0.56	[58]		2.46
23	$2-NH_2-C_6H_4-COOH$	0.57	[8]	-0.84	1.41
		-1.15	[45]		-0.31
24	$2,6-diCH_3-C_5H_4N$	2.30	[62]	2.83	-0.53
25	$2-CH_3-C_6H_4-COOH$	1.76	[45]	1.11	0.65
26	$4-CH_3-C_6H_4-COOH$	1.81	[45]	1.11	0.70
27	$C_6H_5-CH_2-CH_2-NH_2$	1.32	[72]	1.66	-0.34
		1.39	[70]		-0.27
28	$C_6H_5-CH_2-NH-CH_2-C_6H_5$	2.40	[30]	3.57	-1.17
29	$C_6H_5-CO-CH_2-CO-C_6H_5$	5.30	[51]	4.00	1.30
30	$C_6H_5-CH_2-CH(CH_3)-NH-CH_2-C_6H_5$	3.35	[72]	4.69	-1.34

Data on a small number of secondary *f* values are given in Tables VII,8 and VII,9.

In deriving the value of *f*$(CN)_{al.}$, it must be borne in mind that

there is a greater chance of an error. This also appeared to be the case
in the ether — water system, but in the latter it was CH_3-CN that drew
our attention to this possibility and prevented us from making mistakes.
In the chloroform — water system, however, such a guiding structure is
not available.

TABLE VII,7

f VALUES IN THE CHLOROFORM — WATER SYSTEM

No	fragment	f	kn	st.dev.	$\underline{t}$	$f(\text{oct})$
1	C_6H_5	2.348	0	0.049	47.5	1.886
2	C_6H_4	1.995	0	0.047	42.7	1.688
3	CH_3	0.965	0	0.047	20.7	0.702
4	CH_2	0.628	0	0.007	94.8	0.530
5	CH	0.163	0	0.054	3.04	0.235
6	C_5H_4N	0.899	0	0.057	15.8	0.526
7	$(ar)SO_2NH_2$	−2.567		0.058	−44.5	−1.530
8	$(al)COOH$	−2.485	2	0.048	−50.9	−0.954
9	$(al)CO$	−1.136	1	0.042	−26.8	−1.703
10	$(ar)NO_2$	(0.279)		0.060	4.65	−0.078
11	$(al)OH$	−2.367	1	0.050	−47.2	−1.491
12	$(al)NH_2$	−1.935	3	0.057	−33.7	−1.428
13	$(al)NH$	−2.386		0.058	−41.2	−1.825
14	$(al)Br$	−0.134		0.064	−2.09	0.27
15	c_M	0.318		0.015	21.0	0.287

The value in parentheses is suspect (see the section on the
diethylether — water system)

n = 97 (+1);   s = 0.082;

r = 0.997;   F = 2,478;   const. term = 0.000

TABLE VII,8

COMPILATION OF STRUCTURES USED FOR THE DERIVATION OF
SECONDARY f VALUES (CHLOROFORM — WATER SYSTEM)

fragment	compound	log P values			
		obs.	ref.	calc.	res.
$(ar)CONH_2$	$C_5H_4N-3-CONH_2$	−1.37	[8]	−1.31	−0.06
	$C_5H_4N-4-CONH_2$	−1.22	[76]	−1.31	0.09
	$C_6H_5-CONH_2$	0.11	[32]	0.14	−0.03
$(al)CONH_2$	CH_3-CONH_2	−2.00	[8]	−2.01	0.01

TABLE VII,8 (continued)

fragment	compound	obs.	ref.	calc.	res.
		— log P values —			
(ar)O	$CH_3-CH_2-CONH_2$	-1.40	[32]	-1.38	-0.02
	$4-CH_3O-C_6H_4-SO_2NH_2$	0.15	[55]	0.15	0.00
	$4-CH_3O-C_6H_4-COOH$	0.90	[5]	0.86	0.04
	$2-CH_3O-C_6H_4-CH_2CH(CH_3)$ CH_3-NH	3.00	[74]	3.05	-0.05
(al)CN	$NC-CH_2-CN$	-0.53	[77]	-0.68	0.15
	$NC-CH_2-COOH$	-2.17	[8]	-2.04	-0.13
	$NC-CH_2-CH_2-CN$	-0.23	[7]	-0.36	0.13
(al)I	$I-CH_2-COOH$	-0.82	[8]	-0.75	-0.07
		-0.73	[45]		0.02
	$I-CH_2-CH_2-COOH$	-0.40	[45]	-0.44	0.04
		-0.32	[5]		0.12
(al)Cl	$Cl-CH_2-COOH$	-1.67	[48]	-1.54	-0.13
		-1.35	[4]		0.18
		-1.92	[5]		-0.38*
	$Cl-CH_2-CH_2-COOH$	-0.86	[2]	-1.23	0.37*

*) Outliers

TABLE VII,9

SECONDARY *f* VALUES (CHLOROFORM – WATER SYSTEM)

nd	fragment	*f* (chloroform)	*f* (oct)
1	(ar)CONH_2	-2.21	-1.11
2	(al)CONH_2	-2.98	-1.97
3	(ar)O	-0.25	-0.43
4	(al)CN	-1.13	-1.07
5	(al)I	0.16	0.59
6	(al)Cl	-0.63	0.06

BENZENE – WATER SYSTEM

The total number of structures that lend themselves to derivation of
the hydrophobic fragmental constants was 67, involving 127 measured
log P values. Twenty-three structures, totalling 35 log P values, had to
be qualified as outliers; see Table VII,10 for data on the structures
used and Table VII,11 for the *f* values found.

TABLE VII,10

STRUCTURES INTRODUCED INTO STATISTICAL TREATMENT
(BENZENE — WATER SYSTEM)

no	Compound	log P(obs)	ref.	log P(clc)	res.
1	$Br-CH_2-COOH$	−1.41	[45]	−1.24	−0.17
2	$Cl-CH_2-COOH$	−1.45	[105]	−1.49	0.04
		−1.60	[85]		−0.11
3	CH_3-COOH	−1.97	[19]	−1.90	−0.07
		−1.80	[43]		0.10
		−2.00	[47]		−0.10
		−2.05	[78]		−0.15
		−1.74	[11]		0.16
4	CH_3-CH_2-OH	−1.58	[79]	−1.52	−0.06
		−1.49	[80]		0.03
5	$Br-CH_2-CH_2-COOH$	−0.85	[45]	−0.92	0.07
6	$CH_3-CH(Br)-COOH$	−0.62	[45]	−0.70	0.08
7	$Cl-CH_2-CH_2-COOH$	−1.06	[45]	−1.17	0.11
8	$CH_3-CO-CH_3$	−0.04	[19]	−0.03	−0.01
		−0.03	[83]		0.00
		−0.04	[84]		−0.01
		0.00	[85]		0.03
9	CH_3-CH_2-COOH	−1.20	[19]	−1.29	0.09
		−1.16	[86]		0.13
		−1.37	[47]		−0.08
		−1.22	[45]		0.07
10	$CH_3-CH_2-CH_2-OH$	−0.87	[79]	−0.90	0.03
11	$CH_3-CH_2-CH(Br)-COOH$	−0.08	[45]	−0.08	0.00
12	$CH_3-CH_2-CH_2-COOH$	−0.65	[86]	−0.67	0.02
		−0.65	[45]		0.02
13	$(CH_3)_2CH-COOH$	−0.74	[19]	−0.75	0.01
		−0.72	[45]		0.03
		−0.81	[5]		−0.06
14	$CH_3-(CH_2)_3-OH$	−0.19	[87]	−0.28	0.09
		−0.34	[79]		−0.06
		−0.38	[80]		−0.10
15	$(CH_3-CH_2)_2NH$	−0.02	[82]	−0.06	0.04
		−0.05	[24]		0.01
		−0.05	[5]		0.01

Refs. pg 262

TABLE VII,10 (continued)

no	Compound	log P(obs)	ref.	log P(clc)	res.
16	$CH_3-(CH_2)_2-CH(Br)-COOH$	0.50	[45]	0.53	-0.03
17	$CH_3-(CH_2)_3-COOH$	-0.05	[86]	-0.05	0.00
		-0.09	[45]		-0.04
18	$(CH_3)_2CH-CH_2-COOH$	-0.23	[45]	-0.13	-0.10
19	$CH_2\begin{smallmatrix}CH_2-CH_2\\ \\CH_2-CH_2\end{smallmatrix}NH$	-0.06	[88]	-0.10	0.04
20	$CH_3-(CH_2)_4-OH$	0.19	[80]	0.34	-0.15
21	$3-NO_2-C_6H_5-OH$	0.42	[89]	0.33	0.09
		0.38	[32]		0.05
22	$3-Br-C_6H_4-NH_2$	2.20	[90]	2.11	0.09
23	$4-Br-C_6H_4-NH_2$	2.06	[90]	2.11	-0.05
		2.12	[93]		0.01
24	$3-Cl-C_6H_4-NH_2$	1.93	[90]	1.87	0.06
		1.94	[91]		0.07
25	$4-Cl-C_6H_4-NH_2$	1.82	[92]	1.87	-0.05
		1.81	[90]		-0.06
		1.91	[93]		0.04
		1.80	[91]		-0.07
26	$4-NO_2-C_6H_4-NH_2$	0.92	[94]	0.98	-0.06
		0.93	[92]		-0.05
		0.95	[93]		-0.03
27	C_6H_5-OH	0.36	[95]	0.34	0.02
		0.34	[56]		0.00
		0.41	[96]		0.07
		0.40	[36]		0.06
		0.37	[97]		0.03
		0.32	[44]		-0.02
		0.42	[32]		0.08
28	$1,3-diOH-C_6H_4$	-2.11	[32]	-2.05	-0.06
29	$1,4-diOH-C_6H_4$	-2.16	[32]	-2.05	-0.11
30	$C_6H_5-NH_2$	1.03	[89]	0.99	0.04
		1.00	[93]		0.01
31	$1,3-diNH_2-C_6H_4$	-0.79	[98]	-0.76	-0.03
		-0.77	[92]		-0.01
		-0.75	[93]		0.01

TABLE VII,10 (continued)

		-0.75	[32]		0.01
32	$CH_3-(CH_2)_4-COOH$	0.67	[86]	0.57	0.10
		0.63	[45]	0.57	0.06
33	$(CH_3)_2CH-(CH_2)_2-COOH$	0.57	[45]	0.49	0.08
34	$CH_3-(CH_2)_2-NH-(CH_2)_2-CH_3$	1.05	[82]	1.17	-0.12
35	$3-NO_2-C_6H_4-COOH$	0.21	[99]	0.26	-0.05
36	$4-NO_2-C_6H_4-COOH$	0.31	[52]	0.26	0.05
37	C_6H_5-COOH	0.21	[45]	0.27	-0.06
		0.24	[67]		-0.03
		0.18	[100]		-0.09
		0.36	[12]		0.09
		0.12	[102]		-0.15
38	$4-NH_2-C_6H_4-COOH$	-1.46	[89]	-1.47	0.01
39	$3-CH_3-C_6H_4-OH$	0.88	[56]	0.83	0.05
40	$3-CH_3-C_6H_4-NH_2$	1.50	[92]	1.47	0.03
		1.51	[90]		0.04
41	$2-CH_3-C_6H_4-NH_2$	1.53	[92]	1.47	0.06
42	$4-CH_3-C_6H_4-NH_2$	1.43	[92]	1.47	-0.04
		1.49	[90]		0.02
		1.38	[93]		-0.09
43	$C_6H_5-CH_2-COOH$	-0.03	[45]	0.00	-0.03
44	$4-CH_3-CH_2-C_6H_4-OH$	1.44	[56]	1.44	0.00

Outliers:

1	CH_3-COOH	-2.20	[6]	-1.91	-0.29
2	CH_3-CH_2-OH	-0.01	[19]	-1.52	1.51
3	$CH_3-NH-CH_3$	-0.82	[82]	-1.31	0.49
4	$CH_3-CH_2-CH_2-OH$	-0.65	[80]	-0.90	0.25
5	$2-NO_2-C_6H_4-OH$	2.33	[53]	0.32	2.01
6	$4-NO_2-C_6H_4-OH$	0.07	[32]	0.32	-0.25
7	$2-Cl-C_6H_4-NH_2$	2.13	[92]	1.87	0.26
		2.08	[91]		0.21
8	$3-NO_2-C_6H_4-NH_2$	1.31	[94]	0.97	0.34
		1.30	[92]		0.33
		1.36	[93]		0.39
9	$2-NO_2-C_6H_4-NH_2$	1.78	[94]	0.97	0.81
		1.81	[93]		0.84

Refs. pg 262

TABLE VII,10 (continued)

no	Compound	log P(obs)	ref.	log P(clc)	res.
10	1,2-diOH-C_6H_4	-1.19	[32]	-2.06	0.87
11	C_6H_5-NH_2	-0.05	[60]	0.98	-1.03
12	2-OH-C_6H_4-NH_2	-0.84	[92]	-1.41	0.57
13	4-OH-C_6H_4-NH_2	-1.65	[92]	-1.41	-0.24
14	1,2-diNH_2-C_6H_4	-0.28	[98]	-0.76	0.48
		-0.26	[92]		0.50
		-0.26	[93]		0.50
		-0.26	[32]		0.50
15	1,4-diNH_2-C_6H_4	-1.17	[98]	-0.76	-0.41
		-1.17	[92]		-0.41
16	2-NO_2-C_6H_4-COOH	-0.30	[52]	0.26	-0.56
		-0.21	[60]		-0.47
17	C_6H_5-COOH	-0.21	[101]	0.27	-0.48
18	2-OH-C_6H_4-COOH	0.45	[67]	-2.13	2.58
		0.38	[103]		2.51
19	3-NH_2-C_6H_4-COOH	-2.02	[89]	-1.48	-0.54
20	2-NH_2-C_6H_4-COOH	-0.40	[89]	-1.48	1.08
		-0.27	[93]		1.21
21	3-CH_3-C_6H_4-NH_2	1.28	[93]	1.47	-0.19
22	2-CH_3-C_6H_4-NH_2	1.13	[93]	1.47	-0.34
23	4-CH_3-C_6H_4-NH_2	1.70	[104]	1.47	0.23

TABLE VII,11

f VALUES IN THE BENZENE – WATER SYSTEM

No	fragment	*f*	kn	st.dev.	t	*f*(oct)
1	C_6H_5	2.230	0	0.052	42.4	1.886
2	C_6H_4	1.768	0	0.069	25.7	1.688
3	CH_3	0.947	0	0.032	29.4	0.702
4	CH_2	0.619	0	0.010	60.3	0.530
5	CH	0.212	0	0.047	4.55	0.235
6	(al)COOH	-2.898	3	0.045	-64.0	-0.954
7	(al)CO	-1.967		0.066	-29.9	-1.703
8	(ar)NO_2	(0.451)		0.051	8.84	-0.939
9	(al)OH	-3.128	4	0.053	-59.3	-1.491

TABLE VII,11 (continued)

10	(ar)NH$_2$	−1.285		0.046	−27.8	−0.854
11	(al)NH	−3.242		0.065	−50.0	−1.825
12	(al)Cl	−0.147	5	0.062	−2.38	0.061
13	(al)Br	0.094	5	0.059	1.58	0.270
14	c_M	0.298		0.015	19.5	0.287

The value in parentheses is suspect (see the section on the diethylether - water system)

n = 92 (+1);   s = 0.075;

r = 0.998;   F = 1,478;   const. term = 0.044

A few structures could be used for the calculation of secondary f values; see Table VII,12.

TABLE VII,12

SECONDARY f VALUES (BENZENE - WATER SYSTEM)

fragment (f-value)	Compound	log P values			
		obs.	ref.	clc.	res.
(al)I 0.42	I-CH$_2$-COOH	−1.08	[45]	−0.96	−0.12
	I-CH$_2$-CH$_2$-COOH	−0.52	[45]	−0.64	0.12
(al)NH$_2$ −2.40	CH$_3$-NH$_2$	−1.34	[82]	−1.45	0.11
	CH$_3$-(CH$_2$)$_3$-NH$_2$	0.14	[24]	−0.40	−0.26*
	C$_6$H$_5$-CH$_2$-NH$_2$	0.61	[91]	0.45	0.16
(al)COO −1.51	CH$_3$-COO-CH$_3$	0.47	[106]	0.38	0.09
	CH$_3$-CH$_2$-OOC-CH$_3$	1.01	[106]	1.00	0.01
	CH$_3$-COO-(CH$_2$)$_2$-CH$_3$	1.52	[106]	1.62	−0.10

*) Outlier

OIL - WATER SYSTEM

For the organic phase of the oil - water system, a glyceryl triester such as olive oil, cottonseed oil or peanut oil is often used. The partition values found do not appear to be very dependent on the triester employed for the partitioning system, and hence the available values can be used indiscriminately.

There were 163 measurements for 105 structures available, which were divided over 78 aliphatic compounds and 27 phenyl derivatives. Eventually, 59 structures (on which 97 measurements had been carried out) could be

Refs. pg 262

TABLE VII,13

STRUCTURES INTRODUCED INTO STATISTICAL TREATMENT
(OIL – WATER SYSTEM)

no	Compound	log P(obs)	ref.	log P(clc)	res.
1	CH_3-OH	−1.96	[107]	−2.02	0.06
		−2.01	[108]		0.01
		−2.11	[109]		−0.09
		−2.02	[110]		0.00
2	$Br-CH_2-COOH$	−0.72	[111]	−0.57	−0.15
3	CH_3-COOH	−1.30	[111]	−1.41	0.11
		−1.52	[112]		−0.11
		−1.57	[113]		−0.16
4	CH_3-CH_2-Br	1.57	[114]	1.42	0.15
5	CH_3-CH_2-Cl	1.38	[114]	1.19	0.19
6	CH_3-CONH_2	−3.08	[23]	−3.10	0.02
		−3.08	[115]		0.02
7	$CH_3-OOCNH_2$	−1.60	[23]	−1.52	−0.08
		−1.40	[114]		0.12
		−1.40	[116]		0.12
8	CH_3-CH_2-OH	−1.52	[117]	−1.43	−0.09
		−1.45	[107]		−0.02
		−1.45	[108]		−0.02
		−1.49	[109]		−0.06
		−1.33	[118]		0.10
9	$CH_3-CH(Br)-COOH$	−0.18	[111]	−0.07	−0.11
10	$Br-CH_2-CH_2-COOH$	−0.34	[111]	−0.34	0.00
11	$CH_3-CO-CH_2-Cl$	0.03	[107]	−0.07	0.10
12	$Cl-CH_2-CH_2-COOH$	−0.53	[111]	−0.57	0.04
13	$CH_3-CO-CH_3$	−0.70	[117]	−0.67	−0.03
		−0.64	[118]		0.03
14	CH_3-CH_2-COOH	−0.80	[111]	−0.82	0.02
		−0.85	[113]		−0.03
15	$CH_3-CH_2-CONH_2$	−2.44	[119]	−2.51	0.07
16	$CH_3-CH_2-OOCNH_2$	−1.12	[23]	−0.93	−0.19
		−0.92	[107]		0.01
		−0.85	[116]		0.08
		−1.00	[120]		−0.07
17	$CH_3-(CH_2)_2-OH$	−0.85	[107]	−0.84	−0.01

TABLE VII,13 (continued)

		-0.81	[108]	-0.84	0.03
		-0.89	[109]		-0.05
		-0.81	[110]		0.03
18	$(CH_3)_2CH-OH$	-1.05	[23]	-0.92	-0.13
19	$CH_3-O-(CH_2)_2-OH$	-2.25	[23]	-2.25	0.00
20	$CH_3-(CH_2)_2-COOH$	-0.21	[111]	-0.23	0.02
		-0.35	[112]		-0.12
21	$(CH_3)_2CH-COOH$	-0.12	[111]	-0.32	0.20
22	$CH_3-(CH_2)_2-CONH_2$	-2.02	[23]	-1.92	-0.10
		-2.02	[115]		-0.10
23	$CH_3-(CH_2)_3-OH$	-0.28	[107]	-0.25	-0.03
		-0.20	[110]		0.05
24	$(CH_3)_2CH-CH_2-OH$	-0.36	[107]	-0.34	-0.02
		-0.24	[108]		0.10
		-0.26	[110]		0.08
25	$CH_3-CH_2-CH(CH_3)-OH$	-0.42	[110]	-0.34	-0.08
26	$CH_3-CH_2-O-CH_2-CH_3$	0.58	[107]	0.71	-0.17
		0.60	[121]		-0.11
27	$CH_3-CH_2-O-(CH_2)_2-OH$	-1.72	[23]	-1.66	-0.06
28	$CH_3-(CH_2)_2-CH(Br)-CONH_2$	-0.62	[122]	-0.59	-0.03
29	$CH_3-(CH_2)_3-COOH$	0.48	[111]	0.36	0.12
		0.41	[112]		0.05
30	$(CH_3)_2CH-CH_2-COOH$	0.27	[111]	0.27	0.00
31	$CH_3-(CH_2)_3-CONH_2$	-1.15	[79]	-1.33	0.18
32	$CH_3-(CH_2)_4-OH$	0.36	[110]	0.34	0.02
33	$(CH_3)_2CH-(CH_2)_2-OH$	0.26	[107]	0.25	0.01
		0.33	[108]		0.08
		0.32	[110]		0.07
34	$CH_3-(CH_2)_2-CH(CH_3)-OH$	0.17	[110]	0.25	-0.08
35	$(CH_3-CH_2)_2CH-OH$	0.20	[110]	0.25	-0.05
36	$CH_3-O-(CH_2)_2-O-(CH_2)_2-OH$	-2.38	[23]	-2.49	0.11
37	C_6H_5-OH	0.81	[56]	0.69	0.12
		0.78	[107]		0.09
		0.60	[114]		-0.09
		0.75	[123]		0.06
38	$1,4-diOH-C_6H_4$	-0.83	[124]	-0.94	0.11
39	$CH_3-CO-(CH_2)_2-CO-CH_3$	-1.09	[23]	-1.11	0.02

Refs. pg 262

TABLE VII,13 (continued)

no	Compound	log P(obs)	ref.	log P(clc)	res.
40	$CH_3-CH(Br)-(CH_2)_2-CONHCONH_2$	-0.54	[125]	-0.68	0.14
41	$(CH_3)_2CH-CH(Cl)-CONHCONH_2$	-0.11	[122]	0.07	-0.18
42	$CH_3-(CH_2)_3-CONHCONH_2$	-0.31	[122]	-0.36	0.05
43	$CH_3-(CH_2)_4-COOH$	0.83	[111]	0.95	-0.12
44	$(CH_3)_2CH-(CH_2)_2-COOH$	0.90	[112]	0.86	0.04
45	$CH_3-(CH_2)_5-OH$	0.88	[110]	0.93	-0.05
46	$CH_3-CH_2-O-CH(CH_3)-O-CH_2-CH_3$	0.90	[114]	0.74	0.16
47	C_6H_5-COOH	0.66	[126]	0.59	0.07
		0.54	[127]		-0.05
48	$C_6H_5-CONH_2$	-0.51	[107]	-0.40	-0.11
		-0.36	[79]		0.04
		-0.42	[120]		-0.02
		-0.36	[118]		0.04
49	$3-CH_3-C_6H_4-OH$	1.29	[56]	1.31	-0.02
		1.21	[123]		-0.10
50	$2-CH_3-C_6H_4-OH$	1.34	[123]	1.31	0.03
51	$4-CH_3-C_6H_4-OH$	1.21	[123]	1.31	-0.10
52	$CH_3-(CH_2)_5-COOH$	1.69	[112]	1.54	0.15
53	$C_6H_5-CO-CH_2-Cl$	1.99	[107]	2.14	-0.15
54	$4-OH-C_6H_4-CH_2-COOH$	-1.04	[128]	-1.16	0.08
55	$1,4-diCH_3-O-C_6H_4$	2.15	[107]	2.15	0.00
		2.21	[114]		0.06
56	$CH_3-(CH_2)_3-O-(CH_2)_2-O$ $HO-(CH_2)_2$	-0.92	[23]	-0.72	-0.20
57	$4-CH_3-O-C_6H_4-CH_2-COOH$	0.45	[128]	0.38	0.07
58	$4-CH_3-CH_2-C_6H_4-CH_2-COOH$	1.56	[128]	1.68	-0.12
59	$4-CH_3-CH_2-O-C_6H_4-CH_2-COOH$	0.92	[128]	0.97	-0.05

Outliers:

no	Compound	log P(obs)	ref.	log P(clc)	res.
1	$Cl-CH_2-COOH$	-1.10	[111]	-0.84	-0.26
2	$HO-(CH_2)_2-OH$	-3.31	[23]	-3.83	0.52
3	$Cl-CH_2-CO-CH_2-Cl$	-0.28	[107]	0.49	-0.77
4	$H_2NOC-CH_2-CONH_2$	-4.10	[115]	-6.24	2.14
5	$CH_3-CO-CH_3$	-1.10	[107]	-0.67	-0.43
6	$Cl-CH_2-CH(OH)-CH_2-OH$	-1.92	[23]	-3.09	1.17
7	$CH_3-CH_2-OOCNH_2$	-1.52	[79]	-0.93	-0.59

TABLE VII,13 (continued)

8	$CH_3-CH(OH)-CONH_2$	-3.24	[115]	-4.65	1.41
9	$(CH_3)_2CH-OH$	-1.32	[23]	-0.92	-0.40
10	$CH_3-CH(OH)-CH_2-OH$	-2.77	[23]	-3.33	0.56
11	$HO-CH_2-CH(OH)-CH_2-OH$	-4.15	[23]	-5.70	1.55
12	$CH_3-CH_2-CH(Br)-COOH$	0.14	[111]	0.48	-0.34
13	$CH_2^{}\!-\!CH\!-\!O\!-\!CH_3$ $\diagdown CH_2$	0.83	[121]	0.14	0.69
		0.70	[129]		0.56
14	$CH_3-(CH_2)_2-COOH$	-0.46	[113]	-0.23	-0.23
15	$CH_3-CH_2-CH(CH_3)-OH$	-0.60	[23]	-0.34	-0.26
16	$CH_3-CH_2-O-CH_2-CH_3$	0.38	[79]	0.71	-0.33
		0.36	[130]		-0.35
17	$HO-CH_2-CH_2-CH(OH)-CH_3$	-2.37	[23]	-3.45	1.08
18	$HO-(CH_2)_4-OH$	-2.68	[23]	-3.37	0.69
19	$CH_3-CH(OH)-CH(OH)-CH_3$	-2.47	[23]	-2.83	0.36
20	$CH_3-CH_2-O-(CH_2)_2-OH$	-1.15	[107]	-1.66	0.51
21	$CH_3-O-CH_2-CH(OH)-CH_2-OH$	-2.58	[23]	-4.16	1.58
22	$CH_3-CH_2-CH(Br)-CONHCONH_2$	-0.43	[122]	-0.23	-0.20
23	$CH_3-(CH_2)_2-CH(Br)-COOH$	0.55	[111]	1.07	-0.52
24	$(CH_3)_2CH-CH(Br)-CONH_2$	-0.20	[122]	-0.71	0.51
25	$CH_2^{}\!-\!CH\!-\!O\!-\!CH_2\!-\!CH_3$ $\diagdown CH_2$	1.20	[121]	0.73	0.47
26	$CH_3-(CH_2)_3-CONH_2$	-0.50	[131]	-1.33	0.83
		-1.64	[115]		-0.31
27	$(CH_3)_2CH-CH_2-OOCNH_2$	0.73	[132]	0.12	0.61
28	$HO-(CH_2)_5-OH$	-2.21	[23]	-2.78	0.57
29	$CH_3-CH_2-O-CH-CH(OH)-CH_2-OH$	-2.13	[23]	-3.57	1.44
30	$CH_3-(CH_2)_2-CH(Br)-CONHCONH_2$	-0.36	[125]	0.35	-0.71
		-0.19	[122]		-0.54
31	$CH_3-CH_2-CH(Br)-CH_2-CONHCONH_2$	-0.45	[125]	0.00	-0.45
32	$CH_3-CH(Br)-CH(CH_3)-CONHCONH_2$	0.23	[125]	-0.09	0.32
33	$Br-(CH_2)_2-CH(CH_3)-CONHCONH_2$	-0.04	[125]	-0.71	0.67
34	$(CH_3)_2CH-CH_2-CONHCONH_2$	-0.16	[122]	-0.48	0.32
35	$CH_3-CH_2-O-CH(CH_3)-O-CH_2-CH_3$	0.18	[107]	0.74	-0.56
36	$CH_3-CH_2-O-(CH_2)_2-O-(CH_2)_2-OH$	-2.22	[23]	-1.94	-0.28
37	$HO-(CH_2)_3-O-(CH_2)_3-OH$	-2.70	[23]	-4.31	1.61
38	$2-OH-C_6H_4-COOH$	1.00	[107]	-1.08	2.08

Refs. pg 262

TABLE VII,13 (continued)

no	Compound	log P(obs)	ref.	log P(clc)	res.
39	4-OH-C_6H_4-COOH	0.22	[133]	−1.08	1.30
40	C_6H_5-CONH$_2$	−0.66	[132]	−0.40	−0.26
41	2-OH-C_6H_4-CONH$_2$	0.45	[107]	−2.06	2.51
		1.15	[114]		3.21
		0.41	[79]		2.47
		0.34	[132]		2.40
		1.15	[118]		3.21
42	2-OH-C_6H_4-O-CH$_3$	1.48	[114]	0.57	0.91
		0.96	[123]		0.41
43	CH$_3$-CH$_2$-O-CH$_2$-CH(OH)-CH$_2$-O CH$_3$-CH$_2$	−0.96	[23]	−1.44	0.48
44	CH$_3$-(CH$_2$)$_7$-OH	1.77	[110]	2.07	−0.30
45	C_6H_5-CH(CH$_3$)-COOH	0.51	[127]	0.93	−0.42
46	C_6H_5-CH$_2$-CH$_2$-COOH	0.72	[126]	1.02	−0.30
		0.82	[134]		−0.20
47	C_6H_5-CH(CH$_2$-CH$_3$)-COOH	0.74	[127]	1.52	−0.78
		1.16	[135]		−0.36
48	C_6H_5-CH(CH$_3$)-CH$_2$-COOH	1.06	[134]	1.52	−0.46
49	C_6H_5-(CH$_2$)$_3$-COOH	0.92	[126]	1.60	−0.68
		1.17	[134]		−0.43
50	C_6H_5-CH(CH$_2$-CH$_2$-CH$_3$)-COOH	0.84	[127]	2.11	−1.27
		1.41	[135]		−0.70
51	C_6H_5-CH(CH$_3$)-(CH$_2$)$_2$-COOH	1.46	[134]	2.11	−0.65
52	C_6H_5-(CH$_2$)$_4$-COOH	1.03	[126]	2.19	−1.16
53	C_6H_5-(CH$_2$)$_5$-COOH	1.25	[126]	2.78	−1.53

accommodated in the regression analysis, the remaining structures being
qualified as outliers. For pertinent information, see Tables VII,13 and
VII,14.

TABLE VII,14

f VALUES IN THE OIL − WATER SYSTEM

No	fragment	*f*	kn	st.dev.	$\underline{t}$	*f*(oct)
1	C_6H_5	2.091	0	0.063	33.3	1.886
2	C_6H_4	1.906	0	0.066	28.9	1.688
3	CH$_3$	0.808	0	0.042	19.0	0.702

TABLE VII,14 (continued)

4	CH_2	0.589	0	0.010	58.3	0.530
5	CH	0.282	0	0.052	5.37	0.235
6	(al)COOH	−2.253	2	0.049	−46.0	−0.954
7	(al)CO	−2.326	2	0.058	−40.4	−1.703
8	(al)$CONH_2$	−3.945	4	0.058	−68.2	−1.970
9	(al)$OOCNH_2$	−2.363		0.058	−40.8	−1.481
10	(al)$CONHCONH_2$	−2.967		0.079	−17.6	
11	(al)O	−2.124	4	0.049	−43.1	−1.581
12	(al)OH	−2.861	4	0.049	−57.9	−1.491
13	(al)Cl	−0.244		0.064	−3.82	0.061
14	(al)Br	−0.013		0.061	−0.21	0.270
15	c_M	0.355		0.016	22.3	0.287

$n = 97 \; (+2);$ $s = 0.099;$

$r = 0.996;$ $F = 863;$ const. term $= 0.038$

Only one secondary f value could be made available, namely that of the COO group; see Table VII,15.

TABLE VII,15

SECONDARY f VALUE OF THE COO GROUP

Compound	log P value			
	obs.	ref.	clc.	res.
CH_3−COO−CH_3	−0.37	[23]	−0.31	−0.06
Br−CH_2−COO−CH_2−CH_3	1.12	[107]	1.11	0.01
CH_3−CH_2−COO−CH_3	0.40	[23]	0.27	0.13
	0.60	[114]		0.33[*]

[*]) Outlier

$$f \left[COO(al) \right] = -1.93$$

OLEYL ALCOHOL — WATER SYSTEM

Here 85 measurements, referring to 65 structures, which were again aliphatic (12 structures) or phenyl derivatives (53 compounds), were performed. It was possible to accommodate 55 measurements (49 structures) in the regression analysis. For pertinent information, see Tables VII,16 and VII,17.

Refs. pg 262

TABLE VII, 16

STRUCTURES INTRODUCED INTO STATISTICAL TREATMENT
(OLEYLALCOHOL – WATER SYSTEM)

no	Compound	log P(obs)	ref.	log P(clc)	res.
1	CH_3-COOH	−0.66	[136]	−0.60	−0.06
2	CH_3-CH_2-OH	−1.00	[79]	−1.08	0.08
3	$HOOC-CH_2-COOH$	−1.28	[136]	−1.26	−0.02
4	CH_3-CH_2-COOH	−0.09	[136]	−0.07	−0.02
5	$CH_3-(CH_2)_2-OH$	−0.45	[79]	−0.54	0.09
6	$CH_3-(CH_2)_2-COOH$	0.46	[136]	0.47	−0.01
7	$CH_3-(CH_2)_3-OH$	−0.19	[79]	−0.01	−0.18
8	$HOOC-(CH_2)_3-COOH$	−0.96	[136]	−0.97	0.01
9	$CH_3-(CH_2)_4-COOH$	1.65	[136]	1.54	0.11
10	$3-CH_3-C_6H_4-OH$	1.79	[137]	1.70	0.09
11	$2-CH_3-C_6H_4-OH$	1.81	[137]	1.70	0.11
12	$4-CH_3-C_6H_4-OH$	1.80	[137]	1.70	0.10
13	$3-CH_3-O-C_6H_4-OH$	1.15	[137]	1.19	−0.04
14	$2-CH_3-O-C_6H_4-OH$	1.15	[137]	1.19	−0.04
15	$4-CH_3-O-C_6H_4-OH$	1.00	[137]	1.19	−0.19
16	$4-NH_2-C_6H_4-COO-CH_3$	1.09	[138]	1.11	−0.02
17	$4-NH_2-C_6H_4-COO-CH_2-CH_3$	1.61	[138]	1.64	−0.03
		1.58	[139]		−0.06
18	$4-NH_2-C_6H_4-COO-CH(CH_3)_2$	2.05	[139]	2.09	−0.04
19	$4-NH_2-C_6H_4-COO-(CH_2)_2-CH_3$	2.28	[138]	2.18	0.10
		2.17	[139]		−0.01
20	$4-NH_2-C_6H_4-COO-(CH_2)_3-CH_3$	2.77	[138]	2.71	0.06
21	$4-NH_2-C_6H_4-COO-CH_2-CH(CH_3)_2$	2.68	[139]	2.62	0.06
22	$4-NH_2-C_6H_4-COO-CH(CH_3)-CH_2-CH_3$	2.59	[139]	2.62	−0.03
23	$4-NH_2-C_6H_4-COO-(CH_2)_2-N(CH_3)_2$	0.88	[140]	0.80	0.08
24	$4-NH_2-C_6H_4-COO-(CH_2)_4-CH_3$	3.41	[141]	3.25	0.16
25	$4-NH_2-C_6H_4-COO-CH(CH_3)-CH(CH_3)_2$	3.10	[139]	3.06	0.04
26	$4-NH_2-C_6H_4-COO-(CH_2)_4-CH_3$	3.24	[138]	3.25	−0.01
27	$4-NH_2-C_6H_4-COO-(CH_2)_5-CH_3$	3.71	[138]	3.78	−0.07
28	$4-NH_2-C_6H_4-COO-(CH_2)_2-N(CH_2-CH_3)_2$	1.90	[142]	1.87	0.03
		1.79	[143]		−0.08
29	$4-NH_2-C_6H_4-COO-(CH_2)_6-CH_3$	4.26	[138]	4.32	−0.06
30	$4-CH_3-C_6H_4-COO-(CH_2)_2-N(CH_2-CH_3)_2$	3.13	[144]	3.24	−0.11
31	$4-NH_2-C_6H_4-O-(CH_2)_2-N(CH_2-CH_3)_2$	2.32	[143]	2.27	0.05

TABLE VII, 16 (continued)

32	$4-NH_2-C_6H_4-COO-CH(CH_3)-CH_2$ $(CH_3-CH_2)_2N$	2.24	[142]	2.31	-0.07
33	$4-NH_2-C_6H_4-COO-CH_2-CH(CH_3)$ $(CH_3-CH_2)_2N$	2.32	[142]	2.31	0.01
34	$4-NH_2-C_6H_4-(CH_2)_4-N(CH_2-CH_3)_2$	3.26	[143]	3.32	-0.06
35	$4-NH_2-C_6H_4-COO-(CH_2)_7-CH_3$	4.78	[138]	4.85	-0.07
36	$3-CH_3-CH_2-O-C_6H_4-COO-(CH_2)_2$ $(CH_3-CH_2)_2N$	3.35	[145]	3.27	0.08
37	same, $2-C_2H_5O-$ derivative	3.07	[145]	3.27	-0.20
38	same, $4-C_2H_5O-$ derivative	3.40	[145]	3.27	0.13
39	$4-CH_3(CH_2)_2-C_6H_4-COO-(CH_2)_2$ $(CH_3-CH_2)_2N$	4.32	[141]	4.31	0.01
40	same, $3-CH_3-(CH_2)_2-O-$ deriv.	3.92	[145]	3.80	0.12
41	same, $3-(CH_3)_2CH-O-$ deriv.	3.53	[145]	3.71	-0.18
42	same, $4-CH_3-(CH_2)_2-O-$ deriv.	3.92	[145]	3.80	0.12
43	same, $4-(CH_3)_2CH-O-$ deriv.	3.75	[145]	3.71	0.04
44	same, $4-CH_3-(CH_2)_3-$ deriv.	4.72	[146]	4.85	-0.13
45	same, $3-CH_3-(CH_2)_3-O-$ deriv.	4.29	[145]	4.34	-0.05
		4.34	[146]		0.00
46	same, $3-(CH_3)_2CH-CH_2-O-$ deriv.	4.28	[145]	4.24	0.04
47	same, $4-CH_3-(CH_2)_3-O-$ deriv.	4.27	[147]	4.34	-0.07
		4.46	[145]		0.12
		4.29	[146]		-0.05
48	same, $4-(CH_3)_2CH-CH_2-O-$ deriv.	4.35	[145]	4.24	0.11
49	$4-NH_2-C_6H_4-COO-(CH_2)_2-N(n.C_4H_9)_2$	4.07	[140]	4.01	0.06

Outliers:

1	$HO-CH_2-COOH$	-1.70	[136]	-2.27	0.57
2	$HO-CH(CH_3)-COOH$	-1.21	[136]	-1.83	0.62
3	$HOOC-CH(OH)-CH_2-COOH$	-1.74	[136]	-2.74	1.00
4	$4-NH_2-C_6H_4-COO-(CH_2)_2-N(CH_3)_2$	-0.40	[140]	0.80	-1.20
5	$4-HO-C_6H_4-COO-(CH_2)_2-N(CH_2-CH_3)_2$	2.40	[144]	1.57	0.83
		0.70	[144]		-0.87
6	$3-NH_2-C_6H_4-COO-(CH_2)_2-N(C_2H_5)_2$	1.26	[146]	1.87	-0.61
7	same, $2-NH_2-$ derivative	3.04	[146]	1.87	1.17
8	same, $4-NH_2-$ derivative	1.32	[148]	1.87	-0.55
		1.32	[148]		-0.55
		-0.30	[140]		-2.17

Refs. pg 262

TABLE VII,16

no	Compound	log P(obs)	ref.	log P(clc)	res.
		1.18	[146]	1.87	-0.69
9	$4-CH_3-C_6H_4-COO-(CH_2)_2-N(C_2H_5)_2$	1.66	[144]	3.24	-1.58
10	same, $2-CH_3-O-$ derivative	2.46	[145]	2.73	-0.27
11	$4-NH_2-C_6H_4-COO-(CH_2)_3-N(C_2H_5)_2$	-0.10	[140]	1.88	-1.98
12	$4-CH_3-CH_2-O-C_6H_4-COO-(CH_2)_2$	0.90	[144]	3.27	-2.37
	$N(C_2H_5)_2$	3.54	[147]		0.27
		2.46	[148]		-0.81
13	$4-NH_2-C_6H_4-COO-CH(CH_3)-CH(CH_3)$	2.66	[142]	4.15	-1.49
	$(C_2H_5)_2N$				
14	$4-NH_2-C_6H_4-COO-CH_2-CH_2-CH_2-CH_2$	0.48	[140]	3.82	-3.34
	$(C_2H_5)_2N$				
15	$4-NH_2-C_6H_4-COO-CH_2-CH_2$	3.09	[140]	4.34	-1.25
	$(\underline{n}.C_3H_7)_2N$	1.35	[140]		-2.99
16	$2-n.C_3H_7-O-C_6H_4-COO-(CH_2)_2$	3.47	[145]	3.80	-0.33
	$(C_2H_5)_2N$				
17	same, $2-i.C_3H_7-O-$ deriv.	3.43	[145]	3.71	-0.28
18	same, $3-n.C_4H_9-$ deriv.	4.48	[146]	4.85	-0.37
		4.57	[146]		-0.28
19	same, $2-n.C_4H_9-O-$ deriv.	3.98	[145]	4.34	-0.36
		3.81	[146]		-0.51
20	same, $2-i.C_4H_9-O-$ deriv.	3.97	[145]	4.24	-0.27
21	$4-NH_2-C_6H_4-COO-(CH_2)_2-$ N$(C_4H_9)_2$	2.28	[140]	4.01	-1.73

TABLE VII,17

f VALUES IN THE OLEYLALCOHOL – WATER SYSTEM

No	fragment	*f*	kn	st.dev.	$\underline{t}$	*f*(oct)
1	C_6H_4	2.009	0	0.138	14.5	1.688
2	CH_3	0.688	0	0.084	8.14	0.702
3	CH_2	0.535	0	0.010	53.8	0.530
4	CH	0.289	0	0.092	3.13	0.235
5	(al)COOH	-1.289		0.068	-19.0	-0.954
6	(ar)COO	-0.901		0.071	-12.8	-0.431
7	(al)OH	-2.301	5	0.103	-22.3	-1.491
8	(ar)O	-0.512		0.046	-11.1	-0.433
9	(ar)NH$_2$	-0.685		0.104	-6.60	-0.854

TABLE VII,17 (continued)

| 10 | (al)N | -2.590 | 0.130 | -20.0 | -1.012 |
| 11 | c_M | 0.262 | 0.028 | 9.23 | 0.287 |

$n = 55\ (+1);$ $s = 0.097;$

$r = 0.998;$ $F = 1,440;$ const. term $= 0.000$

TOLUENE - WATER SYSTEM

Partitioning data pertinent to the toluene - water system were so limited that the procedure followed for the above mentioned systems had to be abandoned. In order to derive optimum benefit from the available material, it was necessary to consider each compound critically and to build up the f set in such a manner that finally the smallest possible residu total was obtained. Tables VII,18 and VII,19 give the relevant data.

TABLE VII,18

PARTITION VALUES IN THE TOLUENE - WATER SYSTEM

no	Compound	log P(obs)	ref.	log P(clc)	res.
1	CH_3-NH_2	-1.40	[82]	-1.60	0.20
2	$Br-CH_2-COOH$	-1.55	[45]	-1.22	-0.33*
3	CH_3-COOH	-1.90	[85]	-1.86	-0.04
4	$CH_3-NH-CH_3$	-1.08	[82]	-1.22	0.14
		-1.28	[41]		-0.06
5	$CH_3-CH_2-NH_2$	-1.28	[41]	-1.40	0.12
6	$CH_3-CH(Br)-COOH$	-0.80	[45]	-0.79	-0.01
7	$Br-(CH_2)_2-COOH$	-0.97	[45]	-0.96	-0.01
8	CH_3-CH_2-COOH	-1.34	[19]	-1.30	-0.04
		-1.33	[45]		-0.03
9	$CH_3-(CH_2)_2-NH$	-0.65	[41]	-0.48	-0.17
10	$(CH_3)_3N$	-0.36	[82]	-0.80	0.44*
		-0.36	[41]		0.44*
		-0.34	[7]		0.46*
11	$CH_3-CH_2-CH(Br)-COOH$	-0.27	[45]	-0.23	-0.04
12	$CH_3-(CH_2)_2-COOH$	-0.82	[45]	-0.74	-0.08
13	$(CH_3)_2CH-COOH$	-0.86	[19]	-0.87	0.01
		-0.87	[45]		0.00

TABLE VII,18 (continued)

no	Compound	log P(obs)	ref.	log P(clc)	res.
14	$CH_3-(CH_2)_3-NH_2$	0.30	[150]	0.08	0.22
15	$(CH_3-CH_2)_2NH$	-0.09	[82]	-0.10	0.01
		-0.20	[41]		-0.10
		-0.24	[151]		-0.14
16	$CH_3-(CH_2)_2-CH(Br)-COOH$	0.38	[45]	0.33	0.05
17	$CH_3-(CH_2)_3-COOH$	-0.20	[45]	-0.18	-0.02
18	$(CH_3)_2CH-CH_2-COOH$	-0.35	[45]	-0.31	-0.04
19	C_6H_5-OH	0.22	[56]	0.26	-0.04
		0.32	[96]		0.06
		0.23	[85]		-0.03
20	$CH_3-(CH_2)_4-COOH$	0.56	[45]	0.38	0.18
		1.03	[152]		0.65*
21	$(CH_3-CH_2-CH_2)_2NH$	1.16	[41]	1.02	0.14
22	$(CH_3-CH_2)_3N$	1.00	[41]	0.88	0.12
		0.76	[11]		-0.12
		0.92	[153]		0.04
23	C_6H_5-COOH	0.36	[45]	0.40	-0.04
		0.48	[154]		0.08
24	$C_6H_5-CH_2-COOH$	0.09	[48]	-0.04	0.13
		-0.13	[45]		-0.09
25	$2-CH_3-C_6H_4-COOH$	1.10	[45]	0.92	0.18
26	$4-CH_3-C_6H_4-COOH$	0.68	[45]	0.92	-0.24
27	$(CH_3-CH_2-CH_2)_3N$	2.52	[41]	2.56	-0.04

* Outliers

TABLE VII,19

f VALUES IN THE TOLUENE – WATER SYSTEM

no	fragment	*f*(toluene)	*f*(octanol)
1	CH_3	0.74	0.702
2	CH_2	0.56	0.530
3	CH	0.25	0.235
4	C_6H_5	2.00	1.886
5	C_6H_4	1.78	1.688
6	COOH(al)	-2.60	-0.954
7	COOH(ar)	-1.60	-0.093

TABLE VII,19 (continued)

8	OH(ar)	-1.74	-0.343
9	NH_2(al)	-2.34	-1.428
10	NH(al)	-2.70	-1.825
11	N(al)	-3.02	-2.160
12	Br(al)	-0.08	0.270
13	c_M	0.30	0.287

CARBON TETRACHLORIDE - WATER SYSTEM

The discussion of the toluene - water system also applies to the carbon tetrachloride - water system. Partitioning data were extremely scanty, and those which were available had to be handled analogously to the procedure described above for the toluene - water system; see Tables VII,20 and VII,21.

TABLE VII,20

PARTITION VALUES IN THE CARBON TETRACHLORIDE - WATER SYSTEM

no	Compound	log P(obs)	ref.	log P(clc)	res.
1	H-COOH	-3.12	[43]	-2.74^a	-0.38*
2	CH_3-COOH	-1.92	[43]	-2.42	0.50*
		-2.45	[47]		-0.03
3	CH_3-CH_2-COOH	-1.79	[47]	-1.76	-0.03
		-1.62	[48]		0.14
4	$CH_3(CH_2)_2$-COOH	-1.02	[48]	-1.10	0.08
5	$(CH_3)_2$CH-COOH	-1.47	[155]	-1.25	-0.22*
6	$(CH_3)_2$CH-CH_2-COOH	-0.54	[48]	-0.59	0.05
7	CH_3-CH_2-OH	-2.93	[156]	-1.65	-1.28*
		-1.61	[157]		0.04
8	CH_3-$(CH_2)_2$-OH	-0.93	[156]	-0.99	0.06
9	CH_3-$(CH_2)_3$-OH	-0.44	[87]	-0.33	-0.11
		-0.29	[156]		0.04
10	CH_3-$(CH_2)_4$-OH	0.30	[156]	0.33	-0.03
		0.36	[158]		0.03
11	CH_3-$(CH_2)_6$-OH	1.63	[156]	1.65	-0.02
12	CH_3-CON$(C_2H_5)_2$	-0.45	[158]	-0.32	-0.13
13	CH_3-CH_2-CON$(CH_3)_2$	-0.82	[158]	-0.98	0.16
14	CH_3-CH_2-CON$(C_2H_5)_2$	0.32	[158]	0.34	-0.02

Refs. pg 262

TABLE VII,20 (continued)

no	Compound	log P(obs)	ref.	log P(clc)	res.
15	C_6H_5-OH	−0.42	[56]	−0.44	0.02
		−0.50	[96]		−0.06
		−0.36	[36]		0.08
16	$CH_3-COO-CH_3$	0.41	[106]	0.32	0.09
17	$CH_3-COO-CH_2-CH_3$	0.95	[106]	0.98	−0.03
18	$CH_3-COO-CH_2-CH_2-CH_3$	1.59	[106]	1.64	−0.05
19	$C_6H_5-NH_2$	0.25	[36]	0.38	−0.13
		0.60	[89]		0.22*
20	$2-CH_3-C_6H_4-NH_2$	1.18	[92]	1.11	0.07
21	$3-CH_3-C_6H_4-NH_2$	1.15	[92]	1.11	0.04
22	$4-CH_3-C_6H_4-NH_2$	1.14	[36]	1.11	0.03
		1.11	[92]		0.00
23	$3-Cl-C_6H_4-NH_2$	1.37	[92]	1.34	0.03
24	$4-Cl-C_6H_4-NH_2$	1.31	[92]	1.34	−0.03
25	$2-Cl-C_6H_4-NH_2$	1.73	[92]	1.34	0.39*
26	$CH_3-CO-CH_3$	−0.37	[19]	−0.35	−0.02
		−0.35	[46]		0.00
		−0.34	[5]		−0.01

*) Outliers
a) H(neg) = 1.230 × 0.46 = 0.56 (see pg

TABLE VII,21

f VALUES IN THE CARBON TETRACHLORIDE − WATER SYSTEM

no	fragment	*f*(carbon tetrachl.)	*f*(octanol)
1	C_6H_5	2.30	1.886
2	C_6H_4	2.15	1.688
3	CH_3	0.88	0.702
4	CH_2	0.66	0.530
5	CH	0.29	0.235
6	(al)COOH	−3.30	−0.954
7	(al)COO	−1.44	−1.292
8	(ar)OH	−2.74	−0.343
9	(al)OH	−3.19	−1.491
10	(al)CO	−2.11	−1.703
11	(al)CON	−4.28	−2.894

TABLE VII,21 (continued)

12	(ar)NH$_2$)	−1.92	−0.854
13	(ar)Cl	1.11	0.922

XYLENE − WATER SYSTEM

This system also suffered from a lack of sufficient data and was therefore unsuitable for a computerized calculation. The partition values are to be found in Tables VII,22 and VII,23.

TABLE VII,22

PARTITION VALUES IN THE XYLENE − WATER SYSTEM

no	Compound	log P(obs)	ref.	log P(clc)	res.
1	CH_3-NH_2	−1.00	[5]	−1.54	0.54*
2	$Br-CH_2-COOH$	−1.37	[5]	−1.42	0.05
3	CH_3-COOH	−1.92	[85]	−1.94	0.02
4	$CH_3-NH-CH_3$	−0.68	[5]	−1.15	0.47*
5	$CH_3-CH_2-NH_2$	−0.66	[5]	−0.96	0.30*
6	$Br-CH(CH_3)-COOH$	−1.01	[5]	−0.97	−0.04
7	CH_3-CH_2-COOH	−1.32	[48]	−1.36	0.04
		−1.24	[5]		0.12
8	$CH_3-(CH_2)_2-NH_2$	−0.36	[5]	−0.38	0.02
9	$(CH_3)_3N$	−0.44	[82]	−0.76	0.32*
10	$CH_3-(CH_2)_2-COOH$	−0.78	[5]	−0.78	0.00
11	$(CH_3)_2CH-COOH$	−0.80	[5]	−0.91	0.11
12	$CH_3-(CH_2)_3-NH_2$	0.04	[5]	0.20	−0.16
13	$(CH_3)_2CH-CH_2-NH_2$	0.10	[5]	0.07	0.03
14	$(CH_3-CH_2)_2NH$	−0.10	[5]	0.01	−0.11
15	$CH_3-(CH_2)_3-COOH$	−0.33	[5]	−0.20	−0.13
16	$(CH_3)_2CH-CH_2-COOH$	−0.31	[48]	−0.33	0.02
	CH$_2$–CH$_2$ / ...	−0.10	[5]		0.23*
17	CH$_2$ NH / CH$_2$–CH$_2$	0.03	[5]	0.21	−0.18
18	$CH_3-(CH_2)_4-NH_2$	0.44	[5]	0.78	−0.34*
19	C_6H_5-OH	0.13	[56]	0.16	−0.03
		0.18	[85]		0.02
20	$CH_3-(CH_2)_4-COOH$	0.34	[5]	0.38	−0.04

Refs. pg 262

TABLE VII,22 (continued)

no	Compound	log P(obs)	ref.	log P(clc)	res.
21	$(CH_3)_2CH-(CH_2)_2-COOH$	0.18	[5]	0.25	-0.07
22	$CH_3-CH(CH_3)-NH-CH(CH_3)_2$	1.17	[5]	0.91	0.26*
23	$(CH_3-CH_2-CH_2)_2NH$	1.24	[5]	1.17	0.07
24	$CH_3-(CH_2)_5-NH_2$	0.89	[5]	1.36	-0.47*
25	$(CH_3-CH_2)_3N$	1.11	[5]	0.98	0.13
26	$C_6H_5-CH_2-NH_2$	0.30	[5]	0.33	-0.03
27	$(CH_3)_2CH-(CH_2)_3-COOH$	1.24	[5]	0.83	0.41*
28	$CH_3-(CH_2)_6-NH_2$	1.34	[5]	1.94	-0.60*
29	$C_6H_5-CH_2-COOH$	-0.38	[5]	-0.07	-0.31*
30	$C_6H_5-CH_2-NH-CH_3$	1.39	[5]	0.72	0.67*
31	$\begin{matrix} CH_2-NH \\ CH_2 \quad\quad CH-CH_2-CH_2-CH_3 \\ CH_2-CH_2 \end{matrix}$	1.95	[5]	1.82	0.13
32	$C_6H_5-(CH_2)_2-COOH$	0.46	[5]	0.51	-0.05
33	$C_6H_5-CH_2-NH-CH_2-CH_3$	1.72	[5]	1.30	0.42*
34	$C_6H_5-CH_2-CH(CH_3)-NH_2$	0.96	[161]	1.36	-0.40*
35	$C_6H_5-CH_2-CH(NH_2)-CH_2-CH_3$	1.32	[161]	1.94	-0.62*
36	$C_6H_5-CH_2-CH(CH_3)-NH-CH_3$	1.58	[161]	1.75	-0.17
37	$C_6H_5-CH_2-N(CH_3)-CH_2-CH_2-CH_3$	1.81	[161]	2.27	-0.46*

*) Outliers

The number of outliers in the above reported investigations is excessively large. Many of them are present in the series of primary amine structures.

TABLE VII,23

f VALUES IN THE XYLENE - WATER SYSTEM

no	fragment	*f*(xylene)	*f*(octanol)
1	C_6H_5	2.06	1.886
2	CH_3	0.77	0.702
3	CH_2	0.58	0.530
4	CH	0.26	0.235
5	(al)COOH	-2.71	-0.954
6	(ar)OH	-1.90	-0.343
7	(al)NH_2	-2.31	-1.428
8	(al)NH	-2.69	-1.825

9	(al)N	-3.07	-2.160
10	(al)Br	-0.22	0.270
11	c_M	0.31	0.287

CYCLOHEXANE — WATER SYSTEM

There is an obvious trend in the results obtained thus far; in going from the octanol — water system to a system where the organic phase
has less solubility of water, the positive f values tend to become even
more positive and the negative f values even more negative; in other
words, there is a distinctly greater discriminative effect. It may be
expected, therefore, that in the cyclohexane — water system, in which
the solubility of water in the cyclohexane phase is only 0.001 M, there
will be a fairly large discriminative effect.

An initial calculation of an f set in the cyclohexane — water system afforded an incomprehensible result for a number of fragments; we
give the following examples:

f (C_6H_5): 2.334   (octanol): 1.886
f (C_6H_4): 2.251   (octanol): 1.688
f (C_6H_3): 2.009   (octanol): 1.431

It can be seen from these examples that the ratio between f (C_6H_4)
and f (C_6H_3) is in order but the difference between these two f values
is much too small and the difference between f(C_6H_5) and f(C_6H_4) is
even non-significant. Similar peculiarities were observed with other
fragments.

On a careful examination of all the information, it became clear
that these discrepancies had to be ascribed to the same coincidence as
that which had arisen in the diethylether — water partitioning of CN-
-containing structures (see section 2). Circumvention of this problem
appeared possible by a critical and well balanced composition of the
data set based, of course, on experiences with the other solvent systems, as reported above. Relevant material is presented in Table VII,24.
In all, 211 structures (269 measurements) were introduced in the regression analysis; seventy (110 measurements) manifested themselves as
outliers.

While working out the ultimate regression equation, it was deemed

Refs. pg 262

desirable not to introduce a constant c_M. Therefore, $f(OH)_{al}$ occurs side by side with $f(OH)_{ar}$ in Table VII,25.

TABLE VII,24

STRUCTURES INTRODUCED INTO STATISTICAL TREATMENT
(CYCLOHEXANE – WATER SYSTEM)

no	Compound	log P(obs)	ref.	log P(clc)	res.
1	CH_3-CH_2-OH	-1.96	[163]	-2.13	0.17
		-2.10	[156]		0.03
2	$CH_3-(CH_2)_2-OH$	-1.49	[79]	-1.48	-0.01
		-1.60	[156]		-0.12
3	$CH_3-(CH_2)_3-OH$	-0.72	[87]	-0.83	0.11
		-0.95	[156]		-0.12
4	$CH_3-CH_2-NH-CH_2-CH_3$	0.02	[164]	0.14	-0.12
5	$CH_3-NH-CH_2-CH_2-CH_3$	0.17	[164]	0.14	0.03
6	$CH_3-(CH_2)_4-OH$	-0.26	[156]	-0.19	-0.07
7	$CH_3-NH-(CH_2)_3-CH_3$	0.82	[164]	0.78	0.04
8	$4-Br-C_6H_4-OH$	-0.09	[165]	-0.12	0.03
9	$4-Cl-C_6H_4-OH$	-0.26	[165]	-0.18	-0.08
10	$4-F-C_6H_4-OH$	-1.00	[149]	-0.90	-0.10
11	$3-NO_2-C_6H_4-OH$	-1.57	[89]	-1.61	0.04
		-1.52	[32]		0.09
12	$3-Cl-C_6H_4-NH_2$	0.69	[166]	0.68	0.01
13	$4-Cl-C_6H_4-NH_2$	0.69	[92]	0.68	0.01
		0.60	[166]		-0.08
14	C_6H_5-OH	-0.72	[167]	-0.84	0.12
		-0.77	[168]		0.07
		-0.81	[165]		0.03
		-1.00	[149]		-0.16
15	C_6H_5-NH	0.02	[169]	0.01	0.01
		0.05	[166]		0.04
		0.08	[89]		0.07
16	$3-OH-C_6H_4-NH_2$	-3.24	[92]	-3.34	0.10
17	$4-OH-C_6H_4-NH_2$	-3.44	[92]	-3.34	-0.10
18	$1,3-diNH_2-C_6H_4$	-2.44	[92]	-2.48	0.04
19	$CH_3-CH_2-NH-(CH_2)_3-CH_3$	1.52	[164]	1.43	0.09
20	$CH_3-CH(CH_3)-NH-CH(CH_3)_2$	1.29	[164]	1.25	0.04
21	$(CH_3-CH_2-CH_2)_2NH$	1.53	[164]	1.43	0.10

TABLE VII,24 (continued)

22	$CH_3-NH-(CH_2)_4-CH_3$	1.38	[164]	1.43	-0.05
23	$3-Br-C_6H_4-COH$	2.10	[170]	1.97	0.13
24	$4-Br-C_6H_4-COH$	2.03	[170]	1.97	0.06
25	$2-Cl-C_6H_4-COH$	2.02	[170]	1.92	0.10
26	$3-Cl-C_6H_4-COH$	1.92	[170]	1.92	0.00
27	$4-Cl-C_6H_4-COH$	1.82	[170]	1.92	-0.10
28	$3-F-C_6H_4-COH$	1.29	[170]	1.20	0.09
29	$4-F-C_6H_4-COH$	1.16	[170]	1.20	-0.04
30	C_6H_5-COH	1.13	[172]	1.25	-0.12
		1.34	[32]		0.09
31	$C_6H_5-O-CH_3$	2.30	[174]	2.36	-0.06
32	$C_6H_5-CH_2-OH$	-0.62	[172]	-0.62	0.00
33	$3-CH_3-C_6H_4-OH$	-0.15	[167]	-0.23	0.08
		-0.20	[168]		0.03
		-0.10	[175]		0.13
34	$4-CH_3-C_6H_4-OH$	-0.10	[167]	-0.23	0.13
		-0.19	[168]		0.04
35	$4-CH_3-O-C_6H_4-OH$	-1.08	[165]	-0.98	-0.10
36	$3-CH_3-C_6H_4-NH_2$	0.64	[169]	0.63	0.01
		0.58	[92]		-0.05
		0.55	[166]		-0.08
37	$2-CH_3-C_6H_4-NH_2$	0.67	[169]	0.63	0.04
		0.61	[92]		-0.02
		0.65	[166]		0.02
38	$4-CH_3-C_6H_4-NH_2$	0.58	[169]	0.63	-0.05
		0.55	[92]		-0.08
		0.50	[166]		-0.13
39	$3-CH_3-O-C_6H_4-NH_2$	-0.13	[92]	-0.13	0.00
40	$CH_3-(CH_2)_6-OH$	1.12	[156]	1.11	0.01
41	$2-Br-5-NO_2-C_6H_3-CH=CH-NO_2$	1.66	[172]	1.64	0.02
42	$2-Cl-5-NO_2-C_6H_4-CH=CH-NO_2$	1.49	[172]	1.59	-0.10
43	$2,6-diCl-C_6H_3-CH=CH-NO_2$	3.12	[172]	3.02	0.10
44	$2-F-5-NO_2-C_6H_3-CH=CH-NO_2$	0.87	[172]	0.87	0.00
45	$4-Br-C_6H_4-CH=CH-NO_2$	2.33	[172]	2.46	-0.13
46	$2-Br-C_6H_4-CH=CH-NO_2$	2.44	[172]	2.46	-0.02
47	$3-Br-C_6H_4-CH=CH-NO_2$	2.48	[172]	2.46	0.02
48	$3-Cl-C_6H_4-CH=CH-NO_2$	2.33	[172]	2.41	-0.08

Refs. pg 262

TABLE VII,24 (continued)

no	Compound	log P(obs)	ref.	log P(clc)	res.
49	$2-Cl-C_6H_4-CH=CH-NO_2$	2.52	[172]	2.41	0.11
50	$4-F-C_6H_4-CH=CH-NO_2$	1.61	[172]	1.69	−0.08
51	$3-F-C_6H_4-CH=CH-NO_2$	1.74	[172]	1.69	0.05
52	$2-NO_2-C_6H_4-CH=CH-NO_2$	0.89	[172]	0.98	−0.09
53	$3-NO_2-C_6H_4-CH=CH-NO_2$	1.01	[172]	0.98	0.03
54	$C_6H_5-CH=CH-NO_2$	1.80	[172]	1.74	0.06
55	$4-OH-C_6H_4-CH=CH-NO_2$	−1.60	[172]	−1.60	0.00
56	$3-CH_3-C_6H_4-COH$	1.80	[170]	1.87	−0.07
57	$2-CH_3-C_6H_4-COH$	1.86	[170]	1.87	−0.01
58	$3,4-diCH_3-C_6H_3-OH$	0.20	[168]	0.33	−0.13
59	$3,5-diCH_3-C_6H_3-OH$	0.27	[168]	0.33	−0.06
60	$3-CH_3-CH_2-C_6H_4-OH$	0.43	[167]	0.42	0.01
		0.36	[168]		−0.06
61	$4-CH_3-CH_2-C_6H_4-OH$	0.44	[167]	0.42	0.02
		0.36	[168]		−0.06
62	$1,3-diCH_3-O-C_6H_4$	2.32	[174]	2.22	0.10
63	$2,4-diCH_3-C_6H_3-NH_2$	1.23	[169]	1.19	0.04
64	$2,5-diCH_3-C_6H_3-NH_2$	1.22	[169]	1.19	0.03
65	$2,6-diCH_3-C_6H_3-NH_2$	1.35	[169]	1.19	0.16
		1.16	[166]		−0.03
66	$3,5-diCH_3-C_6H_3-NH_2$	1.18	[169]	1.19	−0.01
67	$[(CH_3)_2CH-CH_2]_2NH$	2.57	[164]	2.54	0.03
68	$[CH_3-(CH_2)_3]_2NH$	2.68	[164]	2.72	−0.04
69	$2,4-diCl-C_6H_3-CH=C(CH_3)-NO_2$	3.61	[172]	3.66	−0.05
70	$2-Br-C_6H_4-CH=C(CH_3)-NO_2$	3.01	[172]	3.10	−0.09
71	$3-Br-C_6H_4-CH=C(CH_3)-NO_2$	3.05	[172]	3.10	−0.05
72	$4-Cl-C_6H_4-CH=C(CH_3)-NO_2$	2.97	[172]	3.05	−0.08
73	$4-F-C_6H_4-CH=C(CH_3)-NO_2$	2.47	[172]	2.33	0.14
74	$2-NO_2-C_6H_4-CH=C(CH_3)-NO_2$	1.52	[172]	1.62	−0.10
75	$4-NO_2-C_6H_4-CH=C(CH_3)-NO_2$	1.59	[172]	1.62	−0.03
76	$3-NO_2-C_6H_4-CH=C(CH_3)-NO_2$	1.62	[172]	1.62	0.00
77	$2-CH_3-C_6H_4-CH=CH-NO_2$	2.40	[172]	2.36	0.04
78	$4-CH_3-C_6H_4-CH=CH-NO_2$	2.42	[172]	2.36	0.06
79	$4-CH_3-O-C_6H_4-CH=CH-NO_2$	1.73	[172]	1.61	0.12
80	$2,4-diCH_3-O-C_6H_3-COH$	0.88	[170]	0.92	−0.04
81	$3-CH_3-5-C_2H_5-C_6H_3-OH$	0.89	[167]	0.98	−0.09

TABLE VII,24 (continued)

82	$2-CH_3-3-C_2H_5-C_6H_3-OH$	0.97	[168]	0.98	-0.01
83	$2-CH_3-4-C_2H_5-C_6H_3-OH$	1.01	[168]	0.98	0.03
84	$2-CH_3-5-C_2H_5-C_6H_3-OH$	1.02	[168]	0.98	0.04
85	$3-CH_3-6-C_2H_5-C_6H_3-OH$	1.06	[168]	0.98	0.08
86	$2-CH_3-CH_2-CH_2-C_6H_4-OH$	1.18	[168]	1.06	0.12
87	$2-(CH_3)_2CH-C_6H_4-OH$	1.08	[175]	0.97	0.11
88	$4-(CH_3)_2CH-C_6H_4-OH$	0.77	[168]	0.97	-0.20
89	$3-(CH_3)_2CH-C_6H_4-OH$	0.81	[175]	0.97	-0.16
90	$CH_3-(CH_2)_4-NH-(CH_2)_3-CH_3$	3.25	[164]	3.37	-0.12
91	$3-Cl-C_6H_4-CH=CH-COO-CH_3$	3.09	[176]	3.23	-0.14
92	$4-Cl-C_6H_4-CH=CH-COO-CH_3$	3.21	[176]	3.23	-0.02
93	$3-F-C_6H_4-CH=CH-COO-CH_3$	2.57	[176]	2.51	0.06
94	$2,4-diCl-C_6H_3-CH=C(C_2H_5)-NO_2$	4.40	[172]	4.31	0.09
95	$3,4-diCl-C_6H_3-CH=C(C_2H_5)-NO_2$	4.40	[172]	4.31	0.09
96	$2-Br-6-CH_3O-C_6H_3-CH=C(CH_3)-NO_2$	2.92	[172]	2.91	0.01
97	$4-Cl-C_6H_4-CH=CH-CONH-CH_3$	-0.58	[177]	-0.56	-0.02
98	$C_6H_5-CH=CH-COO-CH_3$	2.44	[176]	2.56	-0.12
99	$4-NO_2-C_6H_4-CH=C(C_2H_5)-NO_2$	2.39	[172]	2.26	0.13
100	$C_6H_5-CH=CH-CONH-CH_3$	-1.27	[177]	-1.23	-0.04
101	$2-CH_3-C_6H_4-CH=C(CH_3)-NO_2$	2.98	[172]	3.00	-0.02
102	$4-CH_3-C_6H_4-CH=C(CH_3)-NO_2$	3.00	[172]	3.00	0.00
103	$4-CH_3O-C_6H_4-CH=C(CH_3)-NO_2$	2.33	[172]	2.25	0.08
104	$4-F-C_6H_4-CH=CH-COO-C_2H_5$	3.07	[176]	3.16	-0.09
105	$C_6H_5-CH=CH-COO-C_2H_5$	3.27	[179]	3.21	0.06
106	$3-CH_3-C_6H_4-CH=CH-COO-CH_3$	3.24	[176]	3.18	0.06
107	$4-CH_3-C_6H_4-CH=CH-COO-CH_3$	3.16	[176]	3.18	-0.02
108	$2-CH_3O-C_6H_4-CH=CH-COO-CH_3$	2.54	[176]	2.42	0.12
109	$3-CH_3O-C_6H_4-CH=CH-COO-CH_3$	2.45	[176]	2.42	0.03
110	$4-CH_3O-C_6H_4-CH=CH-COO-CH_3$	2.35	[176]	2.42	-0.07
111	$C_6H_5-CH=CH-CONH-CH_2-CH_3$	-0.58	[177]	-0.58	0.00
112	$4-CH_3O-C_6H_4-CH=CH-CONH-CH_3$	-1.43	[177]	-1.36	-0.07
113	$2-CH_3-C_6H_4-CH=C(C_2H_5)-NO_2$	3.61	[172]	3.65	-0.04
114	$2-CH_3O-C_6H_4-CH=C(C_2H_5)-NO_2$	2.88	[172]	2.89	-0.01
115	$3-CH_3O-C_6H_4-CH=C(C_2H_5)-NO_2$	2.88	[172]	2.89	-0.01
116	$4-Cl-C_6H_4-CH=CH-CONH-CH(CH_3)_2$	0.79	[177]	0.64	0.15
117	$4-Cl-C_6H_4-CH=CH-CONH-\underline{n}.C_3H_7$	0.74	[177]	0.74	0.00
118	$4-CH_3-C_6H_4-CH=CH-COO-CH_2-CH_3$	3.87	[176]	3.83	0.04

Refs. pg 262

TABLE VII,24 (continued)

no	Compound	log P(obs)	ref.	log P(clc)	res.
119	$3-CH_3O-C_6H_4-CH=CH-COO-C_2H_5$	3.13	[176]	3.07	0.06
120	$2-CH_3O-C_6H_4-CH=CH-COO-C_2H_5$	3.05	[176]	3.07	-0.02
121	$4-CH_3O-C_6H_4-CH=CH-COO-C_2H_5$	3.13	[176]	3.07	0.06
122	$C_6H_5-CH=CH-CONH-CH_2-CH_2-CH_3$	0.01	[177]	0.07	-0.06
123	$C_6H_5-CH=CH-CONH-CH(CH_3)_2$	0.02	[177]	-0.02	0.04
124	$4-CH_3O-C_6H_4-CH=CH-CONH-C_2H_5$	-0.70	[177]	-0.72	0.02
125	$2,4-diCH_3O-C_6H_3-CH=C(C_2H_5)-NO_2$	2.68	[172]	2.70	-0.02
126	$2,5-diCH_3O-C_6H_3-CH=C(C_2H_5)-NO_2$	2.68	[172]	2.70	-0.02
127	$4-Cl-C_6H_4-CH=CH-CONH-\underline{n}.C_4H_9$	1.45	[177]	1.38	0.07
128	$4-Cl-C_6H_4-CH=CH-CONH-CH_2-CH(CH_3)_2$	1.28	[177]	1.29	-0.01
129	$C_6H_5-CH=CH-CONH-\underline{n}.C_4H_9$	0.66	[177]	0.71	-0.05
130	$C_6H_5-CH=CH-CONH-CH_2-CH(CH_3)_2$	0.53	[177]	0.62	-0.09
131	$4-CH_3O-C_6H_4-CH=CH-CONH-\underline{n}.C_3H_7$	-0.13	[177]	-0.07	-0.06
132	$4-CH_3O-C_6H_4-CH=CH-CONH-i.C_3H_7$	-0.12	[177]	-0.07	-0.05
133	$4-Cl-C_6H_4-CH=CH-CONH-\underline{n}.C_5H_{11}$	2.08	[177]	2.03	0.05
134	$4-Cl-C_6H_4-CH=CH-CONH-(CH_2)_2$ $(CH_3)_2CH$	2.01	[177]	1.94	0.07
135	$C_6H_5-CH=CH-CONH-\underline{n}.C_5H_{11}$	1.38	[177]	1.36	0.02
136	$C_6H_5-CH=CH-CONH-(CH_2)_2-CH(CH_3)_2$	1.29	[177]	1.27	0.02
137	$4-CH_3O-C_6H_4-CH=CH-CONH-\underline{n}.C_4H_9$	0.53	[177]	0.58	-0.05
138	$4-CH_3O-C_6H_4-CH=CH-CONH-CH_2$ $(CH_3)_2CH$	0.45	[177]	0.48	-0.03
139	$4-CH_3O-C_6H_4-CH=CH-CONH-\underline{n}.C_5H_{11}$	1.23	[177]	1.22	0.01
140	$4-CH_3O-C_6H_4-CH=CH-CONH-(CH_2)_2$ $(CH_3)_2CH$	1.06	[177]	1.13	-0.07
141	$C_6H_5-CH=CH-CONH-\underline{n}.C_7H_{15}$	2.72	[180]	2.65	0.07

Outliers:

no	Compound	log P(obs)	ref.	log P(clc)	res.
1	CH_3-OH	-1.84	[162]	-2.77	0.93
2	CH_3-CH_2-OH	-2.37	[79]	-2.13	-0.24
3	$CH_3-(CH_2)_3-OH$	-1.12	[79]	-0.83	-0.29
4	$CH_3-CH_2-NH-CH_2-CH_3$	-0.34	[24]	0.14	-0.48
5	(piperidine ring structure)	0.10	[164]	0.46	-0.36
6	$2,6-diCl-C_6H_3-NH_2$	2.04	[166]	1.29	0.75

TABLE VII,24 (continued)

7	2-F-C_6H_4-OH	-0.15	[149]	-0.90	0.75
8	3-F-C_6H_4-OH	-0.70	[149]	-0.90	0.20*
9	2-NO_2-C_6H_4-OH	1.49	[53]	-1.61	3.10
10	4-NO_2-C_6H_4-OH	-1.93	[53]	-1.61	-0.32
		-1.79	[32]	-1.61	-0.18*
11	3-Cl-C_6H_4-NH_2	0.89	[92]	0.68	0.21*
12	2-Cl-C_6H_4-NH_2	1.25	[92]	0.68	0.57
		1.09	[166]		0.41
13	4-NO_2-C_6H_4-NH_2	-1.00	[94]	-0.76	-0.24
		-1.00	[92]		-0.24
14	3-NO_2-C_6H_4-NH_2	-0.42	[94]	-0.76	0.34
		-0.42	[92]		0.34
15	2-NO_2-C_6H_4-NH_2	0.36	[94]	-0.76	1.12
16	2-HO-C_6H_4-NH_2	-2.37	[92]	-3.34	0.97
17	1,2-diNH_2-C_6H_4	-1.65	[92]	-2.48	0.83
18	1,4-diNH_2-C_6H_4	-2.81	[92]	-2.48	-0.33
19	2-Br-C_6H_4-COH	2.21	[170]	1.97	0.24
20	2-HO-C_6H_4-COH	-1.02	[173]	-2.09	1.07
		-0.50	[171]		1.59
21	2-CH_3-C_6H_4-OH	0.13	[167]	-0.23	0.36
		0.10	[168]		0.33
		0.20	[175]		0.43
22	2-CH_3O-C_6H_4-NH_2	0.52	[92]	-0.13	0.65
23	4-CH_3O-C_6H_4-NH_2	-0.41	[92]	-0.13	-0.28
24	3,4-diCl-C_6H_3-CH=CH-NO_2	2.68	[172]	3.02	-0.34
25	2,4-diCl-C_6H_3-CH=CH-NO_2	2.76	[172]	3.02	-0.26
26	3-NO_2-4-F-C_6H_3-CH=CH-NO_2	0.31	[172]	0.87	-0.56
27	2-F-C_6H_4-CH=CH-NO_2	1.94	[172]	1.69	0.25
28	4-NO_2-C_6H_4-CH=CH-NO_2	0.72	[172]	0.98	-0.26
29	3-HO-C_6H_4-CH=CH-NO_2	-1.36	[172]	-1.60	0.24
30	4-CH_3-C_6H_4-COH	1.68	[170]	1.87	-0.19*
31	3-CH_3O-4-HO-C_6H_3-COH	-0.75	[32]	-2.28	1.53
32	2,3-diCH_3-C_6H_3-OH	0.51	[168]	0.33	0.18*
33	2,4-diCH_3-C_6H_3-OH	0.55	[168]	0.33	0.22
		0.76	[167]		0.43
34	2,5-diCH_3-C_6H_3-OH	0.57	[168]	0.33	0.24
		0.77	[167]		0.44

Refs. pg 262

TABLE VII,24 (continued)

no	Compound	log P(obs)	ref.	log P(clc)	res.
35	2,6-diCH$_3$-C$_6$H$_3$-OH	1.28	[167]	0.33	0.95
		0.93	[168]		0.60
36	3,5-diCH$_3$-C$_6$H$_3$-OH	0.54	[167]	0.33	0.21*
37	2-C$_2$H$_5$-C$_6$H$_4$-OH	0.68	[168]	0.42	0.26
		0.83	[167]		0.41
38	C$_6$H$_5$-O-CH$_2$-CH$_3$	2.77	[174]	3.01	-0.24
39	2-Cl-5-NO$_2$-C$_6$H$_3$-CH=C(CH$_3$)-NO$_2$	2.46	[172]	2.23	0.23
40	2,6-diCl-C$_6$H$_3$-CH=C(CH$_3$)-NO$_2$	4.40	[172]	3.66	0.74
41	3,4-diCl-C$_6$H$_3$-CH=C(CH$_3$)-NO$_2$	3.36	[172]	3.66	-0.30
42	3-F-C$_6$H$_4$-CH=C(CH$_3$)-NO$_2$	2.57	[172]	2.33	0.24
43	2-F-C$_6$H$_4$-CH=C(CH$_3$)-NO$_2$	2.67	[172]	2.33	0.34
44	3-Cl-C$_6$H$_4$-CH=C(CH$_3$)-NO$_2$	2.63	[172]	3.05	-0.42
45	2-Cl-C$_6$H$_4$-CH=C(CH$_3$)-NO$_2$	3.31	[172]	3.05	0.26
46	C$_6$H$_5$-CH=C(CH$_3$)-NO$_2$	2.69	[172]	2.38	0.31
47	3-CH$_3$O-C$_6$H$_4$-CH=CH-NO$_2$	1.89	[172]	1.60	0.29
48	2-CH$_3$O-C$_6$H$_4$-CH=CH-NO$_2$	2.15	[172]	1.60	0.55
49	3-CH$_3$O-4-HO-C$_6$H$_3$-CH=CH-NO$_2$	0.04	[172]	-1.79	1.83
50	2,3-diCH$_3$O-C$_6$H$_3$-COH	1.16	[170]	0.92	0.24
51	3,4-diCH$_3$O-C$_6$H$_3$-COH	0.32	[170]	0.92	-0.60
52	3-C$_2$H$_5$O-4-HO-C$_6$H$_3$-COH	0.03	[32]	-1.64	1.67
53	3-CH$_3$-5-C$_2$H$_5$-C$_6$H$_3$-OH	0.73	[168]	0.98	-0.25
54	3-C$_2$H$_5$-4-CH$_3$-C$_6$H$_3$-OH	0.74	[168]	0.98	-0.24
55	3-<u>n</u>.C$_3$H$_7$-C$_6$H$_4$-OH	0.83	[168]	1.06	-0.23
56	4-<u>n</u>.C$_3$H$_7$-C$_6$H$_4$-OH	0.86	[168]	1.06	-0.20*
57	2-Cl-C$_6$H$_4$-CH=CH-COO-CH$_3$	3.44	[175]	3.23	0.21*
58	4-Cl-C$_6$H$_4$-CH=C(C$_2$H$_5$)-NO$_2$	3.19	[172]	3.70	-0.51
59	2-Cl-C$_6$H$_4$-CH=C(C$_2$H$_5$)-NO$_2$	3.23	[172]	3.70	-0.47
60	3-Cl-C$_6$H$_4$-CH=C(C$_2$H$_5$)-NO$_2$	4.40	[172]	3.70	0.70
61	3-Cl-6-CH$_3$O-C$_6$H$_3$-CH=C(CH$_3$)-NO$_2$	3.02	[172]	2.86	0.16*
62	C$_6$H$_5$-CH=C(C$_2$H$_5$)-NO$_2$	3.45	[172]	3.03	0.42
63	2-CH$_3$O-C$_6$H$_4$-CH=C(CH$_3$)-NO$_2$	2.61	[172]	2.25	0.36
64	3-CH$_3$O-C$_6$H$_4$-CH=C(CH$_3$)-NO$_2$	2.63	[172]	2.25	0.38
65	2,5-diCH$_3$O-C$_6$H$_3$-CH=CH-NO$_2$	1.79	[172]	1.25	0.54
66	2,3-diCH$_3$O-C$_6$H$_3$-CH=CH-NO$_2$	1.93	[172]	1.25	0.68
67	2,4-diCH$_3$O-C$_6$H$_3$-CH=CH-NO$_2$	2.04	[172]	1.25	0.79
68	3,4-diCH$_3$O-C$_6$H$_3$-CH=CH-NO$_2$	0.88	[172]	1.25	-0.37

TABLE VII,24 (continued)

69	2-HO-3-CH_3O-C_6H_3-CH=CH-NO_2	0.77	[172]	-1.79	2.56
70	2-CH_3-5-i.C_3H_7-C_6H_3-OH	1.30	[168]	1.54	-0.24
71	2-Cl-C_6H_4-CH=CH-COO-CH_2-CH_3	4.19	[176]	3.88	0.31
72	3-F-C_6H_4-CH=CH-COO-CH_2-CH_3	3.38	[176]	3.16	0.22
73	4-Cl-C_6H_4-CH=CH-CONH-CH_3	0.06	[177]	-0.56	0.62
74	2-CH_3O-4-HO-C_6H_3-CH=CH-COO-CH_3	0.54	[176]	-0.98	1.52
75	4-CH_3-C_6H_4-CH=C(C_2H_5)-NO_2	3.10	[172]	3.65	-0.55
76	4-i.C_3H_7-C_6H_4-CH=CH-NO_2	3.82	[172]	3.56	0.26
77	2-C_2H_5O-C_6H_4-CH=C(CH_3)-NO_2	3.07	[172]	2.89	0.18*
78	4-CH_3O-C_6H_4-CH=C(C_2H_5)-NO_2	2.49	[172]	2.89	-0.40
79	2,5-diCH_3O-C_6H_3-CH=C(CH_3)-NO_2	2.49	[172]	2.05	0.44
80	2,4-diCH_3O-C_6H_3-CH=C(CH_3)-NO_2	2.54	[172]	2.05	0.49
81	2,3-diCH_3O-C_6H_3-CH=C(CH_3)-NO_2	2.57	[172]	2.05	0.52
82	3,4-diCH_3O-C_6H_3-CH=C(CH_3)-NO_2	1.71	[172]	2.05	-0.34
83	3-C_2H_5O-4-HO-C_6H_3-CH=C(CH_3)-NO_2	1.49	[172]	-0.51	2.00
84	3-CH_3O-4-HO-C_6H_3-CH=C(C_2H_5)-NO_2	1.57	[172]	-0.51	2.08
85	3,4-diCH_3O-C_6H_3-CH=CH-COO-CH_3	1.47	[176]	2.23	-0.76
86	3,5-diCH_3O-C_6H_3-CH=CH-COO-CH_3	0.00	[176]	2.23	-2.23
87	2-CH_3O-4-HO-C_6H_3-CH=CH-COO-C_2H_5	1.20	[176]	-0.33	1.55
88	4-i.C_3H_7-C_6H_4-CH=C(CH_3)-NO_2	3.52	[172]	4.20	-0.68
89	2,3-diCH_3O-C_6H_3-CH=C(C_2H_5)-NO_2	3.22	[172]	2.70	0.52
90	3,4-diCH_3O-C_6H_3-CH=C(C_2H_5)-NO_2	2.17	[172]	2.70	-0.53
91	3-C_2H_5O-4-HO-C_6H_3-CH=C(C_2H_5)-NO_2	2.24	[172]	0.14	2.10
92	3-CH_3O-C_6H_4-CH=CH-COO-n.C_3H_7	3.90	[176]	3.72	0.18*
93	3,4-diCH_3O-C_6H_3-CH=CH-COO-C_2H_5	2.16	[176]	2.88	-0.72
94	3,5-diCH_3O-C_6H_3-CH=CH-COO-C_2H_5	0.68	[176]	2.88	-2.20
		3.44	[176]	2.88	0.56
95	2-CH_3O-4-HO-C_6H_3-CH=CH-COO n.C_3H_7	1.88	[176]	0.32	1.56
96	4-i.C_3H_7-C_6H_4-CH=C(C_2H_5)-NO_2	2.89	[172]	4.85	-1.96
97	3,4-diC_2H_5O-C_6H_3-CH=C(CH_3)-NO_2	2.45	[172]	3.34	-0.89
98	3,4-diCH_3O-C_6H_3-CH=CH-COO n.C_3H_7	2.74	[176]	3.53	-0.79

*) The asterisked structures are no longer outliers in the final
regression equation; actually they have to be transferred from
the table of outliers to the table of fitting structures.

Refs. pg 262

TABLE VII,25

f VALUES IN THE CYCLOHEXANE — WATER SYSTEM

no	fragment	*f*	st.dev.	<u>t</u>	*f*(oct)
1	C_6H_5	2.312	0.062	37.4	1.886
2	C_6H_4	2.123	0.089	23.9	1.688
3	C_6H_3	1.881	0.140	13.4	1.431
4	CH_3	0.805	0.062	13.0	0.702
5	CH_2	0.646	0.008	84.8	0.530
6	CH	0.397	0.067	5.89	0.235
7	(ar)COH	−1.058	0.067	−15.8	−0.380[*]
8	(ar)NO_2	−0.578	0.068	−8.56	−0.080
9	(al)OH	−3.578	0.063	−56.8	−1.491
10	(ar)OH	−3.157	0.064	−49.2	−0.343
11	(ar)NH_2	−2.301	0.063	−36.3	−0.854
12	(al)NH	−2.766	0.098	−28.2	−1.825
13	(ar)O	−0.754	0.018	−41.0	−0.433
14	(ar)F	0.139	0.069	2.03	0.399
15	(ar)Cl	0.856	0.064	13.3	0.922
16	(ar)Br	0.909	0.068	13.4	1.131
17	(ar)CH=CH−NO_2	−0.569	0.066	−8.64	0.395[*]
18	(ar)CH=C−NO_2	−0.733	0.041	−17.9	0.220[*]
19	(ar)CH=CH−COO	−0.555	0.042	−13.1	0.042[*]
20	(ar)CH=CH−CONH	−4.343	0.043	−100.	−1.100[*]

[*]) By *f* summation of the constituting fragments

$n = 164 (+1);$ $s = 0.080;$

$r = 0.998;$ $F = 2,802;$ const. term $= 0.000$

PENTANOL — WATER SYSTEM

The solvent systems dealt with in the above sections 2 − 10 all have slopes ⩾ 1.0 in the pattern of LEO <u>et al.</u>'s solvent regression equations, with the apparant exception of the cyclohexane − water system for donor structures, but this might be due to the extremely poor quality of the regression equation. On the whole, these solvent systems (except for the oleyl alcohol − water system) need a differentiation between donor−acceptor or even donor−acceptor−neutral structures.

The remaining equations from the review of LEO <u>et al.</u> are so-called

"sole" equations and do not call for any differentiation. Typically,
nearly all of their slopes are well below 1.0. Of the six systems of
this type that were dealt with by LEO et al., there are only two which
facilitate conversion of the constituent structures into hydrophobic
fragmental constants, namely the pentanol — water and the n. butanol —
— water systems, while the measurements by ARCHIBALD [9] on a few
carboxylic acids in the sec. butanol — water system afforded four valuable constants for a solvent system that can be considered highly interesting because the two phases are not far from being mutually miscible.

The partition data for the pentanol — water system were not suitable for computerized treatment and the procedure described above for
the toluene — water system was followed. All relevant data are listed
in Tables VII,26 and VII,27.

TABLE VII,26

PARTITION VALUES IN THE n. PENTANOL — WATER SYSTEM

no	Compound	log P(obs)	ref.	log P(clc)	res.
1	$H-NH_2$	−0.85	[83]	−0.78	−0.07
2	$H-COOH$	−0.26	[9]	−0.27	0.01
3	CH_3-NH_2	−0.45	[181]	−0.54	0.09
4	CH_3-COOH	−0.02	[9]	−0.03	0.01
		−0.03	[181]		0.00
		−0.03	[83]		0.00
		−0.04	[10]		−0.01
5	$HOOC-CH_2-COOH$	−0.22	[48]	−0.10	−0.12
6	CH_3-CH_2-COOH	0.54	[9]	0.48	0.06
		0.37	[48]		−0.11
7	$CH_3-CH(OH)-COOH$	−0.32	[21]	−0.26	−0.06
		−0.40	[48]		−0.14
8	$HOOC-(CH_2)_2-COOH$	−0.15	[83]	0.14	0.29*
		−0.19	[48]		0.33*
9	$HOOC-CH(OH)-CH(OH)-COOH$	−1.21	[48]	−1.34	0.13
10	$CH_3-(CH_2)_2-COOH$	1.05	[9]	0.99	0.06
		0.97	[181]		−0.02
		1.03	[48]		0.04
11	$(CH_3)_2CH-COOH$	0.99	[48]	0.88	0.11
12	$CH_3-CH_2-CH(OH)-COOH$	0.05	[26]	0.25	−0.20

Refs. pg 262

TABLE VII,26 (continued)

no	Compound	log P(obs)	ref.	log P(clc)	res.
13	$CH_3-(CH_2)_3-COOH$	1.55	[9]	1.50	0.05
		1.40	[181]		-0.10
14	$(CH_3)_2CH-CH_2-COOH$	1.13	[48]	1.39	-0.26*
15	C_6H_5-OH	1.55	[182]	1.42	0.13
		1.21	[83]		-0.21*
		1.50	[56]		0.08
16	$CH_3-(CH_2)_4-COOH$	2.04	[9]	2.01	0.03
17	$C_6H_5-CH_2-COOH$	1.48	[48]	1.53	-0.05
		1.57	[182]		0.04

*) Outliers

TABLE VII,27

f VALUES IN THE n. PENTANOL — WATER SYSTEM

no	fragment	*f*	*f*(oct)
1	C_6H_5	1.73	1.886
2	CH_3	0.68	0.702
3	CH_2	0.51	0.530
4	CH	0.23	0.235
5	H	0.17	0.175
6	(al)COOH	-0.71	-0.954
7	(al)OH	-1.27	-1.491
8	(al)NH_2	-1.22	-1.428
9	(ar)OH	-0.31	-0.343
10	c_M	0.27	0.287

n. BUTANOL — WATER SYSTEM

This system has been used more frequently for partition measurements than the pentanol — water system discussed above, so that the
data obtained possess enough congruity for setting up a normal regression procedure. In all, 68 measurements referring to 51 structures could
be used in our calculation. Seventeen of these measurements, referring
to 13 structures had to be removed as outliers. The data are presented
in Tables VII,28 and VII,29.

TABLE VII,28

STRUCTURES INTRODUCED INTO STATISTICAL TREATMENT
($\underline{n}$. BUTANOL – WATER SYSTEM)

no	Compound	log P(obs)	ref.	log P(clc)	res.
1	CH_3-COOH	0.09	[9]	0.11	-0.02
		0.07	[181]		-0.04
2	$HOOC-CH_2-COOH$	-0.28	[17]	-0.19	-0.09
		-0.11	[48]		0.08
3	CH_3-CH_2-COOH	0.51	[9]	0.52	-0.01
		0.51	[181]		-0.01
		0.43	[48]		-0.09
4	$CH_3-CH(OH)-COOH$	-0.10	[181]	-0.14	0.04
5	$H_2N-CH_2-CH(OH)-CH_2-NH_2$	-0.92	[183]	-0.86	-0.06
6	$HOOC-(CH_2)_2-COOH$	0.00	[17]	0.06	-0.06
		-0.02	[183]		-0.08
7	$HOOC-CH(OH)-CH_2-COOH$	-0.63	[183]	-0.59	-0.04
8	$CH_3-CH_2-CH_2-COOH$	0.95	[9]	0.92	0.03
		0.97	[183]		0.05
		0.96	[181]		0.04
		0.91	[48]		-0.01
9	$(CH_3)_2CH-CH_2-OH$	0.93	[183]	0.98	-0.05
10	$CH_3-(CH_2)_3-NH_2$	0.92	[183]	0.91	0.01
11	$(CH_3-CH_2)_2NH$	0.74	[183]	0.77	-0.03
12	$HO-(CH_2)_2-NH-(CH_2)_2-OH$	-0.70	[183]	-0.68	-0.02
		-0.69	[181]		-0.01
13	$H_2N-(CH_2)_4-NH_2$	-0.12	[183]	-0.26	0.14
14	$HOOC-(CH_2)_3-COOH$	0.21	[17]	0.16	0.05
		0.30	[183]	0.16	0.14
15	$CH_3-(CH_2)_3-COOH$	1.36	[9]	1.33	0.03
		1.39	[181]		0.06
16	$(CH_3)_2CH-CH_2-COOH$	1.30	[183]	1.25	0.05
		1.13	[48]		-0.12
17	$H_2N-(CH_2)_5-NH_2$	0.16	[183]	0.14	0.02
18	$HOOC-CH_2-CH(COOH)-CH_2-COOH$	0.01	[183]	-0.07	0.08
19	$HOOC-(CH_2)_4-COOH$	0.55	[183]	0.56	-0.01
		0.44	[17]		-0.12
20	$CH_3-(CH_2)_4-COOH$	1.87	[183]	1.73	0.14
		1.89	[181]		0.16

Refs. pg 262

TABLE VII,28 (continued)

no	Compound	log P(obs)	ref.	log P(clc)	res.
		1.86	[9]		0.13
21	$CH_3(CH_2)_2$-NH-$(CH_2)_2$-CH_3	1.62	[183]	1.58	0.04
22	$(CH_3$-$CH_2)_3$N	1.32	[183]	1.41	-0.09
23	HO-$(CH_2)_2$-N$(CH_2$-$CH_3)_2$	0.58	[183]	0.68	-0.10
24	HO-CH-CH_2-NH-CH_2-CH-OH | | CH_3 CH_3	-0.15	[183]	-0.03	-0.12
25	$(HO$-CH_2-$CH_2)_3$N	-0.58	[183]	-0.77	0.19
26	C_6H_5-CH_2-NH_2	0.98	[183]	1.01	-0.03
27	HOOC-CH_2-CH(COOH)-$(CH_2)_2$-COOH	0.07	[17]	0.03	0.04
28	HOOC-$(CH_2)_5$-COOH	0.77	[17]	0.97	-0.20
29	C_6H_5-CH_2-COOH	1.43	[183]	1.43	0.00
		1.48	[182]		0.05
30	C_6H_5-CH(OH)-COOH	0.72	[183]	0.77	-0.05
31	$(CH_3)_2$CH-CH_2-NH-CH_2-CH$(CH_3)_2$	2.38	[183]	2.23	0.15
32	CH_3-$(CH_2)_7$-NH_2	2.35	[183]	2.53	-0.18
33	HOOC-$(CH_2)_7$-COOH	1.66	[183]	1.78	-0.12
34	C_6H_5-CH_2-CH$(COOH)_2$	1.48	[183]	1.44	0.04
35	C_6H_5-CH(OH)-CH(CH_3)-NH-CH_3	1.18	[183]	1.19	-0.01

Outliers:

no	Compound	log P(obs)	ref.	log P(clc)	res.
1	CH_3-NH_2	0.00	[181]	-0.30	0.30
2	H_2N-CH_2-COOH	-1.81	[184]	-0.61	1.20
3	H_2N-CH(CH_3)-COOH	-1.60	[184]	-0.28	1.32
4	$(CH_3)_3$N	0.49	[181]	0.19	0.30
5	HOOC-CH(OH)-CH(OH)-COOH	-0.78	[48]	-1.25	0.47
6	CH_3-CH_2-CH(NH_2)-COOH	-1.34	[184]	0.12	-1.46
		-1.79	[48]		-1.91
7	$(CH_3$-$CH_2)_2$NH	0.43	[155]	0.77	-0.36
		0.42	[155]		-0.35
8	CH_3-$(CH_2)_2$-CH(NH_2)-COOH	-0.98	[184]	0.52	-1.50
9	HOOC-CH(NH_2)-CH$(CH_3)_2$	-1.14	[184]	0.44	-1.58
10	CH_3-$(CH_2)_3$-CH(NH_2)-COOH	-0.51	[184]	0.92	-1.43
11	HOOC-CH(NH_2)-CH_2-CH$(CH_3)_2$	-0.74	[184]	0.84	-1.58
		-1.21	[183]		-2.05
12	CH_3-$(CH_2)_5$-NH_2	2.02	[183]	1.72	0.30

TABLE VII,28 (continued)

13	HOOC$-$(CH$_2$)$_6$$-$COOH	0.92	[17]	1.37	$-$0.45
14	H$_2$N$-$CH$-$(CH$_2$)$_3$$-$N(CH$_2$$-CH_3$)$_2$	1.08	[183]	1.36	$-$0.28
	$\quad\;\;$ CH$_3$				

TABLE VII,29

f VALUES IN THE n. BUTANOL $-$ WATER SYSTEM

no	fragment	*f*	st.dev.	t	*f*(oct)
1	C$_6$H$_5$	1.547	0.067	23.2	1.886
2	CH$_3$	0.640	0.054	11.8	0.702
3	CH$_2$	0.404	0.012	33.8	0.530
4	CH	0.087	0.064	1.37	0.235
5	(al)OH	$-$0.794	0.068	$-$11.6	$-$1.491
6	(al)COOH	$-$0.527	0.056	$-$9.34	$-$0.954
7	(al)NH$_2$	$-$0.940	0.064	$-$14.6	$-$1.428
8	(al)NH	$-$1.318	0.069	$-$19.2	$-$1.825
9	(al)N	$-$1.727	0.111	$-$15.6	$-$2.160
10	c$_M$	0.152	0.019	7.81	0.287

$n = 51 \; (+1);$ $\qquad$ $s = 0.097;$

$r = 0.994;$ $\qquad$ $F = 349;$ $\qquad$ const. term $= 0.000$

sec. BUTANOL $-$ WATER SYSTEM

Six aliphatic carboxylic acids were used for the calculation of four fragmental constants. The data obtained are listed in Tables VII,30 and VII,31.

TABLE VII,30

PARTITION VALUES IN THE sec. BUTANOL $-$ WATER SYSTEM

no	Compound	log P(obs)	ref.	log P(clc)	res.
1	H$-$COOH	0.03	[9]	0.03	0.00
2	CH$_3$$-$COOH	0.08	[9]	0.09	$-$0.01
3	CH$_3$$-CH_2$$-$COOH	0.39	[9]	0.41	$-$0.02
4	CH$_3$$-$(CH$_2$)$_2$$-$COOH	0.72	[9]	0.73	$-$0.01
5	CH$_3$$-$(CH$_2$)$_3$$-$COOH	1.06	[9]	1.05	0.01
6	CH$_3$$-$(CH$_2$)$_4$$-$COOH	1.39	[9]	1.37	0.02

Refs. pg 262

TABLE VII,31

f VALUES IN THE sec. BUTANOL - WATER SYSTEM

no	fragment	*f*	*f*(oct)
1	CH_3	0.42	0.702
2	CH_2	0.32	0.530
3	H(neg)	0.36	0.462
4	(al)COOH	−0.33	−0.954

Some comments are called for when all the facts are reviewed as a whole:

(1) All 12 solvent systems give a set of hydrophobic fragmental constants that satisfy the general condition

$$\log P(s) = \sum \underline{a}_n f_n(s) \qquad (VII-7)$$

where s indicates the solvent system used.

The statistical data obtained on application of eqn. VII-7 are, in general, of excellent quality; hence it is concluded that in any solvent system a set of hydrophobic fragmental constants can be defined.

(2) The systems investigated differ rather widely in discriminative power. There seems to be a link between this discrimination and the polarity of the organic phase of the solvent system employed; perhaps the degree of water solubility in the organic phase of the system might be indicative of this discrimination.

REFERENCES

1 R. Collander, Acta Chem. Scand., 3(1949)717.

2 K.B. Sandell, Naturwissensch., 51(1964)336.

3 O.C. Dermer, W.G. Markham and H.M. Trimble,
   J. Amer. Chem. Soc., 63(1941)3524.

4 K.B. Sandell, Monatsh. Chem., 92(1961)1066.

5 H.W. Smith, J. Phys. Chem., 25(1921)204,605.

6 C.P. Brown and A.R. Mathieson, J. Phys. Chem., 58(1954)1057.

7 A. Hantzsch and A. Vagt, Z. Physik. Chem., 38(1901)705.

8 K.B. Sandell, Naturwissensch., 53(1966)330.

9 R.C. Archibald, J. Amer. Chem. Soc., 54(1932)3178.

10 W. Perschke and E. Tschufaroff, Z. Anorg. Allg. Chem.,
   151(1926)121.

11 A. Hantzsch and F. Sebaldt, Z. Physik. Chem., 30(1899)258.

12 N. Schilow and L. Lepin, Z. Physik. Chem., 101(1922)353.

13 E.J. Ross, J. Physiol., 112(1951)229.

14 W. Walter and H.L. Weidemann, Ann. Chem., 685(1965)29.

15 O.C. Dermer and V.H. Dermer, J. Amer. Chem. Soc., 65(1943)1653.

16 E.E. Chandler, J. Amer. Chem. Soc., 30(1908)696.

17 C.S. Marvel and J.C. Richards, Anal. Chem., 21(1949)1480.

18 E.A. Klobbie, Z. Physik. Chem., 24(1897)615.

19 H.J. Almquist, J. Phys. Chem., 37(1933)991.

20 W.U. Behrens, Z. Anal. Chem., 69(1926)97.

21 R. Dietzel and E. Rosenbaum, Biochem. Z., 185(1927)275;
    189(1927)348.

22 J. Pinnow, Z. Anal. Chem., 54(1915)321.

23 R. Collander, Phys. Plant., 7(1954)420.

24 K.B. Sandell, Naturwissensch., 49(1962)12.

25 G.S. Forbes and A.S. Coolidge, J. Amer. Chem. Soc., 41(1918)150.

26 R. Dietzel and P. Schmitt, Z. Untersuch. Lebensm., 63(1932)369.

27 T.E. Gier and J.O. Hougen, J. Ind. Eng. Chem., 45(1953)1362.

28 H. Thies and E. Ermer, Naturwissensch., 49(1962)37.

29 Y. Tuzuki, Bull.Chem. Soc. Japan, 13(1938)337.

30 M.H. Bickel and H.J. Weder, J. Pharm. Pharmacol., 21(1969)160.

31 J.J. Lindberg, Soc. Sci. Fennica, Comm. Phys. Math., 21(1958)1.

32 K.B. Sandell, Monatsh. Chem., 89(1958)36.

33 J. Halmekoski and A. Nissema, Suomen Kemistilehti, 35B(1962)188.

34 J. Pinnow, Z. Anal. Chem., 50(1911)162.

35 R.A. Robinson, J. Chem. Soc., (1952)253.

36 W. Vaubel, Journ. f. Prakt. Chem., 67(1903)473.

37 W. Ruhland and U. Heilmann, Planta, 39(1951)91.

38 A. Sekera, A. Borovansky and C. Vrba, Ann. Pharm. Franc.,
    16(1958)525.

39 F. Auerbach and H. Zeglin, Z. Physik. Chem., 103(1922)200.

40 N.P. Komar and L.S. Manzhelii, Chem. Abstr. 61,3736.

41 T.S. Moore and T.F. Winmill, J. Amer. Chem. Soc., 101(1912)1635.

42 W.A. Felsing and S.E. Buckley, J. Phys. Chem., 37(1933)779.

43 A. Bekturov, J. Gen. Chem., 9(1939)419.

44 V. Rothmund and N.T.M. Wilsmore, Z. Phys. Chem., 40(1902)611.

45 H.W. Smith and T.A. White, J. Phys. Chem., 33(1929)1953.

46 W. Herz and W. Rathman, Z. Elektrochem., 19(1913)552.

47 M. Davies, P. Jones, D. Patnaik and E.A. Moelwyn-Hughes,
    J. Chem. Soc., (1951)1249.

48 N. Kolossowsky, F. Kulikow and A. Bekturow, Bull. Soc. Chim.,
    2(1935)460.

49 A.A. Albert, J. Chem. Soc., (1951)1376.

50 D.B. Stevancevic and V.G. Antonijevic, Bull. Boris. Kidric. Inst.
    Nucl. Sci., 16(1965)11.

51 G. Schweitzer and E. van Willis, in Adv. in Anal. Chem.,
    5(1965)169.

52 W. Kemula and H. Buchowski, J. Phys. Chem., 63(1959)155.

53 W. Kemula, H. Buchowski and J. Teperek, Bull. Acad. Polon. Sci.
    Sec. Sci. Chim., 12(1964)347.

54 C.A.M. Hogben, D.J. Tocco, B.B. Brodie and L.S. Schanker,
    J. Pharm. & Exptl. Ther., 125(1959)275.

55 N. Kakeya, N. Yata, A. Kamada and M. Aoki, Chem. Pharm. Bull.,
    17(1969)2558.

56 A.B. Lindenberg and M. Massin, J. Chim. Phys., 61(1964)112.

57 L.S. Schanker, J. Johnson and J. Jeffrey, Am. J. Physiol.,
    207(1964)503.

58 R. Modin and A. Tilly, Acta Pharm. Suecica, 5(1968)311.

59 I.F. Skidmore and M.W. Whitehouse, Biochem. Pharmacol.,
    14(1965)547.

60 S.E. Mayer, R.P. Maickel and B.B. Brodie, J. Pharm. & Exptl. Ther.,
    127(1959)205.

61 L.S. Schanker, P.A. Nafpliotis and J.M. Johnson, J. Pharm. &
    Exptl. Ther. 133(1961)325.

62 C. Golumbic and M. Orchin, J. Amer. Chem. Soc., 72(1950)4145.

63 T. Morishita, M.Yamazaki, N. Yata and A. Kamada, Chem. Pharm.
    Bull. Japan, 21(1973)2309.

64 T. Koizumi, T. Arita and K. Kakemi, Chem. Pharm. Bull. Japan,
    12(1964)413.

65 D.P. Rall, J.R. Stabenau and C.G. Zubrod, J. Pharm. & Exptl. Ther.,
    125(1959)185.

66 J. Rieder, Arzneim. Forsch., 13(1963)81.

67 W.S. Hendrixson, Z. Anorg. Chem., 13(1897)73.

68 M. Davies and D.M.L. Griffiths, J. Chem. Soc., (1955)132.

69 B. Hök, Svensk Kem. Tidskr., 65(1953)182.

70 R. Modin and M. Johansson, Acta Pharm. Suecica, 8(1971)561.

71 R. Collander, Acta Chem. Scand., 4(1950)1085.

72 T.B. Vree, A.Th.J.M. Muskens and J.M. van Rossum,
    J. Pharm. Pharmacol., 21(1969)774.

73 K.C. Yeh and W.I. Higuchi, J. Pharm. Sci., 10(1972)1648.

74 T.B. Vree, Thesis Katholieke Universiteit, Nijmegen, 1973.

75 M.W. Whitehouse and P.D.G. Dean, Biochem. Pharmacol., 14(1965)557.

76 K.Kakemi, H. Sezaki, H. Okamura and S. Ashida, Chem. Pharm. Bull.,
    17(1969)1332.

77   W. Walter and H. Weidemann, Monatsh. Chem., 93(1962)1235.

78   M. Davies and D.M.L. Griffiths, Z. Phys. Chem., 2(1954)353.

79   K.H. Meyer and H. Hemmi, Biochem. Z., 277(1935)39.

80   E. Hutchinson, J. Phys. Chem., 52(1948)897.

81   J.L.R. Morgan and H.K. Benson, Z. Anorg. Chem., 55(1907)356.

82   W. Herz and E. Stanner, Z. Physik. Chem., 128(1927)399.

83   W. Herz and H. Fischer, Chem. Ber., 37(1904)4746.

84   P. Gross and K. Schwarz, Monatsh. Chem., 55(1930)287.

85   W. Herz and H. Fischer, Chem. Ber., 38(1905)1138.

86   F. Brown and C. Bury, J. Chem. Soc., 123(1923)2430.

87   E. Meeussen and P. Huyskens, J. Chim. Phys., 63(1966)845.

88   G.V. Georgievics, Monatsh. Chem., 36(1915)391.

89   W. Kemula, H. Buchowski and W. Pawlowski, Rocz. Chem.,
     43(1969)1555.

90   B. Flürscheim, J. Chem. Soc., 97(1910)84.

91   G. Williams and F.G. Soper, J. Chem. Soc., (1930)2469.

92   W. Kemula, H. Buchowski and W. Pawlowski, Rocz. Chem.,
     42(1968)1951.

93   R. Farmer and F. Warth, J. Chem. Soc., 85(1904)1713.

94   W. Kemula, H. Buchowski and W. Pawlowski, Bull. Acad. Polon. Sci.
     Ser. Sci. Chim., 12(1964)491.

95   K. Endo, Bull. Chem. Soc., Japan, 1(1926)25.

96   F. Philbrick, J. Amer. Chem. Soc., 56(1934)2581.

97   J.C. Philip and C.H.D. Clark, J. Chem. Soc., 127(1925)1274.

98   J.Mindowicz and I. Uruska, Zeszyty Nauk Politech. Gdansk, Chem.,
     26(1962)79.

99   B. de Sziszkovski, Chem. Abstr. 9,2014.

100 A.K.M. Shamsul Huq and S.A.K. Lodhi, J. Phys. Chem., 70(1966)1354.

101 F.T. Wall, J. Amer. Chem. Soc., 64(1942)472.

102 R. van Duyne, S. Taylor, S. Christian and H. Affsprung,
     J. Phys. Chem., 71(1967)3427.

103 K. Kakemi, H. Sezaki, E. Suzuki and M. Nakano, Chem. Pharm. Bull.,
     17(1969)242.

104 A.B. Taubman, Z. Physik. Chem., 161(1932)141.

105 G.V. Georgievics, Z. Physik. Chem., 90(1915)47.

106 I. Traube, Archiv f. Physiol., 105(1904)541.

107 R. Macy, J. Ind. Hyg. Toxicol. 30(1948)140.

108 B. Wroth and E. Reid, J. Amer. Chem. Soc., 38(1916)2316.

109 S.Y. Gerlsma, J. Biol. Chem., 243(1968)959.

110 B.A. Lindenberg, J. Chim. Phys., 48(1951)350.

111 M. Bodansky and A.V. Meigs, J. Phys. Chem., 36(1932)814.

112 M. Bodansky, J. Biol. Chem., 79(1928)241.

113 N. Gordon and E. Reid, J. Phys. Chem., 26(1922)773.

114 K.H. Meyer and H. Gottlieb-Billroth, Z. Physiol. Chem.,
     112(1921)55.

115 R. Hober and J. Hober, J. Cell. and Physiol., 10(1937)401.

116 F. Baum, Arch. Exp. Path. Pharmakol., 42(1889)119.

117 A. Lindenberg, Compt. Rend. Soc. Biol. 118(1935)1086.

118 H. Meyer, Arch. Exp. Path. Pharmakol., 46(1901)338.

119 J. Büchi and X. Perlia, Arzneim. Forsch., 10(1960)930.

120 E. von Knaffl-Lenz, Arch. Exp. Path. Pharmakol., 84(1919)66.

121 J.C. Krantz Jr., C.J. Carr and W.E. Evans Jr.,
     Anesthesiology, 5(1944)291.

122 A. van der Eeckhout, Arch. Exp. Path. Pharmakol., 57(1907)338.

123 R.C. Doerr and W. Fiddler, J. Agr. Food Chem., 18(1970)936.

124 A. Unmack, Chem. Zentr. (1934II)1862.

125 E. Fourneau and G. Florence, Bull. Soc. Chim., 43(1928)1027.

126 T.C. Daniels and R.E. Lyons, J. Phys. Chem., 35(1931)2049.

127 L.H. Baldinger and J.A. Nieuwland, J. Amer. Pharm. Assoc.,
     22(1933)711.

128 W.A. Bittenbender and E.F. Degering, J. Amer. Pharm. Assoc.,
     28(1939)514.

129 N.M. Cone, S.E. Forman and J.C. Krantz Jr., Proc. Soc. Exp. Biol.
     Med., 48(1941)461.

130 C.D. Leake and M.Y. Chen-Mai, Anesthesia and Analgesia, 10(1931)1.

131 P. Harrass, Arch. Int. Pharmacol. et de Therap. 11(1903)431.

132 R. Bierich, Arch. Physiol., 174(1919)202.

133 A. Seidell, Solubilities of organic compounds, Vol. II,
     D. van Nostrand Comp., New York, 1941.

134 R. H. Goshorn and E.F. Degering, J. Amer. Pharm. Assoc.,
     27(1938)865.

135 G.C. Gross, E.F. Degering and P.A. Tetrault, Proc. Indiana Acad.
     Sci., 49(1939)42.

136 R. Collander, Acta Chem. Scand., 5(1951)774.

137 D.E. Burton, K. Clarke and G.W. Gray, J. Chem. Soc., (1964)1314.

138 J. Büchi, X. Perlia and A. Strässle, Arzneim. Forsch.,
     16(1966)1657.

139 J. Büchi, K.H. Hetterich and X. Perlia, Arzneim. Forsch.,
     18(1968)791.

140 J. Büchi, X. Perlia and S.P. Studach, Arzneim. Forsch.,
     17(1967)1012.

141 J. Büchi, L.T. Oey and X. Perlia, Arzneim. Forsch., 22(1972)1071.

142 J. Büchi, J. Doulakas and X. Perlia, Arzneim. Forsch.,
     19(1969)578.

143 J. Büchi, G. Fischer, M. Mohs and X. Perlia, Arzneim. Forsch.,
     19(1969)1183.

144 J. Büchi, H.K. Bruhin and X. Perlia, Arzneim. Forsch.,
     21(1971)1003.

145 W. Laubender and L. Löbenberg geb. Hunn, Arzneim. Forsch.,
     14(1964)445.

146 J. Büchi, R. Koller and X. Perlia, Arzneim. Forsch., 25(1975)14.

147 C. Rohmann and Th. Eckert, Arch. Pharm., 291(1958)450.

148 J. Büchi, X. Perlia and M. Tinani, Arzneim. Forsch., 21(1971)2074.

149 R.J. Pinney and W. Walters, J. Pharm. Pharmacol., 21(1969)415.

150 R.W. Green and P.W. Alexander, Australian J. Chem., 18(1965)329.

151 V.S. Morello and R.B. Beckmann, Ind. Eng. Chem., 42(1950)1078.

152 Landolt − Börnstein, Zahlenwerte und Funktionen aus Physik,
     Chemie, Astronomie, Geophysik und Technik, Vol. II,2; Springer
     Verlag Berlin, Heidelberg, New York, 1969.

153 N.A. Kolosovskii and S.V. Andryashchenko, J. Gen. Chem.,
     4(1934)1070.

154 K.B. Sandell, Naturwissensch., 49(1962)348.

155 H.W. Smith, J. Phys. Chem., 26(1922)256.

156 I. Hanssens, J. Mullens, C. Deneuter and P. Huyskens,
     Bull. Soc. Chim. France, 10(1968)3942.

157 S. Bugarsky, Z. Physik. Chem., 71(1910)753.

158 M. Nakano and N.K. Patel, J. Pharm. Sci., 59(1970)77.

159 A. Leo, C. Hansch and D. Elkins, Chem. Rev., 71(1971)525.

160 P. Seiler, Eur. J. Med. Chem., 9(1974)473.

161 K.W. Rosenmund, E. Karg and F.K. Marcus, Chem. Ber.,
     75B(1942)1850.

162 E.R. Washburn and H.C. Spencer, J. Amer. Chem. Soc., 56(1934)361.

163 R.D. Vold and E.R. Washburn, J. Amer. Chem. Soc., 54(1932)4217.

164 I.H. Pitman, K. Uekama, T. Higuchi and W.E. Hall,
     J. Amer. Chem. Soc., 94(1972)8147.

165 C. Church, unpublished analysis (see ref. 159).

166 A. Gomez, J. Mullens and P. Huyskens, J. Phys. Chem.,
     76(1972)4011.

167 C. Golumbic, M. Orchin and S. Weller, J. Amer. Chem. Soc.,
     71(1949)2624.

168 N.C. Saha, A. Bhattacharjee, N.G. Basak and A. Lahiri,
     J. Chem. Eng. Data, 8(1963)405.

169 C. Golumbic and G.L. Golbach, J. Amer. Chem. Soc., 73(1951)3966.

170 C.E. Lough, Suffield Memorandum No. 9/70.

171 D.R. Reese, G.M. Irwin, L.W. Dittert, C.W. Chong and J.V. Swintos-

ky, J. Pharm. Sci., 53(1964)591.

172   D.J. Currie, C.E. Lough, R.F. Silver and H.L. Holmes,
      Can. J. Chem., 44(1966)1035.

173   J.J. Banewicz, C.W. Reed and M.E. Levitch, J. Amer. Chem. Soc.,
      79(1957)2693.

174   J.C. McGowan, P.N. Atkinson and L.H. Ruddle, J. Appl. Chem.,
      16(1966)99.

175   J.S. Fritz and C.E. Hedrick, Anal. Chem., 37(1965)1015.

176   C.E. Lough, Suffield Memorandum No. 28/69.

177   A.D. Delaney, D.J. Currie and H.L. Holmes, Can. J. Chem.,
      47(1969)3273.

178   A. Brodin and A. Agren, Acta Pharm. Suecica, 8(1971)609.

179   C.E. Lough, R.F. Silver and F.K. McClusky, Can. J. Chem.,
      46(1968)1943.

180   D.J. Currie and H.L. Holmes, Can. J. Chem., 48(1970)1340.

181   K.F. Gordon, Ind. Eng. Chem., 45(1953)1813.

182   H.M. de Kort and R.F. Rekker, unpublished results.

183   A. England Jr. and E.J. Cohn, J. Amer. Chem. Soc., 57(1935)634.

184   J. Marden, J. Ind. Eng. Chem., 6(1914)315.

185   A. Keston, S. Udenfriend and M. Levy, J. Amer. Chem. Soc.,
      72(1950)748.

CHAPTER VIII

A GENERAL FORMULATION FOR THE SOLVENT REGRESSION EQUATION

DIFFERENTIATION IN THE SOLUTE STRUCTURE; DIETHYLETHER – WATER SYSTEM

As indicated in Chapter VII, a solvent regression equation of the COLLANDER type can be written as

$$\Sigma f_a = \rho \, \Sigma f_b + q \qquad (VIII-1)$$

where Σf_a represents the summation of a series of fragmental constants valid for solvent system a and Σf_b the summation of the same fragmental constants in solvent system b. Also, it was suggested that a further differentiation was possible, which could be accomplished on basis of eqn. VIII-2:

$$\Sigma f_{1(a)} + \Sigma f_{2(a)} = \rho' \, \Sigma f_{1(b)} + \rho'' \, \Sigma f_{2(b)} + q' \qquad (VIII-2)$$

where the subscripts 1 and 2 represent fragment qualifications, 1 denoting, for example, a non-polar and 2 a polar fragment.

In the formulation of eqn. VIII-1, we have related the summations Σf_a and Σf_b to identical collections of fragmental constants. This will enable us to transform eqn. VIII-1 into eqn. VIII-3, which correlates a series of f values in the octanol – water system with a series of f values belonging to a different solvent system:

$$f_a = \rho \, f(\text{octanol}) + q'' \qquad (VIII-3)$$

Regarding the practical elaboration of eqn. VIII-3 for the diethylether – water system, 29 fragments were available (see Table VIII,1 for a complete survey of these fragments). Our calculations resulted in the following equations:

$$f(\text{ether}) = 1.315 \, f(\text{octanol}) - 0.169 \qquad (VIII-4)$$
$$n = 29; \quad r = 0.956; \quad s = 0.492; \quad F = 288; \quad \underline{t} = 17.0$$

$$f(\text{octanol}) = 0.696 \, f(\text{ether}) + 0.071 \qquad (VIII-5)$$
$$n = 29; \quad r = 0.956; \quad s = 0.358; \quad F = 288; \quad \underline{t} = 17.0$$

Relevant data on eqn. VIII-4 are given in column A in Table VIII,1.

It is noteworthy that in an f_a/f_b transformation, it is irrelevant whether the transformed fragment is a donor or an acceptor group. It is

Refs. pg 296

TABLE VIII,1

f(ether) − f(octanol) CONVERSIONS

no	fragment	f octanol	f ether	kn	estimated values and residuals A	B	C
1	C_6H_5	1.886	2.266		2.310 (−0.044)	2.256 (0.009)	2.256 (0.009)
2	C_6H_4	1.688	1.866		2.050 (−0.184)	2.019 (−0.153)	2.020 (−0.154)
3	C_6H_3	1.431	1.677		1.712 (−0.035)	1.712 (−0.035)	1.712 (−0.035)
4	CH_3	0.702	0.919		0.754 (0.164)	0.839 (0.079)	0.841 (0.077)
5	CH_2	0.530	0.531		0.527 (0.003)	0.633 (−0.102)	0.635 (−0.104)
6	CH	0.235	0.277		0.140 (0.136)	0.280 (−0.003)	0.283 (−0.006)
7	H(neg)	0.462	0.59		0.438 (0.151)	0.551 (0.038)	0.554 (0.035)
8	(al)COOH	−0.954	−1.238		−1.422 (0.184)	−1.143 (−0.094)	−1.137 (−0.100)
9	(ar)COOH	−0.93	−0.347	1	−0.291 (−0.055)		−0.371 (0.024)
10	(al)COO	−1.292	−1.553		−1.867 (0.314)	−1.548 (−0.004)	−1.541 (−0.011)
11	(al)CO	−1.703	−1.937		−2.407 (0.470)	−2.040 (0.103)	−2.032 (0.095)
12	(ar)CO	−0.842	−0.94		−1.275 (0.335)	−1.009 (0.069)	−1.003 (0.063)
13	(al)CONH$_2$	−1.970	−3.403	4	−2.758 (−0.644)		−3.403 (0.000)
14	(ar)NHCOO*	−0.795	−3.309	9	−1.213 (−2.095)		−3.312 (0.003)
15	(al)O	−1.581	−1.867		−2.247 (0.380)	−1.893 (0.026)	−1.887 (0.020)
16	(ar)O	−0.433	−0.679	1	−0.738 (0.059)		−0.777 (0.098)
17	(al)OH	−0.343	−0.691		−2.128 (0.249)	−1.786 (−0.092)	−1.779 (−0.099)
18	(ar)OH	−0.343	−0.691	1	−0.619 (−0.071)		−0.670 (−0.020)
19	(al)NH$_2$	−1.428	−2.563	3	−2.046 (−0.516)		−2.492 (−0.070)
20	(ar)NH$_2$	−0.854	−1.375	1	−1.291 (−0.083)		−1.280 (−0.094)

TABLE VIII,1 (continued)

21	(al)NH	−1.825	−2.954	3	−2.568 (−0.385)		−2.967 (0.013)
22	(al)N	−2.160	−3.099	2	−3.008 (−0.090)		−3.104 (0.005)
23	(ar)SO$_2$NH$_2$	−1.530	−1.97	1	−2.180 (0.210)		−2.088 (0.118)
24	(ar)SO$_2$NH	−1.992	−2.50		−2.787 (0.287)	−2.386 (−0.113)	−2.378 (−0.121)
25	(ar)SO$_2$N	−2.454	−2.94		−3.394 (0.454)	−2.939 (−0.000)	−2.930 (−0.009)
26	(al)Cl	0.061	0.176		−0.088 (0.264)	0.071 (0.104)	0.075 (0.100)
27	(al)Br	0.270	0.43		0.186 (0.243)	0.322 (0.107)	0.325 (0.104)
28	(al)I	0.587	0.810		0.602 (0.207)	0.701 (0.108)	0.704 (0.105)
29	c_M	0.287	0.297		0.208 (0.088)	0.342 (−0.045)	0.345 (−0.048)

In parenthesis: residuals

*) Aromatic fragment connected to N

pertinent to ask if a log P(ether) / log P(octanol) transformation is possible with the aid of eqns. VIII-4 and VIII-5. The answer is negative in that in an attempt to do so on the basis of these two equations, the following equations are obtained:

$$\log P(ether) = 1.315 \log P(octanol) - 0.169 \, n \qquad (VIII-6)$$

$$\log P(octanol) = 0.696 \log P(ether) + 0.071 \, n \qquad (VIII-7)$$

from which it appears that the transformation values are dependent on the number of fragments (n) which together constitute the structure.

The table of residuals shows one heavy outlier, viz., $f[(ar)NHCOO]$; its removal from the data set leads to the following regression equation:

$$f(ether) = 1.301 \, f(octanol) - 0.101 \qquad (VIII-8)$$
$$n = 28; \quad r = 0.985; \quad s = 0.276; \quad F = 893; \quad \underline{t} = 29.9$$

or, in its inverted form:

$$f(octanol) = 0.747 \, f(ether) + 0.061 \qquad (VIII-9)$$
$$n = 28; \quad r = 0.985; \quad s = 0.209; \quad F = 893; \quad \underline{t} = 29.9$$

and these are equations that can compete with the solvent regression

Refs. pg 296

equation derived by LEO et al. for the diethylether – water system. The
correlation coefficient is extremely good $(r = 0.985)$, and implies that
as much as 96.8% of the variances are accounted for by eqns. VIII–8 and
9. It must be borne in mind, however, that the equations connect two
sets of parameters which emerge with $r = 0.997$ $[f(\text{ether})$ values$]$ and $r =$
0.995 $[f(\text{octanol})$ values$]$ from their fragment calculations. The s values may likewise be included in these considerations $(s = 0.276$ and $s =$
0.209 in eqns. VIII–8 and 9, respectively, while the original fragment
calculations showed up with $s = 0.097$ for the diethylether – water and
with $s = 0.107$ for the octanol – water system$)$. In this connection, the
question arises of whether the statistical features of eqns. VIII–8 and
9 are really satisfactory and it is worth looking for an additional parameter that could raise the overall statistics to a higher level.

By a rather time-consuming "cleaning" procedure based on trial and
error, data points, first one at a time and next in pairs and so on,
were eliminated from the data set belonging to eqn. VIII–4. By doing
so, we obtained information about the characteristic behaviour of the
various f values in connection with their partners and finally, the regression equations of a "cleaned" series of data points could be composed (see eqn. VIII–10):

$$f(\text{ether}) \;=\; 1.197 \; f(\text{octanol}) \;-\; 0.001 \qquad\qquad (\text{VIII–10})$$
$$n = 19; \quad r = 0.998; \quad s = 0.086; \quad F = 5,786; \quad \underline{t} = 76.1$$

or, in its inverted form:

$$f(\text{octanol}) \;=\; 0.833 \; f(\text{ether}) \;+\; 0.000 \qquad\qquad (\text{VIII–11})$$
$$n = 19; \quad r = 0.998; \quad s = 0.071; \quad F = 5,786; \quad \underline{t} = 76.1$$

Column B in Table VIII,1 indicates which fragments are absent in the
data set belonging to eqns. VIII–10 and 11.

Eqns. VIII–10 and 11 seem to satisfy all our expectations; their
statistical data agree with predictions that can be made from the accu-
racies of the data incorporated in the left- and right-hand terms. In-
terestingly, the intercept can be neglected, the result being that fragment summation can be carried out immediately; in other words, the following can be written without any objection:

$$\Sigma f(\text{ether}) \;=\; 1.197 \; \Sigma f(\text{octanol}) \qquad\qquad (\text{VIII–12})$$

or

$$\log P(\text{ether}) \;=\; 1.197 \; \log P(\text{octanol}) \qquad\qquad (\text{VIII–13})$$

It follows that for the differentiation proposed in eqn. VII-4 in
Chapter VII and presented below as eqn. VIII-14:

$$\Sigma f_{1(a)} + \Sigma f_{2(a)} = \rho' \Sigma f_{1(b)} + \rho'' \Sigma f_{2(b)} + q' \qquad (\text{VIII-14})$$

the 19 fragments from eqn. VIII-10 can be considered as a combination
of $\Sigma f_{1(a)}$ and $\Sigma f_{1(b)}$ and the remaining 10 rejected fragments as a
combination of $\Sigma f_{2(a)}$, $\Sigma f_{2(b)}$ and q'. The implication is that the
diethylether − octanol transformation for the latter 10 fragments, in
contrast to that of the former 19 fragments, cannot be described without an equation that is devoid of an intercept.

There are no doubt several possibilities for solving the problem
in an acceptable manner; our choice was based on a pragmatic approach
in which we attempted to fit the 10 rejected fragments into eqn. VIII-10
using a constant c_M and a suitable set of key numbers. The underlying
thought was that it would be possible to transform eqn. VIII-14 in such
a way that the slopes ρ' and ρ'' could be made equal by the relationship:

$$\rho'' \Sigma f_{2(b)} = \rho' \Sigma f_{2(b)} + c_M \Sigma kn \qquad (\text{VIII-15})$$

so that eqn. VIII-14 would allow the transformation

$$\Sigma f_a = \rho \Sigma f_b + c_M \Sigma kn + q' \qquad (\text{VIII-16})$$

It must be added that we did formally maintain the intercept but, in
fact, we believe that if the constant c_M and the key number are really
meaningful, an intercept would be out of place in this concept.

The fragment conversion that occurs when a particular solvent system is replaced by another can thus be expressed as

$$f_a = \rho f_b + c_M \cdot kn \qquad (\text{VIII-17})$$

Our attempts were highly successful and the results eventually obtained for the diethylether − octanol conversion can be described with
the aid of eqn. VIII-18:

$$f(\text{ether}) = 1.195 \, f(\text{octanol}) - 0.263 \, kn + 0.002 \qquad (\text{VIII-18})$$
$$n = 29; \quad r = 0.998; \quad s = 0.079; \quad F = 5,983; \quad \underline{t}_1 = 91.2; \quad \underline{t}_2 = -31.6$$

or, in its inverted form:

$$f(\text{octanol}) = 0.834 \, f(\text{ether}) + 0.219 \, kn - 0.003 \qquad (\text{VIII-19})$$
$$n = 29; \quad r = 0.998; \quad s = 0.066; \quad F = 4,533; \quad \underline{t}_1 = 91.2; \quad \underline{t}_2 = 27.4$$

The following should be noted:
(1) r, s and F in eqns. VIII-18 and 19 have now reached the same high

level as they had in eqns. VIII-10 and 11;

(2) the regressor values of f(octanol) in eqn. VIII-18 or f(ether) in eqn. VIII-19 have remained virtually unchanged compared with those in eqns. VIII-10 and 11;

(3) the constant terms in eqns. VIII-18 and 19 can be neglected and, as a consequence, these equations may take the forms

$$\Sigma f \text{(ether)} \quad = \quad 1.195 \; \Sigma f \text{(octanol)} \quad - \quad 0.263 \; \Sigma \, kn \qquad \text{(VIII-20)}$$

or

$$\log P \text{(ether)} \quad = \quad 1.195 \quad \log P \text{(octanol)} \quad - \quad 0.263 \; \Sigma kn \quad \text{(VIII-21)}$$

and its inverted form

$$\log P \text{(octanol)} \quad = \quad 0.834 \; \log P \text{(ether)} \quad + \quad 0.219 \; \Sigma \, kn \quad \text{(VIII-22)}$$

where $\Sigma \, kn$ represents the summation for all key numbers relevant for the converted structures. It is eqn. VIII-22 that was used in the preceding chapter (see Table VII,1) for the log P(octanol) calculations indicated under heading III.

The estimates and residuals obtained from eqn. VIII-18 are given in Table VIII,1 under C, and agree with our expectations when compared with the values listed under B and found by using eqn. VIII-10.

In Section 2 of Chapter VII, the deviating character of the CN group was discussed. On the basis of eqn. VIII-18, it can now be predicted for the CN group that $f \text{(CN)}_{\text{ether}} = 1.195 \times (-1.066) - 0.261 \, kn = -1.27$, -1.53, -1.79, -2.04 . . . . . The regression analysis described in that section showed that $f \text{(CN)} = -0.673$, which cannot be fitted in this series of values unless it was done with $kn = -3$, in which event $f \text{(CN)} = -0.675$. Because in our opinion, it is incorrect to take kn to be -3, we prefer that $f \text{(CN)}$ should be taken as -1.27 in the diethylether — water system (eqn. VIII-18 with $kn = 0$), reference being made to the remarks in Chapter VII.

OTHER SOLVENT SYSTEMS

In agreement with the procedure described in the preceding section, modified solvent regressions of the type of eqn. VIII-17 were derived for a number of other solvent systems. The results are given in Tables VIII,2 to VIII,12 in the same order as discussed in Chapter VII. At the foot of each table the relevant regression equations together with some statistically important features are given.

TABLE VIII,2

f(chloroform) $-$ f(octanol) CONVERSIONS

no	fragment	f octanol	f CHCl$_3$	kn	estimated f values and residuals A	B	C
1	C$_6$H$_5$	1.886	2.348		2.131 (0.217)	2.265 (0.083)	2.284 (0.064)
2	C$_6$H$_4$	1.688	1.995		1.879 (0.116)	2.031 (-0.036)	2.046 (-0.051)
3	C$_5$H$_4$N	0.526	0.899	-1	0.399 (0.500)		0.952 (-0.053)
4	CH$_3$	0.702	0.965		0.624 (0.341)	0.865 (0.100)	0.858 (0.107)
5	CH$_2$	0.530	0.628		0.404 (0.223)	0.661 (-0.033)	0.650 (-0.022)
6	CH	0.235	0.163		0.029 (0.134)	0.312 (-0.149)	0.295 (-0.132)
7	(al)COOH	-0.954	-2.485	4	-1.485 (-1.000)		-2.364 (-0.121)
8	(ar)COOH	-0.093	-1.849	6	-0.389 (-1.460)		-1.940 (0.091)
9	(al)CO	-1.703	-1.136	-3	-2.439 (1.303)		-1.120 (-0.016)
10	(ar)CO	-0.842	-0.818	-1	-1.342 (0.524)		-0.696 (-0.122)
11	(al)CONH$_2$	-1.970	-2.98	2	-2.779 (-0.201)		-2.975 (-0.005)
12	(ar)CONH$_2$	-1.110	-2.21	3	-1.684 (-0.526)		-2.246 (0.036)
13	(ar)O	-0.433	-0.25	-1	-0.822 (0.572)		-0.203 (-0.047)
14	(al)OH	-1.491	-2.367	2	-2.169 (-0.198)		-2.398 (0.031)
15	(ar)OH	-0.343	-2.049	5	-0.707 (-1.342)		-1.935 (-0.114)
16	(al)NH$_2$	-1.428	-1.935	1	-2.089 (0.154)		-2.015 (0.080)
17	(ar)NH$_2$	-0.854	-0.981		-1.358 (0.377)	-0.976 (-0.005)	-1.017 (0.036)
18	(al)NH	-1.825	-2.386	1	-2.594 (0.208)		-2.494 (0.108)
19	(ar)SO$_2$NH$_2$	-1.530	-2.567	2	-2.218 (-0.348)		-2.445 (-0.122)
20	(ar)NO$_2$	-0.078	0.279	-1	-0.370 (0.649)		0.224 (0.054)

Refs. pg 296

TABLE VIII,2 (continued)

no	fragment						
21	(al)CN	−1.066	−1.13		−1.628 (0.498)	−1.226 (0.096)	−1.272 (0.142)
22	(al)Cl	0.061	−0.63	2	−0.193 (−0.437)		−0.528 (−0.102)
23	(al)Br	0.270	−0.134	2	0.073 (−0.207)		−0.276 (0.142)
24	(al)I	0.587	0.16	2	0.477 (−0.317)		0.106 (0.054)
25	c_M	0.287	0.318		0.095 (0.223)	0.374 (−0.056)	0.358 (−0.040)

In parenthesis: residuals

$$f(\text{chloroform}) = 1.273\, f(\text{octanol}) - 0.270 \qquad (VIII-23)$$

$n = 25; \quad r = 0.905; \quad s = 0.641; \quad F = 105; \quad \underline{t} = 10.2$

Inverted form of eqn. VIII-23:

$$f(\text{octanol}) = 0.644\, f(\text{chloroform}) + 0.110 \qquad (VIII-24)$$

$n = 25; \quad r = 0.905; \quad s = 0.456; \quad F = 105; \quad \underline{t} = 10.2$

$$f(\text{chloroform}) = 1.183\, f(\text{octanol}) + 0.034 \qquad (VIII-25)$$

$n = 8; \quad r = 0.997; \quad s = 0.094; \quad F = 1{,}200; \quad \underline{t} = 34.6$

Inverted form of eqn. VIII-25:

$$f(\text{octanol}) = 0.841\, f(\text{chloroform}) - 0.027 \qquad (VIII-26)$$

$n = 8; \quad r = 0.997; \quad s = 0.080; \quad F = 1{,}200; \quad \underline{t} = 34.6$

$$f(\text{chloroform}) = 1.205\, f(\text{octanol}) - 0.307\, kn + 0.012 \quad (VIII-27)$$

$n = 25; \quad r = 0.998; \quad s = 0.092; \quad F = 3{,}077; \quad \underline{t}_1 = 66.9; \quad \underline{t}_2 = -33.0$

Inverted form of eqn. VIII-27:

$$f(\text{octanol}) = 0.826\, f(\text{chloroform}) + 0.253\, kn - 0.011 \quad (VIII-28)$$

$n = 25; \quad r = 0.997; \quad s = 0.076; \quad F = 2{,}266; \quad \underline{t}_1 = 66.9; \quad \underline{t}_2 = 28.3$

TABLE VIII,3

$f(\text{benzene}) - f(\text{octanol})$ CONVERSIONS

no	fragment	f octanol	f benzene	kn	estimated f values and residuals		
					A	B	C
1	C_6H_5	1.886	2.230		2.388 (−0.158)	2.154 (0.076)	2.168 (0.062)
2	C_6H_4	1.688	1.768		2.098 (−0.330)	1.927 (−0.159)	1.941 (−0.173)

TABLE VIII,3 (continued)

3	CH$_3$	0.702	0.947		0.651 (0.296)	0.796 (0.151)	0.807 (0.140)
4	CH$_2$	0.530	0.619		0.398 (0.220)	0.598 (0.021)	0.610 (0.009)
5	CH	0.235	0.212		−0.034 (0.246)	0.260 (−0.048)	0.271 (−0.059)
6	(al)COOH	−0.954	−2.898	6	−1.779 (−1.119)		−2.803 (−0.095)
7	(ar)COOH	−0.093	−2.004	7	−0.516 (−1.488)		−2.099 (0.095)
8	(al)COO	−1.292	−1.51		−2.275 (0.765	−1.493 (−0.017)	−1.484 (−0.026)
9	(al)CO	−1.703	−1.967		−2.878 (0.911)	−1.964 (−0.003)	−1.957 (−0.010)
10	(al)OH	−1.491	−3.128	5	−2.567 (−0.561)		−3.136 (0.008)
11	(ar)OH	−0.343	−1.936	5	−0.883 (−1.053)		−1.817 (−0.119)
12	(al)NH$_2$	−1.428	−2.40	3	−2.475 (0.075)		−2.494 (0.094)
13	(ar)NH$_2$	−0.854	−1.285	1	−1.632 (0.348)		−1.265 (−0.020)
14	(al)NH	−1.825	−3.242	4	−3.057 (−0.185)		−3.235 (−0.007)
15	(ar)NO$_2$	−0.078	0.451	−2	−0.494 (0.945)		0.480 (−0.029)
16	(al)Cl	0.061	−0.147	1	−0.290 (0.143)		−0.214 (0.067)
17	(ar)Cl	0.922	1.343	−1	0.974 (0.369)		1.345 (−0.002)
18	(al)Br	0.270	0.094	1	0.017 (0.077)		0.026 (0.067)
19	(ar)Br	1.131	1.584	−1	1.280 (0.304)		1.585 (−0.001)
20	(al)I	0.587	0.42	1	0.482 (−0.062)		0.391 (0.029)
21	c$_M$	0.287	0.298		0.042 (0.256)	0.319 (−0.021)	0.330 (−0.032)

In parenthesis: residuals

f(benzene) = 1.467 f(octanol) − 0.379   (VIII-29)

n = 21;   r = 0.929;   s = 0.649;   F = 121; $\underline{t}$ = 11.0

Inverted form of eqn. VIII-29:

Refs. pg 296

$$f(\text{octanol}) = 0.589 \, f(\text{benzene}) + 0.212 \qquad\qquad (\text{VIII-30})$$
$$n = 21; \quad r = 0.929; \quad s = 0.411; \quad F = 121; \quad \underline{t} = 11.0$$

$$f(\text{benzene}) = 1.148 \, f(\text{octanol}) - 0.010 \qquad\qquad (\text{VIII-31})$$
$$n = 8; \quad r = 0.998; \quad s = 0.097; \quad F = 1{,}541; \quad \underline{t} = 39.3$$

Inverted form of eqn. VIII-31:

$$f(\text{octanol}) = 0.868 \, f(\text{benzene}) + 0.010 \qquad\qquad (\text{VIII-32})$$
$$n = 8; \quad r = 0.998; \quad s = 0.085; \quad F = 1{,}541; \quad \underline{t} = 39.3$$

$$f(\text{benzene}) = 1.149 \, f(\text{octanol}) - 0.285 \, kn + 0.001 \qquad (\text{VIII-33})$$
$$n = 21; \quad r = 0.999; \quad s = 0.078; \quad F = 4{,}783; \quad \underline{t}_1 = 62.4; \quad \underline{t}_2 = -35.8$$

Inverted form of eqn. VIII-33:

$$f(\text{octanol}) = 0.866 \, f(\text{benzene}) + 0.246 \, kn + 0.000 \qquad (\text{VIII-34})$$
$$n = 21; \quad r = 0.998; \quad s = 0.068; \quad F = 2{,}543; \quad \underline{t}_1 = 62.4; \quad \underline{t}_2 = 26.0$$

TABLE VIII,4

$f(\text{oil}) - f(\text{octanol})$ CONVERSIONS

no	fragment	f octanol	f oil	kn	estimated f values and residuals A	B	C
1	C_6H_5	1.886	2.091		2.280 (-0.189)	2.106 (-0.015)	2.147 (-0.056)
2	C_6H_4	1.688	1.906		1.995 (-0.088)	1.888 (0.018)	1.921 (-0.015)
3	CH_3	0.702	0.808		0.571 (0.237)	0.800 (0.008)	0.798 (0.010)
4	CH_2	0.530	0.589		0.323 (0.266)	0.610 (-0.021)	0.602 (-0.013)
5	CH	0.235	0.282		-0.103 (0.385)	0.285 (-0.003)	0.265 (0.017)
6	(al)COOH	-0.954	-2.253	3	-1.819 (-0.434)		-2.131 (-0.122)
7	(ar)COOH	-0.093	-1.543	4	-0.576 (-0.967)		-1.497 (-0.046)
8	(al)COO	-1.292	-1.93	1	-2.307 (0.377)		-1.822 (-0.108)
9	(al)CO	-1.703	-2.326	1	-2.900 (0.574)		-2.291 (-0.035)
10	(ar)CO	-0.842	-1.616	2	-1.657 (0.041)		-1.656 (0.040)
11	(al)$CONH_2$	-1.970	-3.945	5	-3.286 (-0.659)		-3.983 (0.038)

TABLE VIII,4 (continued)

12	$(ar)CONH_2$	−1.109	−2.525	4	−2.043 (−0.482)		−2.655 (0.130)
13	$(al)OOCNH_2$	−1.481	−2.363	2	−2.580 (0.217)		−2.385 (0.022)
14	$(al)O$	−1.581	−2.124	1	−2.724 (0.600)		−2.152 (0.028)
15	$(ar)O$	−0.433	−0.704	1	−1.067 (0.363)		−0.843 (0.139)
16	$(al)OH$	−1.491	−2.861	3	−2.549 (−0.267)		−2.743 (−0.118)
17	$(ar)OH$	−0.343	−1.441	3	−0.937 (−0.504)		−1.435 (−0.006)
18	$(al)Cl$	0.061	−0.244	1	−0.354 (0.110)		−0.280 (0.036)
19	$(al)Br$	0.270	−0.013	1	−0.052 (0.039)		−0.042 (0.029)
20	c_M	0.287	0.355		−0.028 (0.383)	0.342 (0.013)	0.325 (0.030)

In parenthesis: residuals

$$f(oil) = 1.443\, f(octanol) - 0.442 \qquad (VIII-35)$$
$$n = 20; \quad r = 0.964; \quad s = 0.447; \quad F = 239; \quad \underline{t} = 15.5$$

Inverted form of eqn. VIII-35:

$$f(octanol) = 0.644\, f(oil) + 0.258 \qquad (VIII-36)$$
$$n = 20; \quad r = 0.964; \quad s = 0.299; \quad F = 239; \quad \underline{t} = 15.5$$

$$f(oil) = 1.103\, f(octanol) + 0.026 \qquad (VIII-37)$$
$$n = 6; \quad r = 0.999; \quad s = 0.017; \quad F = 10{,}030; \quad \underline{t} = 100$$

Inverted form of eqn. VIII-37:

$$f(octanol) = 0.906\, f(oil) - 0.023 \qquad (VIII-38)$$
$$n = 6; \quad r = 0.999; \quad s = 0.016; \quad F = 10{,}030; \quad \underline{t} = 100$$

$$f(oil) = 1.140\, f(octanol) - 0.347\, kn - 0.003 \qquad (VIII-39)$$
$$n = 20; \quad r = 0.999; \quad s = 0.073; \quad F = 4{,}790; \quad \underline{t}_1 = 58.9; \quad \underline{t}_2 = -25.6$$

Inverted form of eqn. VIII-39:

$$f(octanol) = 0.873\, f(oil) + 0.301\, kn + 0.004 \qquad (VIII-40)$$
$$n = 20; \quad r = 0.998; \quad s = 0.064; \quad F = 2{,}787; \quad \underline{t}_1 = 58.9; \quad \underline{t}_2 = 19.3$$

TABLE VIII,5

f(oleyl alcohol) − f(octanol) CONVERSIONS

no	fragment	f octanol	f ol.alc.	kn	estimated f values and residuals A	B	C
1	C_6H_4	1.688	2.009	−1	1.952 (0.057)		2.010 (−0.001)
2	CH_3	0.702	0.688		0.743 (−0.055)	0.718 (−0.030)	0.726 (−0.038)
3	CH_2	0.530	0.535		0.533 (0.002)	0.535 (0.000)	0.542 (−0.007)
4	CH	0.235	0.289		0.171 (0.118)	0.221 (0.068)	0.226 (0.063)
5	(al)COOH	−0.954	−1.289	1	−1.287 (−0.002)		−1.276 (−0.013)
6	(ar)COO	−0.431	−0.901	2	−0.646 (−0.255)		−0.943 (0.042)
7	(ar)O	−0.433	−0.512		−0.648 (0.136)	−0.490 (−0.022)	−0.490 (−0.022)
8	(al)OH	−1.491	−2.301	3	−1.945 (−0.356)		−2.306 (0.005)
9	(ar)OH	−0.343	−0.866	2	−0.538 (−0.328)		−0.848 (−0.018)
10	(ar)NH_2	−0.854	−0.685	−1	−1.164 (0.479)		−0.714 (0.029)
11	(al)N	−2.160	−2.590	1	−2.766 (0.176)		−2.569 (−0.021)
12	c_M	0.287	0.262		0.235 (0.027)	0.277 (−0.015)	0.281 (−0.019)

In parenthesis: residuals

f(oleyl alc.) = 1.226 f(octanol) − 0.117   (VIII−41)

n = 12; r = 0.983; s = 0.244; F = 299; $\underline{t}$ = 17.3

Inverted form of eqn. VIII−41:

f(octanol) = 0.789 f(oleyl alc.) + 0.084   (VIII−42)

n = 12; r = 0.983; s = 0.196; F = 299; $\underline{t}$ = 17.3

f(oleyl alc.) = 1.064 f(octanol) − 0.029   (VIII−43)

n = 5; r = 0.996; s = 0.045; F = 410; $\underline{t}$ = 20.2

Inverted form of eqn. VIII−43:

f(octanol) = 0.933 f(oleyl alc.) + 0.029   (VIII−44)

n = 5; r = 0.996; s = 0.042; F = 410; $\underline{t}$ = 20.2

f(oleyl alc.) = 1.072 f(octanol) - 0.227 kn - 0.026 (VIII-45)

n = 12;   r = 0.999;   s = 0.033;   F = 8,378; $\underline{t}_1$ = 91.5; $\underline{t}_2$ = -23.1

Inverted form of eqn. VIII-45:

f(octanol) = 0.932 f(oleyl alc.) + 0.211 kn + 0.024 (VIII-46)

n = 12;   r = 0.999;   s = 0.030;   F = 6,200; $\underline{t}_1$ = 91.5; $\underline{t}_2$ = 19.8

TABLE VIII,6

f(toluene) - f(octanol) CONVERSIONS

no	fragment	f octanol	f toluene	kn	estimated f values and residuals		
					A	B	C
1	C_6H_5	1.886	2.00		2.04 (-0.04)	2.00 (0.00)	1.95 (0.05)
2	C_6H_4	1.688	1.78		1.77 (0.01)	1.79 (-0.01)	1.75 (0.03)
3	CH_3	0.702	.0.74		0.43 (0.31)	0.74 (0.00)	0.75 (-0.01)
4	CH_2	0.530	0.56		0.20 (0.36)	0.56 (0.00)	0.57 (-0.01)
5	CH	0.235	0.25		-0.20 (0.45)	0.25 (0.00)	0.27 (-0.02)
6	(al)COOH	-0.954	-2.60	5	-1.82 (-0.78)		-2.47 (-0.13)
7	(ar)COOH	-0.093	-1.60	5	-0.65 (-0.95)		-1.59 (-0.01)
8	(ar)OH	-0.343	-1.74	5	-0.99 (-0.75)		-1.84 (0.10)
9	(al)NH_2	-1.428	-2.34	3	-2.47 (0.13)		-2.34 (0.00)
10	(al)NH	-1.825	-2.70	3	-3.01 (0.31)		-2.74 (0.04)
11	(al)N	-2.160	-3.02	3	-3.46 (0.44)		-3.08 (0.06)
12	(al)Br	0.269	-0.08	1	-0.16 (0.08)		0.00 (-0.08)
13	c_M	0.287	0.30		-0.13 (0.43)	0.30 (0.00)	0.32 (-0.02)

In parenthesis: residuals

f(toluene) = 1.360 f(octanol) - 0.524   (VIII-47)

n = 13;   r = 0.958;   s = 0.523;   F = 124; $\underline{t}$ = 11.1

Inverted form of eqn. VIII-47:

Refs. pg 296

$$f(\text{octanol}) = 0.676\ f(\text{toluene}) + 0.347 \qquad (\text{VIII--48})$$
$$n = 13; \quad r = 0.958; \quad s = 0.368; \quad F = 124; \quad \underline{t} = 11.1$$

$$f(\text{toluene}) = 1.059\ f(\text{octanol}) - 0.002 \qquad (\text{VIII--49})$$
$$n = 6;_{,}\quad r = 0.999; \quad s = 0.004; \quad F > 10,000; \quad \underline{t} = 407$$

Inverted form of eqn. VIII-49:

$$f(\text{octanol}) = 0.944\ f(\text{toluene}) + 0.002 \qquad (\text{VIII--50})$$
$$n = 6; \quad r = 0.999; \quad s = 0.003; \quad F > 10,000; \quad \underline{t} = 407$$

$$f(\text{toluene}) = 1.017\ f(\text{octanol}) - 0.305\ \text{kn} + 0.031 \qquad (\text{VIII--51})$$
$$n = 13; \quad r = 0.999; \quad s = 0.067; \quad F = 4,050; \quad \underline{t}_1 = 49.1; \quad \underline{t}_2 = -25.5$$

Inverted form of eqn. VIII-51:

$$f(\text{octanol}) = 0.979\ f(\text{toluene}) + 0.297\ \text{kn} - 0.028 \qquad (\text{VIII--52})$$
$$n = 13; \quad r = 0.998; \quad s = 0.066; \quad F = 2,087; \quad \underline{t}_1 = 49.1; \quad \underline{t}_2 = 18.2$$

TABLE VIII,7

f(carbon tetrachloride) $-$ f(octanol) CONVERSIONS

no	fragments	f octanol	f CCl$_4$	kn	estimated f values and residual A	B	C
1	C$_6$H$_5$	1.886	2.30		2.27 (0.03)	2.32 (−0.02)	2.34 (−0.04)
2	C$_6$H$_4$	1.688	2.15		1.98 (0.17)	2.08 (0.07)	2.10 (0.05)
3	CH$_3$	0.702	0.88		0.54 (0.34)	0.88 (0.00)	0.89 (−0.01)
4	CH$_2$	0.530	0.66		0.29 (0.37)	0.67 (−0.01)	0.68 (−0.02)
5	CH	0.235	0.29		−0.15 (0.44)	0.32 (−0.03)	0.31 (−0.02)
6	(al)COOH	−0.954	−3.30	6	−1.89 (−1.41)		−3.20 (−0.10)
7	(al)COO	−1.292	−1.44		−2.38 (0.94)	−1.54 (0.10)	−1.56 (0.12)
8	(al)CO	−1.703	−2.11		−2.98 (0.87)	−2.05 (−0.06)	−2.07 (−0.04)
9	(al)CON	−2.894	−4.28	2	−4.73 (0.45)		−4.22 (−0.06)
10	(al)OH	−1.491	−3.19	4	−2.67 (−0.52)		−3.18 (−0.01)
11	(ar)OH	−0.343	−2.74	7	−0.99 (−1.75)		−2.79 (0.05)

TABLE VIII,7 (continued)

bo	fragment						
12	(ar)NH$_2$	−0.854	−1.92	3	−1.74 (−0.18)		−2.05 (0.13)
13	(ar)Cl	0.922	1.11		0.86 (0.25)	1.15 (−0.04)	1.16 (−0.05)

In parenthesis: residuals

$$f(CCl_4) = 1.463\ f(octanol) - 0.490 \qquad (VIII\text{-}53)$$

n = 13; r = 0.932; s = 0.838; F = 73.2; $\underline{t}$ = 8.55

Inverted form of eqn. VIII-53:

$$f(octanol) = 0.594\ f(CCl_4) + 0.255 \qquad (VIII\text{-}54)$$

n = 13; r = 0.932; s = 0.534; F = 73.2; $\underline{t}$ = 8.55

$$f(CCl_4) = 1.217\ f(octanol) + 0.029 \qquad (VIII\text{-}55)$$

n = 8; r = 0.999; s = 0.061; F = 4,529; $\underline{t}$ = 67.3

Inverted form of eqn. VIII-55:

$$f(octanol) = 0.821\ f(CCl_4) - 0.023 \qquad (VIII\text{-}56)$$

n = 8; r = 0.999; s = 0.050; F= 4,529; $\underline{t}$ = 67.3

$$f(CCl_4) = 1.230\ f(octanol) - 0.342\ kn + 0.025 \qquad (VIII\text{-}57)$$

n = 13; r = o.999; s = 0.076; F = 5,058; $\underline{t}_1$ = 72.9; $\underline{t}_2$ = −36.2

Inverted form of eqn. VIII-58:

$$f(octanol) = 0.812\ f(CCl_4) + 0.277\ kn - 0.020 \qquad (VIII\text{-}58)$$

n = 13; r = 0.999; s = 0.062; F = 3,109; $\underline{t}_1$ = 72.9; $\underline{t}_2$ = 28.4

TABLE VIII,8

f(xylene) − f(octanol) CONVERSIONS

bo	fragment	f octanol	f xylene	kn	estimated f values and residuals		
					A	B	C
1	C$_6$H$_5$	1.896	2.06		2.15 (−0.09)	2.06 (0.00)	2.03 (0.03)
2	CH$_3$	0.702	0.77		0.51 (0.26)	0.77 (0.00)	0.74 (0.03)
3	CH$_2$	0.530	0.58		0.28 (0.30)	0.58 (0.00)	0.55 (0.03)
4	CH	0.235	0.26		−0.13 (0.39)	0.26 (0.00)	0.23 (0.03)
5	(al)COOH	−0.954	−2.71	5	−1.78 (−0.93)		−2.82 (0.11)

Refs. pg 296

TABLE VIII,8 (continued)

6	(ar)OH	−0.343	−1.90	4	−0.93 (−0.97)		−1.80 (−0.10)
7	(al)NH$_2$	−1.428	−2.31	2	−2.43 (0.12)		−2.28 (−0.03)
8	(al)NH	−1.825	−2.69	2	−2.99 (0.30)		−2.71 (0.02)
9	(al)N	−2.160	−3.07	2	−3.45 (0.38)		−3.07 (0.00)
10	(al)Br	0.270	−0.22	1	−0.08 (−0.14)		−0.08 (−0.14)
11	c_M	0.287	0.31		−0.06 (0.37)	0.31 (0.00)	0.29 (0.02)

In parenthesis: residuals

$$f(\text{xylene}) = 1.385\, f(\text{octanol}) - 0.458 \qquad \text{(VIII-59)}$$
$$n = 11; \quad r = 0.958; \quad s = 0.529; \quad F = 102; \quad \underline{t} = 10.1$$

Inverted form of eqn. VIII-59:

$$f(\text{octanol}) = 0.663\, f(\text{xylene}) + 0.283 \qquad \text{(VIII-60)}$$
$$n = 11; \quad r = 0.958; \quad s = 0.366; \quad F = 102; \quad \underline{t} = 10.1$$

$$f(\text{xylene}) = 1.092\, f(\text{octanol}) + 0.001 \qquad \text{(VIII-61)}$$
$$n = 5; \quad r = 0.999; \quad s = 0.003; \quad F \quad 10,000; \quad \underline{t} = 465$$

Inverted form of eqn. VIII-61:

$$f(\text{octanol}) = 0.916\, f(\text{xylene}) - 0.001 \qquad \text{(VIII-62)}$$
$$n = 5; \quad r = 0.999; \quad s = 0.002; \quad F \quad 10,000; \quad \underline{t} = 465$$

$$f(\text{xylene}) = 1.087\, f(\text{octanol}) - 0.352 \text{ kn} - 0.022 \qquad \text{(VIII-63)}$$
$$n = 11; \quad r = 0.999; \quad s = 0.075; \quad f = 2,709; \quad \underline{t}_1 = 44.7; \quad \underline{t}_2 = -20.8$$

Inverted form of eqn. VIII-63:

$$f(\text{octanol}) = 0.916\, f(\text{xylene}) + 0.321 \text{ kn} + 0.022 \qquad \text{(VIII-64)}$$
$$n = 11; \quad r = 0.998; \quad s = 0.069; \quad F = 1,537; \quad \underline{t}_1 = 44.7; \quad \underline{t}_2 = 15.6$$

TABLE VIII,9

f(cyclohexane) $-$ f(octanol) CONVERSIONS

no	fragment	f octanol	f c.hex.	kn	estimated f values and residuals A	B	C
1	C_6H_5	1.886	2.312		2.513 (−0.201)	2.353 (−0.041)	2.353 (−0.041)
2	C_6H_4	1.688	2.123		2.167 (−0.044)	2.110 (0.013)	2.105 (0.018)
3	C_6H_3	1.431	1.881		1.717 (0.164)	1.794 (0.087)	1.783 (0.098)
4	CH_3	0.702	0.805		0.443 (0.362)	0.898 (−0.093)	0.869 (−0.064)
5	CH_2	0.530	0.646		0.142 (0.504)	0.686 (−0.040)	0.653 (−0.007)
6	CH	0.235	0.397		−0.374 (0.771)	0.323 (0.074)	0.283 (0.114)
7	$(ar)NO_2$	−0.080	−0.578	1	−0.925 (0.347)		−0.444 (−0.134)
8	$(ar)CH=CH-NO_2$	0.395*	−0.569	3	−0.094 (−0.475)		−0.513 (−0.056)
9	$(ar)CH=C-NO_2$	0.220*	−0.733	3	−0.400 (−0.333)		−0.732 (−0.001)
10	$(ar)CH=CH-COO$	0.042*	−0.555	2	−0.712 (0.157)		−0.623 (0.068)
11	$(ar)CH=CH-CONH$	−1.100*	−4.343	9	−2.709 (−1.634)		−4.380 (0.037)
12	$(al)OH$	−1.491	−3.578	5	−3.392 (−0.186)		−3.542 (−0.036)
13	$(ar)OH$	−0.343	−3.157	8	−1.385 (−1.772)		−3.099 (−0.058)
14	$(ar)O$	−0.433	−0.754	1	−1.542 (0.778)		−0.887 (0.133)
15	$(ar)COH$	−0.380*	−1.058	2	−1.450 (0.392)		−1.152 (0.094)
16	$(ar)NH_2$	−0.854	−2.301	4	−2.278 (−0.023)		−2.411 (0.110)
17	$(al)NH$	−1.825	−2.766	1	−3.977 (1.211)		−2.632 (−0.134)
18	$(ar)F$	0.399	0.139	1	−0.087 (0.226)		0.157 (−0.018)
19	$(ar)Cl$	0.922	0.856	1	0.827 (0.029)		0.812 (0.044)
20	$(ar)Br$	1.131	0.909	1	1.193 (−0.284)		1.074 (−0.165)

*) Calculated from the constituent fragments
In parenthesis: residuals

Refs. pg 296

$$f(\text{c.hexane}) \;=\; 1.749\; f(\text{octanol}) \;-\; 0.785 \qquad\qquad \text{(VIII-65)}$$
$$n = 20; \quad r = 0.926; \quad s = 0.735; \quad F = 109; \quad \underline{t} = 10.4$$

Inverted form of eqn. VIII-65:
$$f(\text{octanol}) \;=\; 0.491\; f(\text{c.hexane}) \;+\; 0.407 \qquad\qquad \text{(VIII-66)}$$
$$n = 20; \quad r = 0.926; \quad s = 0.389; \quad F = 109; \quad \underline{t} = 10.4$$

$$f(\text{c.hexane}) \;=\; 1.230\; f(\text{octanol}) \;+\; 0.034 \qquad\qquad \text{(VIII-67)}$$
$$n = 6; \quad r = 0.996; \quad s = 0.079; \quad F = 555; \quad \underline{t} = 23.6$$

Inverted form of eqn. VIII-67:
$$f(\text{octanol}) \;=\; 0.807\; f(\text{c.hexane}) \;-\; 0.020 \qquad\qquad \text{(VIII-68)}$$
$$n = 6; \quad r = 0.996; \quad s = 0.064; \quad F = 555; \quad \underline{t} = 23.6$$

$$f(\text{c.hexane}) \;=\; 1.254\; f(\text{octanol}) \;-\; 0.332\; kn \;-\; 0.012 \qquad \text{(VIII-69)}$$
$$n = 20; \quad r = 0.998; \quad s = 0.092; \quad F = 3{,}967; \quad \underline{t}_1 = 48.5; \quad \underline{t}_2 = -33.4$$

Inverted form of eqn. VIII-69:
$$f(\text{octanol}) \;=\; 0.792\; f(\text{c.hexane}) \;+\; 0.261\; kn \;+\; 0.014 \qquad \text{(VIII-70)}$$
$$n = 20; \quad r = 0.997; \quad s = 0.073; \quad F = 1{,}757; \quad \underline{t}_1 = 48.5; \quad \underline{t}_2 = 22.0$$

TABLE VIII,10

$f(\text{pentanol}) - f(\text{octanol})$ CONVERSIONS

no	fragment	f octanol	f pentanol	estimated f values and residuals
1	C_6H_5	1.885	1.73	1.70 (0.03)
2	CH_3	0.702	0.68	0.66 (0.02)
3	CH_2	0.530	0.51	0.51 (0.00)
4	CH	0.235	0.23	0.25 (−0.02)
5	H	0.175	0.17	0.20 (−0.03)
6	(al)COOH	−0.954	−0.71	−0.80 (0.09)
7	(al)OH	−1.491	−1.27	−1.27 (0.00)
8	(ar)OH	−0.343	−0.31	−0.26 (−0.05)
9	(al)NH_2	−1.428	−1.22	−1.21 (−0.01)
10	c_M	0.287	0.27	0.30 (−0.03)

In parenthesis: residuals

$$f(\text{pentanol}) \;=\; 0.879\; f(\text{octanol}) \;+\; 0.043 \qquad\qquad (\text{VIII--71})$$

$$n = 10; \quad r = 0.999; \quad s = 0.040; \quad F = 4,640; \quad \underline{t} = 68.1$$

Inverted form of eqn. VIII-71:

$$f(\text{octanol}) \;=\; 1.135\; f(\text{pentanol}) \;-\; 0.049 \qquad\qquad (\text{VIII--72})$$

$$n = 10; \quad r = 0.999; \quad s = 0.045; \quad F = 4,640; \quad \underline{t} = 68.1$$

TABLE VIII,11

$f(\underline{n}.\text{butanol}) - f(\text{octanol})$ CONVERSIONS

no	fragment	f octanol	f $\underline{n}$.but	kn	estimated f values and residuals A	B	C
1	C_6H_5	1.886	1.547		1.465 (0.082)	1.512 (0.035)	1.517 (0.030)
2	CH_3	0.702	0.640		0.586 (0.054)	0.553 (0.087)	0.555 (0.085)
3	CH_2	0.530	0.404		0.459 (−0.055)	0.413 (−0.009)	0.416 (−0.012)
4	CH	0.235	0.087		0.240 (−0.153)	0.174 (−0.087)	0.176 (−0.089)
5	(al)COOH	−0.954	−0.527	1	−0.642 (0.115)		−0.570 (0.043)
6	(al)OH	−1.491	−0.794	2	−1.041 (0.247)		−0.785 (−0.009)
7	(al)NH_2	−1.428	−0.940	1	−0.994 (0.054)		−0.955 (0.015)
8	(al)NH	−1.825	−1.318	1	−1.289 (−0.029)		−1.278 (−0.040)
9	(al)N	−2.160	−1.727		−1.537 (−0.190)	−1.766 (0.039)	−1.770 (0.043)
10	c_M	0.287	0.152		0.278 (−0.126)	0.216 −(0.064)	0.218 (−0.066)

In parenthesis: residuals

$$f(\underline{n}.\text{but}) \;=\; 0.742\; f(\text{octanol}) \;+\; 0.065 \qquad\qquad (\text{VIII--73})$$

$$n = 10; \quad r = 0.990; \quad s = 0.144; \quad F = 421; \quad \underline{t} = 20.5$$

Inverted form of eqn. VIII-73:

$$f(\text{octanol}) \;=\; 1.323\; f(\underline{n}.\text{but}) \;-\; 0.094 \qquad\qquad (\text{VIII--74})$$

$$n = 10; \quad r = 0.990; \quad s = 0.192; \quad F = 421; \quad \underline{t} = 20.5$$

$$f(\underline{n}.\text{but}) \;=\; 0.810\; f(\text{octanol}) \;-\; 0.016 \qquad\qquad (\text{VIII--75})$$

$$n = 6; \quad r = 0.998; \quad s = 0.074; \quad F = 1,035; \quad \underline{t} = 32.2$$

Refs. pg 296

Inverted form of eqn. VIII-75:

$$f(\text{octanol}) = 1.230\, f(\underline{n}.\text{but}) + 0.021 \qquad (\text{VIII-76})$$

$n = 6;\quad r = 0.998;\quad s = 0.091;\quad F = 1{,}035;\quad \underline{t} = 32.2$

$$f(\underline{n}.\text{but}) = 0.813\, f(\text{octanol}) + 0.221\, kn - 0.015 \qquad (\text{VIII-77})$$

$n = 10;\quad r = 0.998;\quad s = 0.061;\quad F = 1{,}190;\quad \underline{t}_1 = 42.2;\quad \underline{t}_2 = 6.13$

Inverted form of eqn. VIII-77:

$$f(\text{octanol}) = 1.226\, f(\underline{n}.\text{but}) - 0.275\, kn + 0.019 \qquad (\text{VIII-78})$$

$n = 10;\quad r = 0.998;\quad s = 0.074;\quad F = 1{,}407;\quad \underline{t}_1 = 42.2;\quad \underline{t}_2 = -6.76$

TABLE VIII,12

$f(\underline{\text{sec}}.\text{butanol}) - f(\text{octanol})$ CONVERSIONS

no	fragment	f octanol	f $\underline{\text{sec}}$.but	kn	estimated f values and residuals A	C
1	CH_3	0.702	0.42		0.43 (-0.01)	0.46 (-0.04)
2	CH_2	0.530	0.32		0.35 (-0.03)	0.35 (-0.03)
3	H(neg)	0.47	0.36		0.32 (0.04)	0.31 (0.05)
4	(al)COOH	-0.954	-0.33	2	-0.33 (0.00)	-0.25 (-0.08)

In parenthesis: residuals

$$f(\underline{\text{sec}}.\text{but}) = 0.46\, f(\text{octanol}) + 0.11 \qquad (\text{VIII-79})$$

Inverted form of eqn. VIII-79:

$$f(\text{octanol}) = 2.18\, f(\underline{\text{sec}}.\text{but}) - 0.23 \qquad (\text{VIII-80})$$

$$f(\underline{\text{sec}}.\text{but}) = 0.66\, f(\text{octanol}) + 0.19\, kn \qquad (\text{VIII-81})$$

Inverted form of eqn. VIII-81:

$$f(\text{octanol}) = 1.52\, f(\underline{\text{sec}}.\text{but}) - 0.29\, kn \qquad (\text{VIII-82})$$

PROPOSAL FOR A GENERALLY APPLICABLE SOLVENT REGRESSION EQUATION

Solvent — water to octanol — water conversion

For the conversion of partitioning data from any solvent system into values pertaining to the octanol — water system, eqn. VIII-83 can be used:

$$f(\text{octanol}) \ = \ \rho' f_a \ \pm \ kn.c_M(\text{oct}) \qquad\qquad (\text{VIII-83})$$

as far as the hydrophobic fragmental constants are concerned. The symbol ρ' in this equation represents the multiplication factor which renders f_a quantifiable in multiples of the constant $c_M(\text{oct})$. The plus
sign in eqn. VIII-83 should be used for solvent systems with an organic
phase in which water is less soluble than it is in octanol, and the
minus sign for solvent systems where the opposite is true.

Table VIII,13 provides a compilation of salient features in the
solvent regressions designed in the preceding sections.

TABLE VIII,13

COMPILATION OF ρ VALUES AND c_M CONSTANTS

Organic phase of solvent system[*]	ρ'	$c_M(\text{oct})$	ρ''	$c_M(a)$
cyclohexane	0.792	0.261	1.254	0.332
carbon tetrachloride	0.812	0.277	1.230	0.342
xylene	0.916	0.321	1.087	0.352
toluene	0.979	0.297	1.017	0.305
benzene	0.866	0.246	1.149	0.285
chloroform	0.826	0.253	1.205	0.307
oil	0.873	0.301	1.140	0.347
diethylether	0.834	0.219	1.195	0.263
oleyl alcohol	0.932	0.211	1.072	0.227
octanol	(1.000)	0.268[a]	(1.000)	
n.pentanol	1.135	b	0.879	c
n.butanol	1.226	0.275	0.813	0.221
sec.butanol	1.52	0.29	0.66	0.19

[*] the organic phases are tabulated in the order of increasing
     solving power for water
[a] average value of the 11 data given in this column
[b] not determined; the value 0.27 is advised
[c] not determined; the value 0.24 is advised (obtained via eqn.
     VIII-89)

The mean of $c_M(\text{oct})$ was established as 0.268 with a standard deviation of 0.035 and this value agrees well with the set of c_M values presented in Part I (Chapter IV). The introduction of the value of 0.268
into eqn. VIII-83 and changing the f summations into log P values will
transform eqn. VIII-83 into eqn. VIII-84:

$$\log P(\text{octanol}) \ = \ \rho' \log P(a) \ \pm \ 0.268 \ \Sigma \ kn \qquad (\text{VIII-84})$$

Refs. pg 296

and this equation offers the possibility of transforming partition data
from any system into those of the octanol – water system. For this procedure to be allowed it is essential that :

(1) correct information about the structures be available;

(2) the key numbers of the various fragments for solvent – octanol con-
     version be known.

As will be clear, a detailed foreknowledge is needed for a fruitful
application of equation VIII-84. This is the logical result of the differentiation in the solute structure, as proposed before and elaborated
in the preceding sections. When using the solvent regression equations
of LEO et al., even such a simple distinction as that of donor and acceptor groups in the structure under conversion may be regarded as sufficient, and sometimes even irrelevant, for instance when "sole" equations are applicable.

Octanol – water to solvent – water conversion

The equation that permits the conversion of f values from the octanol – water system to values appropriate to any solvent system can be
easily derived from eqn. VIII-85:

$$f_a = 1/\rho' \, f(\text{octanol}) \; \mp \; 1/\rho' \, kn.c_M(\text{oct})$$

or

$$f_a = \rho'' \, f(\text{octanol}) \; \mp \; \rho'' \, kn.c_M(\text{oct}) \qquad \text{(VIII-85)}$$

where ρ'' $(= 1/\rho')$ represents the multiplication factor, which renders
$f(\text{octanol})$ quantifiable in multiples of $c_M(a)$.

It is, of course, also possible to write

$$f_a = \rho'' \, f(\text{octanol}) \; \mp \; kn.c_M(a) \qquad \text{(VIII-86)}$$

which implies that

$$c_M(a) = \rho'' \, c_M(\text{oct}) \qquad \text{(VIII-87)}$$

f Summation and transition to log P changes eqn. VIII-86 into

$$\log P(a) = \rho'' \, \log P(\text{octanol}) \; \mp \; 0.268 \, \rho'' \, \Sigma \, kn \qquad \text{(VIII-88)}$$

where the factor 0.268 represents the mean $c_M(\text{oct})$ value established from
Table VIII,13 (column 3).

The implication laid down in eqn. VIII-87 is open for verification;
via the data available from Table VIII,13 we obtain the following regression equation:

$$c_M(a) = 0.229\ \rho'' + 0.042 \qquad (VIII-89)$$
$$n = 12; \quad r = 0.765; \quad s = 0.037; \quad F = 14.1; \quad \underline{t} = 3.80$$

which can be considered to be a reasonable result, even though the correlation coefficient is not as satisfactory as one would wish; s, F and
$\underline{t}$ are acceptable, there are no outliers. One should bear in mind that
the material available for preparing the f sets in the previous chapter
was sometimes unsatisfactory, especially regarding mutual congruity.

It seems that the discriminative power of a solvent system can now
be correctly described as follows on the basis of eqns. VIII-83 and 85
or their respective log P analogues.

A solvent system is isodiscriminative towards the octanol − water
system when $\rho' = \rho'' = 1$ and, consequently, the Σ kn term is absent;
it is hyperdiscriminative towards the octanol − water system when $\rho' < 1$
$(\rho'' > 1)$, in which event the two upper algebraic signs of eqns. VIII-83
and 85 operate simultaneously; hypodiscrimination towards the octanol −
water system is indicated by $\rho' > 1$ $(\rho'' < 1)$, in which event the two
lower algebraic signs in eqns. VIII-83 and 85 operate simultaneously.

In fact, everything seems to depend on the quantifiability of log P
and relatedly on the sequence of "holes" that the solvent system has to
offer to the partitioning solute; the essential role of the Σ kn term
is to account for the difference between one functional group and another when transferred from the water phase to the lipoidic partner of
the solvent system and vice versa; for one functional group this transfer is much easier than for another depending on local dehydration −
solvation possibilities.

It follows that high key numbers will have to be associated with
large differences in polarity between the two organic phases concerned
in the equation, and this applies with even greater force to groups li-
able to strong hydration, such as $CONH_2$.

If membrane lipids are to be included in a range of organic phases,
one evidently needs information in terms of hyper-, iso- and hypodiscrimination of these lipids (combined with water to a solvent system)
towards a reference solvent system (for instance octanol − water).

Definition of hydro- and lipophilicity

In Chapter III, consideration was given to hydro- and lipophilicity,
and the question arose of how to define these two concepts in the most
adequate way. It became clear that structures and fragments had to be

differentiated from each other; it is possible for a structure, which
in its entirety has to be regarded as lipophilic, to possess several
more or less hydrophilic fragments whose behaviour is counterbalanced
by that of a number of lipophilic fragments.

The great diversity of material, especially the consideration of
additional solvent systems as displayed in the previous chapter and in
the above sections of this chapter, has given a broader insight and in-
vites some more discussion on hydro- and lipophilicity.

On reviewing all currently available fragmental constants, the following points may be noted:

(1) There are fragments which, on comparison of any given solvent system with the octanol – water system, can always be fitted into the
following conversion equation:

$$f_a = \rho \, f \text{(octanol)} \qquad\qquad\qquad \text{(VIII-90)}$$

These fragments are C_6H_5, C_6H_4, C_6H_3, CH_3, CH_2 and CH, and it seems
obvious that this set can be extended to all fragments derived from
saturated, unsaturated, aromatic and mixed hydrocarbons; we have in
mind fragments such as $CH_2=CH$, $CH\equiv C$, C and H. It should be noted
that these four fragments have been studied only in the octanol –
water system and could not be included in the other solvent systems
examined.

(2) There are fragments that can be accommodated in eqn. VIII-90 only by
using an extra term:

$$f_a = \rho \, f \text{(octanol)} + q \qquad\qquad \text{(VIII-91)}$$

where this term (q) can be described as a multiple of the constant
c_M. There are many examples, ranging from OH, O, COOH, NH_2, NH and
N, which may occur alone or in combinations, to more complex func-
tional groups, such as $CONH_2$.

(3) A number of other fragments may sometimes seem to belong to class
(1) but, if sufficient allowance is made for the difference between
the polarity of the selected solvent system and that of the octanol
– water system, they will actually belong to class (2). Examples
are F, Cl, Br and I[*].

Whether a given fragment occurs in class (2,3) or in class (1) ap-

---

[*]) Most investigators in the field of SAR qualify the halogen functions
as being lipophilic. Years ago, however, they were classified as
having hydrophilic character by BOOY and BUNGENBERG DE JONG [1].

pears to be connected with the presence or absence of pairs of lone
electrons, and this in itself presents a criterion for qualifying a
fragment as hydrophilic or lipophilic.

Hydrophilic fragments possess one or more pairs of lone electrons
which can be involved in hydrogen bonding with, for example, H_2O molecules. Complete or partial withdrawal of these electrons, which prevents
or reduces this hydrogen bonding, will be accompanied by an increase in
the lipophilicity of the fragment; this is clearly illustrated in an
aromatic system where conjugative coupling is accompanied by the con-
current rise in lipophilicity of (3 ± 1) c_M.

strong H-bonding   weak or absent
                                    H-bonding

It is clear that the proximity effect should also be examined in
the light of this concept and that it should be attributed to a decrease in the acceptor capacity of a group containing pairs of free
electrons and this, in turn, is the result of the −I effect of another
negative group one or two carbon atoms away. When viewed in this way,
the absence of a proximity effect in many benzene derivatives ortho-
substituted by two negative groups is no longer surprising; the lone
pairs have been promoted to π-electrons by the aromatic system and
this supposedly "protects" them from being affected along inductive
routes.

As lipophilic fragments lack lone pairs of electron pairs, they are
not sensitive to the effects referred to above, in either aromatic or
aliphatic substitution patterns.

KEY NUMBERS AND STEIN's N VALUES FOR DESCRIBING HYDROGEN-BONDING EFFECTS

In 1933 COLLANDER and BARLUND [2] examined the variation of the
permeability constant with the chemical structure of the penetrating
species on the plant cell Chara ceratophylla. The results of this investigation, together with some other permeability values, were considered
further by COLLANDER in 1949 [3]. There appeared to be a fairly good

Refs. pg 296

correlation between $PM^{\frac{1}{2}}$ values (where P represents the permeability constant in centimetre-hour^{-1} and M the molecular weight of the penetrating
substance) and the olive oil — water partition coefficient, both taken
logarithmically. As there was evidence to suggest that small molecules
such as water, methanol, ethanol, cyanamide, formamide, acetamide, ethylene glycol and urea would penetrate more quickly than one might expect
from their oil — water partitioning, COLLANDER assigned sieve-like qual-
ities to the membrane in addition to its lipidic nature.

STEIN [4] reconsidered the problem of membrane permeability on the
basis of a novel concept: diffusion in a liquid is identical with a
movement within a lattice. The original COLLANDER data, extended with a
few other values taken from the work of DAINTY and GINZBURG on the alga
Chara australis [5], were incorporated into this re-examination.

> STEIN's work did not come to our attention until Chapter IV had
> already been completed. In the concept he proposes, there are ob-
> vious arguments in favour of our "hole" theorem for understanding
> the quantifiability of f and log P values. For example, STEIN re-
> gards the diffusion process as a movement of the solute through a
> lattice-like structure, representable in a two-dimensional projec-
> tion as a move of the solute molecules across a chess board, the
> squares representing the lattice spaces. The solute molecules can
> have dimensions such that for diffusion they require more vacant
> lattice spaces than would be expected from their volumes alone. In
> terms of a model, this concept is virtually the same as that which
> we proposed in Fig.IV,2 with a ball pushed through the masonry of
> a brick wall.

STEIN suggests that the possibility of a solute crossing the phase
boundary at a lipid — water interface is clearly related to the number
($\underline{N}$) of hydrogen bonds that the permeant makes in water and that, as a
consequence, there must be a functional relationship between $PM^{\frac{1}{2}}$ and $\underline{N}$.

Establishing the correct $\underline{N}$ values (which should be considered as
weighted numerical values that are proportional to apparent hydrogen-
-bonding strengths rather than as an exact representation of the number
of hydrogen bonds) is not easy.

For water, STEIN chose the value $\underline{N} = 4$, following ideas of PAULING
[6]. Hydroxyl groups, regardless of whether they appear in alcoholic
structures or in the COOH group of carboxylic acids, are assigned a val-
ue of $\underline{N} = 2$. $\underline{N} = 1$ is assigned to a C=O group in aldehydes, amides and
carboxylic acids, and $\underline{N} = \frac{1}{2}$ to a C=O group in eaters; $\underline{N}$ is taken 0
for ethers, and $\underline{N} = 1$ is assigned to a C≡N group.

As regards the amines, an NH_2 group, which may reside either in a
primary amine or in a carbonamide, is always assigned $\underline{N} = 2$, and $\underline{N} = 1$

is assigned to a secondary amide.

We derived the regression equation from the 34 data points comprised by the log $PM^{\frac{1}{2}}$ versus N plot in STEIN's monograph and found

$$\log PM^{\frac{1}{2}} = 0.684\ \underline{N} - 1.458 \qquad\qquad (VIII-92)$$
$$n = 34;\quad r = 0.900;\quad s = 0.470;\quad F = 137;\quad \underline{t} = 11.7$$

With a residual of 1.33, H_2O is a severe outlier. On removal of H_2O from the series, the following equation is obtained:

$$\log PM^{\frac{1}{2}} = 0.690\ \underline{N} - 1.477 \qquad\qquad (VIII-92)$$
$$n = 33;\quad r = 0.923;\quad s = 0.412;\quad F = 181;\quad \underline{t} = 13.4$$

This equation leaves no more than 15% of the variances unaccounted for, and permits the conclusion that STEIN's principle of N values is sound. However, it is pertinent to ask if, in assigning such values, no other rules have to be advised. Fig.VIII,1, for example, where three amine patterns are presented in the light of present knowledge about hydrogen bonding (see CONDON [7]), shows weighted hydrogen bonding values of 3, 2 and 1 for primary, secondary and tertiary amines, respectively. In the concept proposed in this figure, an aromatic amine would occur with a value of no more than 2 in accordance with the supposition that the free electron pair of the nitrogen atom will be incorporated in the aniline resonance.

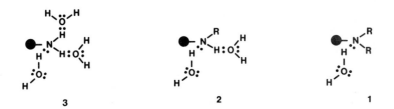

Fig.VIII,1: Primary, secondary and tertiary aliphatic amines, showing hydrogen bonding in accordance with weighting factors of 3, 2 and 1, respectively.

The reason for quoting STEIN's N values for the description of hydrogen bonding was that, in our opinion, there must be some connection between these values and the solvent system differences as they emerged from this chapter and which can apparently be expressed as $kn.c_M$ values. It might be tentatively suggested that a relationship could possibly exist between N and kn, but we prefer to leave any further speculation to the reader.

REFERENCES

1 H.L. Booy and H.G. Bungenberg de Jong, Biocolloids and Their
   Interactions, with special Reference to Coacervates and related
   Systems; Vol. I,2 in Protoplasmatologia, Handbuch der Protoplasma-
   forschung (Editors: L.V. Heilbrunn and F. Weber), Springer Verlag,
   Wien 1956.

2 R. Collander and H. Barlund, Acta Botan. Fennica, 11(1933)1.

3 R. Collander, Physiol. Plant., 2(1949)300.

4 W.D. Stein, The Movement of Molecules across Cell Membranes;
   Vol. 5 in Theoretical and Experimental Biology (Editor: J.F.
   Danielli), Academic Press, New York, London 1967.

5 J. Dainty and B.Z. Ginzburg, Biochem. Biophys. Acta, 79(1964)122.

6 L. Pauling, The Nature of the Chemical Bond, Conell Univ. Press,
   Ithaca, New York 1948.

7 F.E. Condon, J. Amer. Chem. Soc., 87(1965)4481,4485.

CHAPTER IX

MEMBRANES AND THEIR CHARACTERIZATION

INTRODUCTION

It is not the aim of this present chapter to deal exhaustively with the great many features of membrane structures and membrane passages, as there are plenty of other sources for both a cursory orientation and a searching study in this field.

Based on the material in the preceding chapters, the following outlines are devoted exclusively to matters that are clearly related to further characterization of the membrane lipid system that apparently plays an important role in passive transport phenomena.

Interested readers are referred to various sources for additional information, but we again wish to stress that we make no claim to be exhaustive in this bibliographic presentation.

Although some 20 years old, the treatise by BOOY and BUNGENBERG DE JONG [1] on "Biocolloids and Their Interactions", with a highly exploratory chapter on "Biological Membranes", is still worthy of perusal. It is remarkable that several of the concepts then introduced were so unambiguously substantiated in retrospect.

Going back further, one comes across publications such as those of COLLANDER [2], COLLANDER and BARLUND [3], DANIELLI and HARVEY [4] and DAVSON and DANIELLI [5]. As regards more recent developments, several workers have written surveys that are ideally suited to a first, serious acquaintance. Thus, CAPALDI [6] and LEE [7] provide primarily descriptive discussions, whereas the physical aspects of membrane processes are dealt with in greater detail by SCHORR [8]. The volume by FINEAN, COLEMAN and MICHELL [9] also lends itself particularly well to a greater insight into modern developments. More specialized information is to be found in STEIN's monograph [10] and in the volume written by CEREIJIDO and ROTUNNO [11].

Finally, attention is called to the proceedings of the Conference on Membrane Structure and its Biological Applications, held in New York in 1971 [12], and to a book recently published by HARRISON and LUNT [13], which, although intended for advanced undergraduates in biology and bio-

Refs. pg 322

chemistry, is certainly of interest to other readers also.

MEMBRANE FUNCTION

Membrane systems may be thought of as having essentially a protective function. This usually consists of preventing the passage of certain molecules from external to internal bio-phases or *vice versa* and may entail the regulation of definite vital processes. It is evident that a great diversity exists between such "floodgates" and that the nature and capacity of each depends on its location in the bio-system.

For our purposes the following sequence of protective membrane systems may be distinguished:

a. buccal and sub-lingual membranes

b. gastrointestinal membranes

c. various cell membranes

d. blood-brain barrier.

This order corresponds to that of an increasing need for protection of vital parts in the bio-apparatus; the deeper the penetration of a permeant into the bio-system, the sharper is the selection and hence the protection by the "floodgate" system.

In this concept, the buccal membrane system is the least selective. On average, a permeant remains in the buccal cavity for a relatively short period of time, while absorption is usually rather rapid. This may be partly associated with the main function of the cavity, that of taste sensation, which may lead either to a stimulation or to a refusal of intake[*].

Compared with its stay in the buccal cavity, the permeant is present in the gastrointestinal tract for a considerable time whereas its possibilities for permeation are markedly reduced.

The literature data available are insufficient for a rewarding approach to these problems. This is also due to the fact that no experimental planning has ever been directed at obtaining information to derive a membrane differentiation as set out above.

The blood-brain barrier presents a problem in its own right. Although no longer the mystery it once was, it is still not possible to design pertinent experiments that would afford the information needed to

---

[*] The actual situation is even more complex. Absorption into the tongue supposedly differs from absorption into the buccal mucosa (DASTOLI et al. [14]).

classify the lipid system operating in this barrier. KURZ's suggestion
[15] that the blood-brain barrier has a greater discrimination in favour
of more lipid-soluble substances than the cell barrier (as present in
liver cells) is substantiated by the findings of BICKEL and WEDER [16]
that several metabolites of medium polarity, which do not enter the
brain, can be traced in the liver and other cellular tissues. These results seem to be in favour of the above proposed sequence a – b – c – d.

A recent survey on the blood-brain barrier was compiled by OLDEN-
DORF [17].

MEMBRANE STRUCTURE

Current ideas about the biological membrane have been collected by
SINGER and NICHOLSON in their so-called fluid-mosaic model [18] that re-
placed the DAVSON and DANIELLI model [5] frequently used since 1952. The
SINGER and NICHOLSON model is shown schematically in Fig. IX,1.

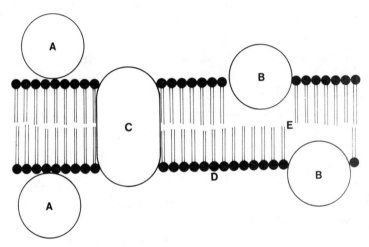

Fig.IX,1: Schematic SINGER and NICHOLSON model for the
biological membrane. A and B: globularly shaped proteins
adjoining the hydrophilic heads of the phospholipidic
part of the bilayer (A) or penetrating partially the bilayer (B). C: ovally shaped protein with head and tail
exposed to the external and internal aqueous phases of
the bio-system and the central part immersed in the lipid
region of the bilayer. D: hydrophilic heads and E: twin
hydrophobic tails of the phospholipid molecules.

The heart of the membrane structure consists of a bilayer of lipid
molecules with structure:

$$X - O - \overset{\overset{O}{\|}}{\underset{\underset{O}{|}}{P}} - O - CH_2 - \overset{\overset{CH_2 - OOC - R'}{|}}{CH} - OOC - R''$$

where X represents a nitrogenous base, usually choline, ethanolamine or
serine and R' and R" are the extended chains of fatty acids. Molecules
of this type, combining a strongly hydrophilic part (the phosphoric acid
ester—base unit) with a strongly hydrophobic part (the twin hydrocarbon
tails), are called "amphiphatic". The fatty acid chains usually consist
of 16 - 20 carbon atoms and sometimes double bonds will occur. Stearate
and oleate units are among the ones often present in these chains. The
cis—oleate double bond produces a bend in the hydrocarbon chain, which
will distort stiff packing in the system. Cholesterol may also be found
in the fatty acid chain region of the bilayer, sometimes in an amount
as high as 60% of the phospholipid fraction.

It is true that the depicted bilayer looks fairly stable but it is
constantly involved in a series of movements. This motional pattern
consists of

(a) rapid rotation about the long axis of the phospholipids;

(b) lateral diffusion of the lipids in the bilayer plane (jumping over);

(c) transversal diffusion of a lipid molecule from one side of the bi-
    layer to the other (flip-flop).

The other membrane constituents are proteins of a globular shape
(A and B in Fig. IX,1) or of a more extended form (C in Fig. IX,1),
with the possibility of spanning the bilayer by itself. A distinction
is made between extrinsic proteins that lie at or near the membrane
surfaces and intrinsic proteins that are partly exposed to water and
partly immersed in the lipid region of the bilayer. These differences
in accommodation on or in the membrane, together with their capacity to
pass through it, are thought to be related to the distribution of the
amino acids over the protein structure. Some explanatory details are
given in Table IX,1. For the 21 amino acids, the contributions of the
substituent R to lipophilicity relative to H (glycine) are listed in
column 4. The figures make it clear whether any of these amino acids
are lipophilic or hydrophilic compared with glycine. In a protein structure made up by glycine units alone, the amount of lipophilicity per
unit may be arrived at as follows: $f(CH_2) + f(CONH) + 1$ p.e.$(1) = f(CH_2)$
$+ f(CON) + f(H_{neg.}) + 4\ c_M = 0.53 + (-2.89) + 0.18 + 1.15 = -1.03.$

When, for comparison, this amount is brought together with the data

TABLE IX,1

RELATIVE LIPOPHILICITY VALUES OF PROTEIN - FRAGMENTS

no	amino acids	R	Σf (rel.)	L/H
1	tryptophane	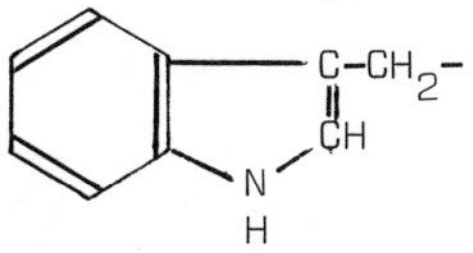 (indole ring)$-C-CH_2-$	2.31	L
2	phenylalanine	$C_6H_5-CH_2-$	2.24	L
3	leucine	$(CH_3)_2CH-CH_2-$	1.99	L
4	isoleucine	$CH_3-CH_2-CH(CH_3)-$	1.99	L
5	tyrosine	$4-OH-C_6H_4-CH_2-$	1.70	L
6	valine	$(CH_3)_2CH-$	1.46	L
7	cystine	$\frac{1}{2}(-CH_2-S-S-CH_2-)$	1.11	$\sim$
8	methionine	$CH_3-S-(CH_2)_2-$	1.08	$\sim$
9	proline	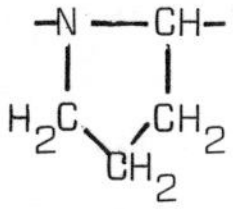 $-N-CH-$ ring with H_2C, CH_2, CH_2	1.01	$\sim$
10	cysteine	$HS-CH_2-$	0.93	$\sim$
11	alanine	CH_3-	0.53	H
12	lysine	$NH_2-(CH_2)_4-$	0.52	H
13	glycine	$H-$	(0.00)	H
14	aspartic acid	$HOOC-CH_2-$	-0.02	H
15	glutamic acid	$HOOC-(CH_2)_2-$	-0.07	H
16	histidine	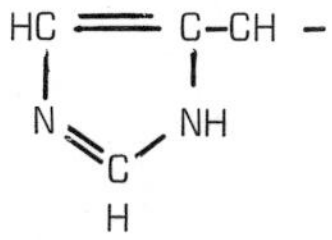 $HC == C-CH-$ imidazole ring N, NH, C, H	-0.23	H
17	threonine	$CH_3-CH(OH)-$	-0.26	H
18	serine	$HO-CH_2-$	-0.56	H
19	asparagine	$NH_2-CO-CH_2-$	-1.05	H
20	glutamine	$NH_2-CO-(CH_2)_2-$	-1.09	H
21	arginine	$NH_2-(C=NH)-NH-(CH_2)_3-$	?	H

H = hydrophilic; L = lipophilic;
$\sim$: the value Σf(rel.) practically counterbalances the lipophilicity of one glycine unit;
[a] f(S-S) = 0.37 was derived from log P = 1.77 measured for dimethyl-disulfide [53].

Refs. pg 322

from Table IX,1, the 21 amino acids can be classified as hydrophilic (H) or lipophilic (L), depending on the sign of the net remainder when the amino acid has replaced one glycine unit.

Obviously, a large variety of protein structures is possible, including both overtly hydrophilic and lipophilic types and a great many intermediate types. It seems that the question of whether the protein molecule in a given membrane system will behave intrinsically or extrinsically depends on the distribution of hydrophobicity and lipophilicity in the protein pattern. Thus, MARCHESI et al. [19] established that the major glycoprotein from the membrane of red blood cells consists basically of three regions with a distribution pattern H - L - H. The two fragments indicated by H are exposed to water at the outer surface of the cell membrane and the aqueous interior of the cell, respectively; the central part (L), which consists of 30 amino acids, is highly hydrophobic and prefers a location in the heart of the bilayer. Table IX,2 shows how the various relationships may be broadly judged in several instances.

TABLE IX,2

SOME PROTEIN LIPOPHILICITIES

source	ratio hydrophilic / lipophilic	
inner membranes of mitochondria	20 / 80	
FOLCH - LEES protein from myelin of sciatic nerve	29 / 71	Av. 30 / 70
rhodopsin	36 / 64	
cytochrome	37 / 63	
cytoplasmic and extrinsic proteins	47 / 53	

The above discussion suggests strongly that membrane proteins are probably responsible for:

(1) a modification of the original partitional behaviour of the lipid heart of the membrane;

(2) the creation of a certain differentiation in the whole complex of membrane systems.

The first claim may hardly raise objections. a passing reference being made to the membrane cholesterol, which is mentioned above and may be presumed to be of some importance in so far as, with its highly hydrophilic OH groups (the f value in a hydrocarbon-like phase such as that of cyclohexane in combination with water amounts to -3.5), it creates hydratable islets at regular intervals.

Regarding the second claim, at present we can only speculate. It is certain that the total amount of protein in a membrane system is subject to large variations; myelin, which is found around nerves, contains no more than 18% protein and plasmamembrane has 50% of protein, whereas this percentage can be as high as 75% for the internal membranes of mitochondria and chloroplasts. Were it possible to couple these variable percentages of protein to a correctly fluctuating amino acid pattern expressed in terms of hydrophilicity and lipophilicity, we could provide a strong argument for the second postulate.

BUCCAL MEMBRANE SYSTEMS

In recent years, a number of interesting papers have been published on buccal absorption.

The buccal absorption test was introduced by BECKETT et al. [20, 21] and applied by other workers, although slightly adjusted according to personal experience. Some of these experiments will be considered below.

Buccal absorption of imipramine and its metabolites

BICKEL and WEDER [22] investigated the buccal absorption of imipramine and nine of its derivatives that had previously been demonstrated to occur in rats as metabolites of imipramine [16]. Table IX,3 includes them as well as (in the second column) the human buccal absorptions expressed as percentages of the total amount in which the compounds disappeared from isotonic phosphate buffer solutions (pH 7.4) after mouth-rinsing for 5 min.

Of the ten compounds, partition coefficients were measured in four different systems (the organic phases being chloroform, n.hexane, diethylether and 1,2-dichloroethane), the aqueous phase always consisting of a phosphate buffer (pH 7.4). BICKEL and WEDER believe that buccal absorption data agree best with the partition coefficients in the hexane - buffer system and state that the other three solvent systems fit much less with their buccal absorption data. With regression analyses based on BICKEL and WEDER's original data, we wish to demonstrate the real quality of all of these fits.

Refs. pg 322

TABLE IX,3

BUCCAL ABSORPTIONS AND PARTITION VALUES OF IMIPRAMINE AND
ITS METABOLITES (pH 7.4)

compound	buccal absorption (%)	solubility in organic phase (%)			
		$CHCl_3$	n.hexane	ether	DClE
imipramine	57 (64.2)	99.8	99.4	99.3	98.5
desimipramine	45 (47.2)	98.5	65.1	88.4	97.8
desdimethylimipramine	36 (49.9)	99.1	70.5		
imipramine-N-oxide	18 (19.8)	97.2	10.0	7.0	71.0
iminodibenzyl	69 (64.2)	99.6	99.2	93.6	
2-OH-imipramine	34 (27.4)	97.4	25.2	84.2	71.0
2-OH-desimipramine	2 (17.9)	55.2	6.3	55.6	50.2
10-OH-imipramine	32	92.5			
10-OH-desimipramine	30	77.0			
2-OH-iminodibenzyl	67 (37.1)	98.0	44.8	97.5	

DClE: 1,2-dichloroethane
In parenthesis: estimated buccal absorption % derived from
n.hexane - buffer partition (eqn. IX-5)

Chloroform - buffer

$$\% \text{ Abs.} = 1.017 \ (\% \text{ org.ph.}) - 53.95 \qquad (IX-1)$$
$$n = 10; \quad r = 0.695; \quad s = 16.11; \quad F = 7.49; \quad \underline{t} = 2.74$$

$$\% \text{ Abs.} = 1.069 \ (\% \text{ org.ph.}) - 58.53 \qquad (IX-2)$$
$$n = 8; \quad r = 0.698; \quad s = 18.13; \quad F = 5.73; \quad \underline{t} = 2.39$$

$$\% \text{ Abs.} = 1.118 \ (\% \text{ org.ph.}) - 61.40 \qquad (IX-3)$$
$$n = 7; \quad r = 0.724; \quad s = 19.08; \quad F = 5.52; \quad \underline{t} = 2.35$$

$$\% \text{ Abs.} = 0.883 \ (\% \text{ org.ph.}) - 47.97 \qquad (IX-4)$$
$$n = 5; \quad r = 0.782; \quad s = 15.62; \quad F = 4.75; \quad \underline{t} = 2.18$$

In none of eqns. IX-1 to IX-4 did outliers occur.

n.Hexane - buffer

$$\% \text{ Abs.} = 0.498 \ (\% \text{ org.ph.}) + 14.80 \qquad (IX-5)$$
$$n = 8; \quad r = 0.786; \quad s = 15.64; \quad F = 9.74; \quad \underline{t} = 3.12$$

There are no outliers, although 2-hydroxy-iminodibenzyl is marginal
with its residual of 29.9%.

$$\% \text{ Abs.} = 0.529 \ (\% \text{ org.ph.}) + 15.25 \qquad (IX-6)$$
$$n = 7; \quad r = 0.822; \quad s = 15.74; \quad F = 10.45; \quad \underline{t} = 3.23$$

There are no outliers.

$$\% \text{ Abs.} = 0.502 \,(\% \text{ org.ph.}) + 10.51 \qquad\qquad (IX\text{-}7)$$
$$n = 5; \quad r = 0.924; \quad s = 9.57; \quad F = 17.65; \quad \underline{t} = 4.20$$

There are no outliers. The strong improvement of eqn. IX-7 compared with eqn. IX-6 is largely attributable to the absence of 2-hydroxy-iminodi-benzyl in the former equation.

Diethylether – buffer

$$\% \text{ Abs.} = 0.567 \,(\% \text{ org.ph.}) - 0.870 \qquad\qquad (IX\text{-}8)$$
$$n = 7; \quad r = 0.750; \quad s = 18.28; \quad F = 6.46; \quad \underline{t} = 2.54$$

$$\% \text{ Abs.} = 0.416 \,(\% \text{ org.ph.}) + 3.340 \qquad\qquad (IX\text{-}9)$$
$$n = 5; \quad r = 0.712; \quad s = 17.63; \quad F = 3.08; \quad \underline{t} = 1.76$$

In neither of these two equations did outliers occur.

1,2-Dichloroethane – buffer

$$\% \text{ Abs.} = 1.002 \,(\% \text{ org.ph.}) - 46.69 \qquad\qquad (IX\text{-}10)$$
$$n = 5; \quad r = 0.945; \quad s = 8.17; \quad F = 25.4; \quad \underline{t} = 5.04$$

There are no outliers.

Calculations were designed so that the comparison covered the larg-est possible number of structures in the four solvent systems.

A correlation for ten structures can be realized only in the chloroform – buffer system (eqn. IX-1). The results are not completely satisfactory, as 52% of the variances cannot be accounted for. For seven structures, the comparison can be extended to the diethylether – buffer system. The $\underline{n}$.hexane – buffer regression then is far the best (only 32% of the variances cannot be explained) but the diethylether – buffer re-gression, with its 44% of unexplained variances, likewise gives a satis-factory result (eqn. IX-8).

A complete comparison of all four systems is possible for no more than five structures. It then appears that the diethylether – buffer re-gression is no longer significant and that the 1,2-dichloroethane – buf-fer system has the best regression. Unfortunately, the paper cited leaves a number of 1,2-dichloroethane partition coefficients unmentioned so that the possible merits of the latter system cannot be appraised sufficiently, and also lacking are the relevant data in the octanol – buffer system, which in our opinion are indispensable for a correct comparison of the solvent systems investigated.

The effect of alkyl substitution in acids on buccal absorption

BECKETT and MOFFAT studied the passage of a number of acids into the buccal membrane of man [24]. The tested compounds included normal fatty acids ranging from butyric to dodecanoic acid and a series of benzoic acids with one, two, three and four methyl substituents.

The acids could be broadly divided into three groups:

(1) acids with the same pK value and different lipid solubilities;

(2) acids with different pK values and the same lipid solubility;

(3) acids with different pK values and different lipid solubilities.

Some explanatory notes are given below.

(1) By decreasing the pH, the concentration of unionized acid species is increased, and because the normal fatty acids hardly differ in pK value, it would be expected that the rate of entry into the biological membrane would increase by a constant amount for each CH_2 group by which the chain is lengthened; experimental work substantiated this expectation. n.Dodecanoic and n.undecanoic acids are different in this respect, as they still show an absorption of 20% and 47%, respectively, at pH 9.0, where the concentration of any unionized acid species should be negligible. BECKETT and MOFFAT believe that the ionized C_{11} and C_{12} acids, with their increased lipid solubilities, are responsible.

(2) The three toluic acids have the same buccal absorptions except for a displacement from each other by approximately the differences in their pK values (pK_a value of o-, p- and m-toluic acid = 3.92, 4.24 and 4.33, respectively).

(3) As expected, mono- and 2,4-dimethyl-substituted benzoic acids display an increase in absorption over benzoic acid, and BECKETT and MOFFAT relate this mainly to the lipid – water partition characteristics of the unionized species. 2,4,6-Trimethyl- and 2,3,5,6-tetramethylbenzoic acid are not in line with their partners in so far that their absorption by the buccal membrane is clearly lower than that of benzoic acid while their lipophilic characters would suggest the contrary. As to the nature of the partitioning system of choice, BECKETT and MOFFAT agree with BICKEL and WEDER, although on practical grounds they prefer the n.heptane – buffer system.

Before commenting on these choices, it is worth considering the partitioning characteristics of n.hexane and n.heptane in more detail.

Partitioning characteristics of the n.hexane - water
and n.heptane - water systems

Scrutiny of the scanty information on the partitioning characteristics in the n.hexane - water system indicates that the most reliable ρ value of this system _versus_ the octanol - water system is 1.15. Based on this value, Table IX,4 gives a survey of f values for those fragments which had previously been qualified as hydrophobic.

TABLE IX,4

HYDROPHOBIC f VALUES IN THE n.HEXANE - WATER SYSTEM

fragment	f(n.hexane)	f(c.hexane)
CH_3	0.81^a	0.805
CH_2	0.61^a	0.646
C_6H_5	2.19^a	2.312
H	0.20^a	0.22^d
C_6H_4	1.94^a	2.123
C_6H_3	1.65^a	1.881
(c_M)	0.33^b	0.36^c

a: from f(n.hexane) = 1.15 f(octanol)
b: from c_M(n.hexane) = 1.15 c_M(octanol)
c: from c_M(c.hexane) = 1.254 c_M(octanol)
d: from f(c.hexane) = 1.254 f(octanol)

The f(cyclohexane) values are included for comparison, and the differences between the various f values are rather small. It therefore appears appropriate to use the f(cyclohexane) analogue in the calculation of n.hexane - water partitioning data when a correct f(n.hexane) value is lacking. For further inspection, a number of structures are listed in Table IX,5, log P(n.hexane) being calculated with the application of an f(cyclohexane) value, when necessary.

The n.heptane - water partitioning system is much too incongruous to permit the derivation of its own set of f values. Table IX,5 gives some log P values to demonstrate that the n.heptane - water and n.hexane - water systems often show such small differences, that there is no objection to the indiscriminate use of the partition data of the two systems and to extend them, when necessary, by f(cyclohexane) values.

At present there seems to be one exception, viz. the COOH group. BRUNZELL [42] measured the partition values of six long-chain fatty acids in n.heptane - water, and from his measurements we derived f(COOH)

TABLE IX,5

CALCULATIONS OF log P($\underline{n}$.hexane) AND log P($\underline{n}$.heptane) VALUES,
APPLYING f(cyclohexane) CONTRIBUTIONS

compound	log P (obs)	I/II	ref.	log P (calc)	res.
C_6H_6	2.45	I	[27]	2.39	0.06
	2.26	II	[28]		−0.13
$C_6H_5-CH_3$	2.99	I	[27]	3.00	−0.01
	2.85	II	[28]		−0.15
1,2-diCH$_3$-C$_6$H$_4$	3.39	II	[28]	3.56	−0.17
1,3-diCH$_3$-C$_6$H$_4$	3.54	II	[28]		−0.02
1,4-diCH$_3$-C$_6$H$_4$	3.45	II	[28]		−0.10
CH$_3$-CH$_2$-OH	−2.26	I	[29]	−2.16	−0.10
CH$_3$-(CH$_2$)$_2$-OH	−1.48	I	[29]	−1.55	0.07
CH$_3$-(CH$_2$)$_3$-OH	−0.78	I	[29]	−0.94	0.16
	−0.86	II	[30]		0.08
C$_6$H$_5$-CH$_2$-OH	−0.76	I	[31]	−0.78	0.02
C$_6$H$_5$-CH$_2$-CH$_2$-OH	−0.39	I	[31]	−0.17	−0.22*
C$_6$H$_5$-OH	−0.96	I	[32]	−0.97	0.01
	−0.70	I	[33]		0.27*
	−0.89	I	[27]		0.08
	−0.92	II	[28]		0.05
	−0.82	II	[34]		0.15
	−1.06	II	[35]		−0.09
2-CH$_3$-C$_6$H$_4$-OH	−0.05	II	[28]	−0.41	0.36*
3-CH$_3$-C$_6$H$_4$-OH	−0.35	II	[28]	−0.41	0.06
4-CH$_3$-C$_6$H$_4$-OH	−0.35	II	[28]	−0.41	0.06
2,3-diCH$_3$-C$_6$H$_3$-OH	0.43	II	[28]	0.11	0.32*
2,4-diCH$_3$-C$_6$H$_3$-OH	0.40	II	[28]	0.11	0.29*
2,5-diCH$_3$-C$_6$H$_3$-OH	0.52	II	[28]	0.11	0.41*
2,6-diCH$_3$-C$_6$H$_3$-OH	0.82	II	[28]	0.11	0.71*
3,4-diCH$_3$-C$_6$H$_3$-OH	0.10	II	[28]	0.11	−0.01
3,5-diCH$_3$-C$_6$H$_3$-OH	0.21	II	[28]	0.11	0.10
C$_6$H$_5$-O-CH$_3$	2.10	II	[36]	2.25	−0.15
C$_6$H$_5$-NH$_2$	−0.02	I	[33]	−0.11	0.09
	−0.05	I	[27]		0.06
	−0.11	II	[37]		0.00
	−0.03	II	[28]		0.08

TABLE IX,5 (continued)

	0.04	II	[34]		0.15
	−0.26	II	[38]		−0.15
	0.04	II	[39]		0.15
$2\text{-}CH_3\text{-}C_6H_4\text{-}\underline{NH_2}$	0.55	II	[28]	0.45	0.10
	0.47	II	[40]		0.02
$3\text{-}CH_3\text{-}C_6H_4\text{-}\underline{NH_2}$	0.54	II	[28]	0.45	0.09
	0.45	II	[35]		0.00
$4\text{-}CH_3\text{-}C_6H_4\text{-}\underline{NH_2}$	0.41	I	[40]	0.45	−0.04
	0.54	I	[41]		0.09
	0.48	II	[28]		0.03
	0.51	II	[34]		0.06
	0.44	II	[40]		−0.01
	0.27	II	[35]		−0.18
$C_6H_5\text{-}\underline{NO_2}$	1.49	I	[27]	1.61	−0.12
	1.43	II	[35]		−0.18
$1,3\text{-}di\underline{NH_2}\text{-}C_6H_4$	−2.60	II	[40]	−2.66	0.06
$C_6H_5\text{-}\underline{F}$	2.47	I	[27]	2.33	0.14
$C_6H_5\text{-}\underline{Cl}$	2.98	I	[27]	3.05	−0.07
$C_6H_5\text{-}\underline{Br}$	3.08	I	[27]	3.10	−0.02

*) Outliers
I $\underline{n}$.hexane − water values
II $\underline{n}$.heptane − water values
Underlined functional groups: f(cyclohexane) values used

= −3.23, which is an apparantly reliable value, considering the observed standard deviation of 0.12. For comparative purposes, the literature unfortunately does not report suitable cyclohexane − water partition data, and so we decided to measure the partition coefficient of a
representative acid in this solvent system. We chose phenylacetic acid
and found log P = −1.23 [43], which, in turn, gave a value of −4.19 for
f(COOH).

Such a striking difference is difficult to interpret and the problem can only be solved by performing partition measurements of a selected series of carboxylic acids in the above hydrocarbon − water systems.
The comparison in Table IX,6 (three log P values from the BECKETT −
MOFFAT series $\underline{versus}$ calculated log P values) clearly demonstrates the
lack of sufficient information.

TABLE IX,6

COMPARISON OF THREE log P VALUES MEASURED BY BECKETT AND MOFFAT,
WITH CALCULATED log P VALUES

compound	log P(obs) (n.heptane)	log P(calc) $f(COOH) = -3.23$	$f(COOH) = -4.19$
C_4H_9COOH	−1.40	−0.59 (−0.99)	−1.55 (0.15)
$C_6H_{13}COOH$	−0.96	0.63 (−1.59)	−0.33 (−0.63)
$C_{10}H_{21}COOH$	1.09	3.07 (−1.98)	2.11 (−1.12)

In parenthesis: residuals

Subsequently, we decided to analyze the whole range of aliphatic and
aromatic carboxylic acid lipophilicities determined by BECKETT and
MOFFAT. For the material employed, see Table IX,7. We aimed at the

TABLE IX,7

PARTITION VALUES OF ALIPHATIC AND AROMATIC CARBOXYLIC ACIDS,
MEASURED IN n.HEPTANE − 0.1 N HYDROCHLORIC ACID [26]

compound	log P (obs)	log P (calc)	res.	log P (calc) res. kn II
C_4H_9-COOH	−1.40	−2.22	0.82*	
$C_6H_{13}-COOH$	−0.96	−1.00	0.04	
$C_7H_{15}-COOH$	−0.37	−0.39	0.02	
$C_8H_{17}-COOH$	0.29	0.22	0.07	
$C_9H_{19}-COOH$	0.70	0.83	−0.13	
$C_{10}H_{21}-COOH$	1.09	1.44	−0.35*	
$C_{11}H_{23}-COOH$	1.27	2.05	−0.78*	
$C_{12}H_{25}-COOH$	1.98	2.66	−0.68*	
C_6H_5-COOH	−0.96	−1.10	0.14	
$2-CH_3-C_6H_4-COOH$	−0.60	−0.52	−0.08	
$3-CH_3-C_6H_4-COOH$	−0.51	−0.52	0.01	
$4-CH_3-C_6H_4-COOH$	−0.64	−0.52	−0.12	
$2,4-diCH_3-C_6H_3-COOH$	−0.06	−0.02	−0.04	
$2,6-diCH_3-C_6H_3-COOH$	−1.00	−0.02	−0.98*	−1.01 (0.01) 3
$2,4,6-triCH_3-C_6H_2-COOH$	−0.46	0.59	−1.05*	0.40(−0.06) 3
$2,3,5,6-tetraCH_3-C_6H-COOH$	−0.43	1.20	−1.63*	−0.45 (0.02) 5

*) Outliers

greatest possible consistency in the whole array of data on the basis
of the equation

$$f(COOH)_{ar} = f(COOH)_{al} + kn.c_M \qquad (IX-11)$$

This condition is valid for the octanol — water system with $kn = 3$ and
$c_M = 0.287$. As regards the n.heptane — water system, c_M can be taken
0.33 (see Table IX,4) and kn is at least 3 but rather 4 or 5; the correct value of kn should be established by trial and error. In our hands,
$kn = 5$ appeared to be most useful, and the best overall fit in the aliphatic series emerged with $f(COOH) = -4.86$, so that the value for
$f(COOH)_{ar}$ is $-4.86 + 5 \times 0.33 = -3.27 (\pm$ some standard deviation). As a
result, benzoic acid, the three toluic acids and 2,4-dimethylbenzoic
acid can be fitted rather well into this equation, optimal consistency
being attained at $f = -3.29$.

The calculations further show that 2,6-dimethyl-, 2,4,6-trimethyl-
and 2,3,5,6-tetramethylbenzoic acid are outliers with residuals of
-0.98, -1.05 and -1.63, respectively. In terms of key numbers, these
residuals are found to be 3, 3 and 5, respectively, taking $c_M = 0.33$.
The implication for 2,6-dimethyl- and 2,4,6-trimethylbenzoic acid is an
advanced but still incomplete de-coupling of the $C_6H_5 \longrightarrow COOH$ resonance
and for 2,3,5,6-tetramethylbenzoic acid such an advanced de-coupling
that the COOH group ends up with an f value equivalent to that of an
aliphatic COOH group. The buttressing effect of the two meta-CH_3 groups
is believed to be responsible for this. As the partition measurements
provide a pattern that, besides being fairly consistent, is in harmony
with our line of argument in Chapter V for 2,6-dimethylbenzoic acid, it
can be concluded that the numerical data reported by BECKETT and MOFFAT
are — in their mutual connexion — correct (it should be noted that in
Table IX,7 four outliers remain). Thus, BECKETT and MOFFAT stated, with
reason, that the absorption of 2,4,6-trimethyl- and 2,3,5,6-tetramethyl-
benzoic acid by the buccal membrane system does not entirely fulfil the
prognosis based on lipophilicity values. We wish to emphasize, however,
that this detracts considerably from the general conclusion on the
n.heptane — water system as being ideally suited in describing buccal
membrane passage.

Buccal absorption of phenylacetic acids

As an extension to their studies on the buccal absorption of aliphatic and aromatic carboxylic acids, BECKETT and MOFFAT performed sim-

Refs. pg 322

ilar experiments with several phenylacetic acids [25,26] . There appeared
to be a good deal of agreement between the results. For details see
Table IX,8, second column, which lists the percentage buccal absorption
of each phenylacetic acid tested at pH 6.9. Again, they noted connecting
links between the process of buccal absorption and the partition coefficients of the acids concerned in the n.heptane — water system.

TABLE IX,8

PERCENTAGE BUCCAL ABSORPTIONS AT pH 6.0 OF para-SUBSTITUTED
PHENYLACETIC ACIDS AS RELATED TO f(n.heptane) AND f(octanol) VALUES
OF THE SUBSTITUENT GROUPS

substituents	% b. abs.	f values n.heptane [*]	octanol
(H)	1	0.20	0.175
NO_2	1	−0.578	−0.080
F	1.5	0.139	0.399
CH_3O	3	0.06	0.269
CH_3	7	0.81	0.702
Cl	7	0.856	0.922
Br	8	0.909	1.131
n.C_3H_7O	10	1.28	1.008
C_2H_5	10	1.42	1.232
I	10	1.48	1.448
C_3H_7	25	2.03	1.762
t.C_4H_9	25	2.60	2.250
C_4H_9	34	2.64	2.292
t.C_5H_{11}	30	3.21	2.780
cyclo C_5H_9	30	2.85	2.360
C_5H_{11}	49	3.25	2.822
cyclo C_6H_{11}	44	3.46	2.890
C_6H_{13}	61	3.86	3.352

[*]) Values are taken or derived from Tables IV,4 and VII,25

By correlating the absorption data reported by BECKETT and MOFFAT
with both the f values of the substituents in the n.heptane — water system and those in the octanol — water system, we obtained the following
equations:

$$\log \% \text{ Abs.} = 0.406\, f\text{(heptane)} + 0.349 \qquad \text{(IX-12)}$$

$$n = 18; \quad r = 0.954; \quad s = 0.175; \quad F = 163; \quad \underline{t} = 12.8$$

$$\log \% \text{ Abs.} = 0.512 \, f(\text{octanol}) + 0.250 \qquad \text{(IX-13)}$$

$$n = 18; \quad r = 0.949; \quad s = 0.183; \quad F = 148; \quad \underline{t} = 12.2$$

The quality of the regression equation utilizing $f(\text{octanol})$ values is slightly less than that in which $f(\underline{n}.\text{heptane})$ values have been incorporated but, in our opinion, the conclusion that the $\underline{n}$.heptane – water system would be exclusively ideal in describing buccal absorptional behaviour is highly debatable.

Physical model approach by HO et al. for the analysis of buccal absorption

The physical model developed by SUZUKI et al.[44,45] for the understanding of drug transport involving the intestinal, gastric and rectal tracts appears to lend itself particularly well to buccal drug absorption studies (HO and HIGUCHI [46], VORA et al. [47]). As used for the description of buccal absorption processes, the model consists of two compartments, viz. the mucosal side, which consists of the bulk phase of the aqueous drug solution and a diffusion layer with given thickness, and, in series with this compartment, a second one build up by a homogeneous lipid phase, also having a given thickness. It is assumed
(1) that there is a perfect sink on the serosal side after passage of the lipid phase and
(2) that only non-ionized molecules pass the lipid membrane.
The authors approximated the steady-state rate of buccal absorption by a first-order process and derived the following equations (for details, see the original papers [46,47]):

$$K_u = B_1 \, f(T) \qquad \text{(IX-14)}$$

$$n = \frac{K_{i+1}}{K_i} = \frac{P_{o,2,i+1}}{P_{o,2,i}} = \frac{T_i}{T_{i+1}} \qquad \text{(IX-15)}$$

where

$$B_1 = A P_{w,1}/V$$

$$f(T) = \frac{1}{(1 + K_a/[H^+])T + 1}$$

and

$$T = P_{w,1}/P_{o,2}$$

The meaning of the various symbols is as follows:

K_u = absorption rate constant

Refs. pg 322

B_1 = constant $\left(\text{time}^{-1}\right)$

$P_{w,1}$ = permeability coefficient of the drug in the aqueous diffu-
           sion layer; it is described by B_1

A = surface area

V = volume of drug solution

$f(T)$ = dimensionless parameter: $0 < f(T) \leqslant 1$

T = diffusion efficiency coefficient

$P_{o,2}$ = permeability coefficient of the drug in the lipid membrane

K_a = dissociation constant

$\left[H^+\right]$ = hydrogen-ion concentration

n = lipid — aqueous incremental partition constant

K_i and K_{i+1} are partition coefficients, the subscripts 1 and i+1
           denoting the compound and modified compound, respective-
           ly; these subscripts are also used to distinguish be-
           tween the connected permeability coefficients $\left(P_{o,2}\right)$
           and the connected diffusion efficiency coefficient (T).

VORA et al. subjected the buccal absorption data of BECKETT and
MOFFAT to a further analysis by means of this physical model. A feature
of the procedure that deserves special attention is the determination
of the value of n, the lipid — aqueous incremental partition constant.
This implies that the effect of a substituent group on the buccal absorption of an unsubstituted structure can be established rather easily.
The values given in Table IX,9 were obtained from a fitting procedure
at maximum absorption levels, B_1 ranging from 0.393 to 0.416 min^{-1}.
VORA et al. concluded that these n values are such that, in terms of
lipophilicity, buccal absorption is ideally approximated by the isobutanol — water system. The n values are basically the partition ratios
in eqn. IX-15, and transformed into logarithms they are simply f values
representing the hydrophobic fragmental effects on buccal absorption
(see the third column in Table IX,9). For $f(CH_2)$, an average is given,
excluding the f value that links o-toluic and benzoic acids. The last
four columns of Table IX,9 list the f values operating in the octanol —
water, sec. butanol — water, n.butanol — water and cyclohexane — water
systems, respectively. Where these values could not be extracted directly from the sets presented in Chapter VIII, they were calculated from
f(octanol) values using the appropriate solvent regression equation.

It is clear that further data would be useful for a definite judg-
ment and it is therefore unfortunate that for the sec. butanol — water

TABLE IX,9

COMPARISON OF INCREMENTAL PARTITION CONSTANTS FOR A NUMBER
OF SUBSTITUENT GROUPS IN A SERIES OF ACIDS

acids		n	$\log n$ $(f_{b.a.})$	f values OCT	SBU	PBU	CH
p.alkylphenylacetic	$(CH_2)_{al}$:	2.22	0.346				
phenylacetic / p.methylphenylacetic	$(CH_3-H)_{ar}$:	2.22	0.346				
benzoic / o.toluic	$(CH_3-H)_{ar}$:	1.05	(0.021)	0.53	0.32	0.40	0.65
/ m.toluic	$(CH_3-H)_{ar}$:	1.40	0.146				
/ p.toluic	$(CH_3-H)_{ar}$:	2.74	0.438				
n.alkanoic	$(CH_2)_{al}$:	2.33	0.367				
phenylacetic / p.F-phenylacetic	$(F)_{ar}$:	1.47	0.17	0.40	0.26	0.32	0.14
p.Cl-phenylacetic	$(Cl)_{ar}$:	3.65	0.56	0.92	0.61	0.75	0.86
p.Br-phenylacetic	$(Br)_{ar}$:	4.29	0.63	1.13	0.74	0.91	0.91
p.I-phenylacetic	$(I)_{ar}$:	5.94	0.77	1.45	0.95	1.17	1.58

$f(CH_2)_{aver.} = 0.33$

b.a. = buccal absorption
OCT = octanol – water system
SBU = sec. butanol – water system
PBU = n.butanol – water system
CH = cyclohexanol – water system

system only a very limited set of structures are available. Yet, the data now at our disposal, seem to indicate that the sec. butanol – water system is fairly representative in describing the buccal absorption process, and it seems preferable to the n.butanol – water system.

The observation that o-toluic acid, with its minor increment over benzoic acid, is an outlier on comparison with the di-o-methyl-substituted acids already mentioned in this chapter is not so surprising. No explanation can be given, however, for the low $f_{b.a.}$ value of the CH_2 contribution of m-toluic acid.

Buccal absorption of some analgesically active p-substituted acetanilides

DEARDEN and TOMLINSON [48,49] examined the buccal absorption of a series of p-substituted acetanilides and correlated the results with their analgesic activities. They noted a significantly correct parabolic relationship with a better fit than that of the analgesic activities and the octanol – water partition data of the investigated structures. For

details, see the original paper, which is also of interest for other
reasons; thus, they proposed a three-compartment instead of the common-
ly used two-compartment model for buccal absorption studies. We only
wish to draw special attention to their view that the octanol — water
system may be regarded as an acceptable model for buccal absorption
studies.

DISCRIMINATIVE POWER OF THE BUCCAL MEMBRANE

From the foregoing considerations about buccal absorption phenomena
two completely different viewpoints emerge.

On the one hand, BICKEL and WEDER, BECKETT and MOFFAT and DEARDEN
and TOMLINSON opt for n.hexane — water, n.heptane — water and octanol —
water, respectively, in describing buccal absorption; in other words,
the buccal membrane is supposed to be either hyperdiscriminative (in being modeled by a hydrocarbon — water system) or isodiscriminative (in
being modeled by the octanol — water system).

On the other hand, VORA et al. regard the isobutanol — water system
as ideally suited in characterizing buccal absorption; in other words,
they prefer a solvent system that has a remarkably hypodiscriminative
power. Curiously, VORA et al. used the same data set as BECKETT and
MOFFAT and yet came to an opposite conclusion. In considering VORA et
al.'s results together with our sets of f values, the buccal membrane
seems to be even more hypodiscriminative (in being modeled slightly better by sec. butanol — water than by isobutanol — water).

Does all this mean that there is a serious controversy about buccal
absorption, or could there be a question of a certain dualistic behav-
ioural pattern in the buccal membrane with respect to its discriminative
properties? The previous chapter presented the following equation, which
connects hydrophobic fragmental values in two different solvent systems,
one of which represents octanol — water:

$$f_a \; = \; \rho'' \, f(\text{octanol}) \; \mp \; \rho'' \; kn.c_M(\text{oct}) \qquad\qquad (\text{IX--16})$$

where f_a denotes the f value in any given solvent system and ρ'' the multiplication factor which renders f(octanol) quantifiable in terms of
$c_M(a)$ multiples. Hypodiscrimination of the solvent system over the octanol — water system was coupled to $\rho'' < 1$ and the use of the plus sign
in eqn. IX-16, while with hyperdiscrimination it was the opposite: $\rho'' > 1$
and the minus sign in eqn. IX-16.

The first term of the right-hand member of eqn. IX-16 is always present, regardless of whether the fragment is lipo- or hydrophilic; the presence of a second term is restricted to hydrophilic fragments and seems to be coupled with hydration/dehydration processes.

On inspection of all of the solvent systems investigated, a clear trend is seen for eqn. IX-16; strong hyperdiscrimination (extra-large $\rho"$) is associated with an extra-large second term in the right-hand member of eqn. IX-16 because extra-high key numbers occur, while the situation is the opposite in the case of strong hypodiscrimination.

As regards the f values to be used in a membrane system, an equation similar to eqn. IX-16 should be valid:

$$f(\text{membrane}) = \rho" \, f(\text{octanol}) \mp \rho" \, kn.c_M(\text{oct}) \qquad (IX-17)$$

Is it a priori reasoning that the trend noted for a true solvent system will be merely repeated in a membrane system, or could it be that the membrane system shows a less pronounced coupling of $\rho"$ and kn magnitudes? As regards the buccal membrane system, we would be out of difficulty if the latter were true; from the fact that $\rho" \sim 0.66$ for the buccal membrane, we could decide on its hypodiscriminative properties (suggestively located in the protein part of the system?) in complete accordance with VORA et al.'s findings, which are based exclusively on the lipophilic parts of the structures; and, taking into account that a hydrophilic group may show up with a kn value that is relatively much greater than may be expected in connection with $\rho"$ 0.66 (suggestively induced by the hydrocarbon-like character of the heart of the membrane?), lipophilicity as a whole can acquire the nature of a drifting factor.

Obviously, any future observations in this area should be collected so as to lend themselves to an efficient verification on the basis of the above postulates. An initial contribution is already at hand and is presented in the following section.

HUMAN ERYTHROCYTE MEMBRANE LIPID

Thanks to SEEMAN et al. [50,51,52], there is an interesting collection of partition coefficients, which were determined in a system that consisted, on the one hand, of purified lipid material from human erythrocyte membranes and, on the other hand, of an aqueous buffer of pH 7.2. Table IX,10 presents a summary of the available data.

Refs. pg 322

Some attention has already been paid to these values in Chapter VI and a systematic coverage is given below. Special consideration will be given to the question of the extent to which it is possible to develop an equation of the IX-17 type and thus to obtain pertinent information on the discriminative power of the membrane lipid.

Nos. 1 - 7 were incorporated into a first regression; log P(calc) values from the octanol - water system were chosen as the dependent parameter. The following equation was obtained:

$$\log P(HEML) = 0.905 \log P(oct) - 0.513 \qquad (IX-18)$$
$$n = 7; \quad r = 0.991; \quad s = 0.149; \quad F = 273; \quad \underline{t} = 16.5$$

The quality of this regression equation is excellent; there are no outliers, but it must be admitted that No. 7 (benzyl alcohol), with a residual of 0.23, differs considerably from the others. On removal of this compound, the equation can be given as:

$$\log P(HEML) = 0.984 \log P(oct) - 0.763 \qquad (IX-19)$$
$$n = 6; \quad r = 0.998; \quad s = 0.072; \quad F = 902; \quad \underline{t} = 30.0$$

For the estimates and the respective residuals of this regression equation, see Table IX,10.

Eqn. IX-19 is, in fact, a summation of several equations of the IX-17 type. The only fragment that plays a role in the second term of the right-hand member is the aliphatic hydroxyl group, and hence:

$$\rho'' \; kn(OH)_{al} \cdot c_M(oct) = 0.763 \qquad (IX-20)$$

With $\rho'' = 0.984$ and $c_M(oct) = 0.268$, this provides a $kn(OH)_{al}$ value of 3; and a value of kn = 0 for ethylurethane will adapt this structure with perfect correctness. The three phenolic structures (Nos. 8, 9 and 10) need a value of kn = 2 in order to fit them correctly into eqn. IX-19.

Thus, the regression equation for the first 11 structures is, with due observation of the correct key numbers:

$$\log P(HEML) = 0.914 \log P(oct) - 0.144 \, \Sigma \, kn - 0.103 \qquad (IX-21)$$
$$n = 11; \quad r = 0.994; \quad s = 0.123; \quad F = 340; \quad \underline{t}\left[\log P(oct)\right] = 22.5;$$
$$\underline{t}(kn) = -2.78$$

With a residual of 0.24, benzyl alcohol is very near to being an outlier. Its removal from the data series and the simultaneous application of one zero card in the data set to achieve some forcing gives the following equation:

TABLE IX,10

PARTITION VALUES IN THE HUMAN ERYTHROCYTE MEMBRANE
LIPID − BUFFER SYSTEM (SEEMAN)

no	compound	log P (oct)	obs.	log P (HEML) clc.I	kn	clc.II
1	$CH_3-(CH_2)_4-OH$	1.37	0.56	0.59 (−0.03)	3	0.58 (−0.02)
2	$CH_3-(CH_2)_5-OH$	1.90	1.11	1.11 (0.00)	3	1.10 (0.01)
3	$CH_3-(CH_2)_6-OH$	2.42	1.60	1.62 (−0.02)	3	1.61 (−0.01)
4	$CH_3-(CH_2)_7-OH$	2.95	2.18	2.14 (0.04)	3	2.13 (0.05)
5	$CH_3-(CH_2)_8-OH$	3.48	2.76	2.66 (0.10)	3	2.65 (0.11)
6	$CH_3-(CH_2)_9-OH$	4.01	3.09	3.18 (−0.09)	3	3.17 (−0.08)
7	$C_6H_5-CH_2-OH$	0.98	0.60	0.20* (0.40)	3	0.20* (0.40)
8	C_6H_5-OH	1.54	0.93		2	0.98 (−0.05)
9	$4-Br-C_6H_4-OH$	2.53	1.96		2	1.96 (0.00)
10	$3-CH_3-4-Cl-C_6H_3-OH$	2.73	2.12		2	2.15 (−0.03)
11	$NH_2-COO-CH_2-CH_3$	−0.16	−0.24		0	−0.21 (−0.03)
12	diphenylhydantoin	2.34[a]	1.60		3	
13	barbital	0.68[b]	−0.17		4	
14	phenobarbital	1.45[b]	0.81		3	
15	pentobarbital	2.07[b]	0.98		5	
16	methadone	3.93[c]	1.51		10	
17	morphine	0.73[d]	0.43		1	
18	chlorpromazine	5.30[e]	3.20		9	
19	procain	1.90[f]	0.65		5	
20	butyrate-ion	0.81	−0.19		4	
21	valerate-ion	1.34	0.28		4	

*) Outlier

a) From log P(5,5-diphenylhydantoin) = 2.47 [54] and log P(5-ethyl-5-phenylhydantoin) = 1.53 [54], it follows that f = −1.44 is an acceptable value for the hydantoin skeleton; hence the value log P = 2.34 was preferred

b) Averaged values from literature (see Table X,1)

Refs. pg 322

c) Ref. [55]
d) Averaged from log P = 0.76 [56] and log P = 0.70 [55]
e) Averaged value from literature (see Chapter I, pg 19)
f) Averaged from log P = 1.87 [54] and log P = 1.92 [54]

$$\log P(\text{HEML}) = 0.982 \log P(\text{oct}) - 0.237 \sum \text{kn} - 0.055 \qquad (IX\text{-}22)$$

$$n = 10(+1); \quad r = 0.999; \quad s = 0.060; \quad F = 1,615; \quad \underline{t}[\log P(\text{oct})] = 40.6$$
$$\underline{t}(\text{kn}) = -8.70$$

The value of ρ" 0.982 means that the HEML is, in its behaviour towards lipophilic groups, not far from <u>isodiscriminative</u>. As regards two
functional groups studied, OH(al) and OH(ar), their kn values are 3 and
2, respectively, which shows unambiguously that the HEML is <u>actually</u>
<u>not similar</u> to the octanol – water system. Here, too, the dual character
as noted for the buccal membrane system occurs.

The partition values of the remaining structures listed in Table
IX-10 can now be interpreted on the basis of eqn. IX-22. The framed
part of the Table gives the key numbers required to correlate the relevant log P(oct) values with the log P(HEML) values by way of this equation. In instances where two species might exist, we opted for the non-
-ionized species to be preferentially taken up by the HEML.

It appears that with compounds such as methadone and chlorpromazine,
the key numbers to be applied can mount up considerably. Another interesting feature of Table IX,10 is the picture presented by the butyrate
and valerate ions. Their partition values (-3.20 [23] and -2.69*) are
not included in the log P(oct) column but the table does present the
log P values of the non-ionized species. The reason is that in the latter instance key numbers are found to fulfil expectations, whereas key
numbers pertaining to the ionic species would be completely out of pro-
portion (-13). The pK_a value of these acids is approximately 4.9, so
that at pH 7.2 only 0.6% is available as free acid. It is pertinent to
ask if this low percentage is sufficient to permit membrane passage.

CONCLUSION

It is clear that for a closely reasoned characterization of membrane
systems, much more material is needed, obtained not only on purified
membrane lipids (experiments of the SEEMAN type) but also on more intact

*)This value has not been reported in the literature; it was deter-
mined by interpolation between the partition values of butyrate
(-3.20) and hexanoate (-2.17 [23]).

systems (experiments of the BECKETT type).

Yet the results discussed above make it necessary to underline
the hypotheses advanced in the previous sections.

The biological membrane seems dualistic in its partitioning aspects
in comparison with true solvent systems, seeing that its behaviour towards lipophilic groups is radically different from that towards hydrophilic groups.

Chapter VI gave a schematic outline of the discriminative behaviour
of a number of solvent systems (see Fig.VI,1). Also included in this
illustration was the HEML, as it presented itself in the light of current knowledge. On reconsideration, the following will emerge (see Fig.
IX,2).

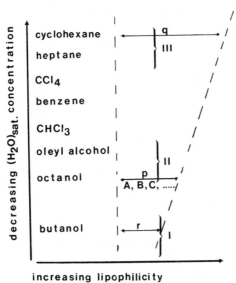

Fig.IX,2: Graphical representation of the discriminating properties of a number of solvent systems
(only indicated by their organic phases) as a function of the water-saturation concentration of the
organic phases; p, q and r denote the discriminating
powers of the octanol – water, cyclohexane – water
and butanol – water systems, respectively, towards
a set of partitioned structures A, B, C ....
I, II and III denote approximate locations of three
membrane systems.

I represents a membrane system whose ρ values are unequivocally
lower than zero, while the key numbers it impresses on hydrophilic

Refs. pg 322

groups seem to belong to a less hypodiscriminative system. The buccal membrane could be regarded as an examole. II is a membrane system with $\rho \sim 1$; key numbers of hydrophilic groups will likewise tend to become increased; the gastrointestinal tract could perhaps be regarded as an example. Finally, III denotes a membrane system with ρ values largely in excess of 1, along with high key numbers for hydrophilic groups, the total picture being perhaps suggestive of the blood-brain barrier.

REFERENCES

1 H.L. Booy and H.G. Bungenberg de Jong, Biocolloids and Their Interactions, with special Reference to Coacervates and related systems; Vol. I,2 in Protoplasmatologia, Handbuch der Protoplasmaforschung (Editors: L.V. Heilbrunn and F. Weber), Springer Verlag, Wien 1956.

2 R. Collander, Soc. Sci. Fennica Comm. Biol., II, No. 6 (1926).

3 R. Collander and H. Barlund, Acta Botan. Fennica, 11(1933)1.

4 J.F. Danielli and E.N. Harvey, J. cellul. comp. Physiol., 5(1935)483.

5 H. Davson and J.F. Danielli, The Permeability of Natural Membranes, Cambridge Univ. Press, Cambdidge, New York 1943.

6 R.A. Capaldi, Scient. Am. 230(1974)26.

7 A.G. Lee, Endeavour, 35(1975)67.

8 H. Schorr, Biologie in unserer Zeit, 5(1975)97.

9 J.B. Finean, R. Coleman and R.M. Michell, Membranes and Their Cellular Functions, Blackwell Sc. Publ., Oxford, London, Edinburgh, Melbourne 1974.

10 W.D. Stein, The Movement of Molecules across Cell Membranes; Vol. 5 in Theoretical and Experimental Biology (Editor: J.F. Danielli), Academic Press, New York London 1967.

11 M. Cerijido and C.A. Rotunno, Introduction to the Study of Biological Membranes, Gordon and Breach, New York 1970.

12 Membrane Structure and its Biological Applications, Proceedings of a Conference held by the New York Academy of Sciences on June 2, 3 and 4, 1971 (Editor: D.E. Green), Annals of the New York Academy of Sciences 195(1972).

13 R. Harrison and G.G. Lunt, Biological Membranes, Their Structure and Function, Blackie & Son Ltd., Glasgow 1976.

14 F.R. Dastoli, D.V. Lopiekes and S. Price, Biochem., 7(1968)1160.

15 H. Kurz, Arch. exp. Path. Pharmak., 247(1964)164.

16 M.H. Bickel and H.J. Weder, Archs. int. Pharmacodyn. Thér., 173(1968)433.

17 W.H. Oldendorf, Ann. Rev. of Pharmacol., 14(1974)239.

18 S.J. Singer and G.L. Nicholson, Science, 175(1972)720.

19 V.T. Marchesi, R.L. Jackson, J.P. Segrest, I. Kahane, Fed. Proc. Proc. Fed. Amer. Soc. Exp. Biol., 32(1973)1833.

20 A.H. Beckett and E.J. Triggs, J. Pharm. Pharmac., 19S(1967)31.

21 A.H. Beckett, R.N. Boyes and E.J. Triggs, J. Pharm. Pharmac., 20(1968)92.

22 M.H. Bickel and H.J. Weder, J. Pharm. Pharmac., 21(1969)160.

23 C. Church, unpublished analysis (see ref. [57]).

24 A.H. Beckett and A.C. Moffat, J. Pharm. Pharmac., 20S(1968)239.

25 A.H. Beckett and A.C. Moffat, J. Pharm. Pharmac., 21S(1969)139.

26 A.H. Beckett and A.C. Moffat, J. Pharm. Pharmac., 21S(1969)144.

27 T. Sekine, Y. Suzuki and N. Ihara, Bull. Chem. Soc., Japan, 46(1973)995.

28 C.L. de Ligny, H. Kreutzer and G.F. Visserman, Rec. trav. chim., Pays Bas, 85(1966)5.

29 K.H. Meyer and H. Hemmi, Biochem. Z., 277(1935)39.

30 N.W. Weisbrodt, M. Kienzle and A.R. Cooke, Proc. Soc. Exp. and Biol. Med., 142(1973)450.

31 A.C. McCulloch and B.H. Stock, Australian J. Pharm., 48S(1966)14.

32 A.B. Lindenberg and M. Massin, J. Chim. Phys., 61(1964)112.

33 K. Nasim, M.C. Meyer and J. Autian, J. Pharm. Sci., 61(1972)1775.

34 C.A.M. Hogben, D.J. Tocco and B.B. Brodie, J. Pharmacol. Exptl. Therap., 125(1959)275.

35 K. Kim and C. Hansch, unpublished results (see ref. [57]).

36 P. Jow and C. Hansch, unpublished results (see ref. [57]).

37 W. Kemula, H. Buchowsky and W. Pawlowski, Rocz. Chem., 43(1969)1555.

38 S. Mayer, R.P. Maickel and B.B. Brodie, J. Pharm. Exptl. Ther., 127(1959)205.

39 B.B. Brodie, H. Kurz and L.S. Schanker, J. Pharm. Exptl. Ther., 130(1960)20.

40 W. Kemula, H. Buchowsky and W. Pawlowski, Rocz. Chem., 42(1968)1951.

41 A. Taubman, Z. Physik. Chem., A161(1932)141.

42 A. Brunzell, J. Pharm. Pharmacol., 8(1956)329.

43 H.M. de Kort and R.F. Rekker, unpublished results.

44 A. Suzuki, W.I. Higuchi and N.F.H. Ho, J. Pharm. Sci., 59(1970)644.

45 A. Suzuki, W.I. Higuchi and N.F.H. Ho, J. Pharm. Sci., 59(1970)651.

46 N.F.H. Ho and W.I. Higuchi, J. Pharm. Sci., 60(1971)537.

47 K.R.M. Vora, W.I. Higuchi and N.F.H. Ho, J. Pharm. Sci., 61(1972)1785.

48 J.C. Dearden and E. Tomlinson, J. Pharm. Pharmacol., 23S(1971)68.

49 J.C. Dearden and E. Tomlinson, J. Pharm. Pharmacol., 23S(1971)73.

50 P. Seeman, Pharmacol. Rev., 24(1972)583.

51 S. Roth and P. Seeman, Biochem. Biophys. Acta, 255(1972)207.

52 P. Seeman, S. Roth and H. Schneider, Biochem. Biophys. Acta, 225(1971)171.

53 D. Nikaitani and C. Hansch, unpublished analysis (see ref. [57]).

54 S. Anderson, unpublished results (see ref. [57]).

55 D. Benson, J. Kaufman and H. Koski, see ref. [57].

56 C. Hansch and S. Anderson, J. Org. Chem., 32(1967)2583.

57 C. Hansch and A. Leo, Pomona College Medicinal Chemistry Project, Issue 6, January 1975.

CHAPTER X

A FEW APPLICATIONS OF f ; PROPOSAL FOR FURTHER ELABORATION OF THE HANSCH APPROACH TO QSAR

In this final chapter, we shall first pay some attention to the application of the hydrophobic fragmental constant and then we shall discuss how a better understanding of partition phenomena could be of value in adjusting, and possibly improving, the HANSCH structure − activity model.

Log P VALUES OF STRUCTURES OF THE UREIDE TYPE

The ureide group, −NH−CO−NH−, is extremely important in medicinal chemistry.

Firstly, there are the barbiturates, many of which have been developed into useful medicines in the last decenniums. These structures usually pass the blood − brain barrier with great ease, thus reaching the CNS level where they display a varied pattern of activities: hypnotic, sedative, anticonvulsive.

"Open" structures such as phenylacetylurea can show a similar variety of activities, while structures of a different type, such as the benzdiazepines, have an activity pattern which, although certainly dissimilar in several respects, could be related to that of the barbiturates.

An attempt will be made to introduce some consistency in the log P values for all of these structures, many of which are apparently not measured so as to yield the most workable results. We shall, of course, not deal with the SARs of barbiturates etc. themselves, as this subject would certainly lie beyond the scope of this monograph.

Barbiturates

Barbiturates have been tested for lipophilicity in various solvent systems in recent years. A survey of the available material is given in Table X,1.

$$\begin{array}{c}
> C \begin{array}{l} CO - NH \\ \\ CO - NH \end{array} > CO \qquad (A)
\end{array}$$

For any one of the solvent systems employed, a skeleton f value (f_A) with the

TABLE X,1

LIPOPHILIC BEHAVIOUR OF BARBITURATES

no	Substituents	log P values			
		obs.	ref.	calc.	res.
	Octanol — Water				
1	none	−1.41	[1]	−1.43	0.02
		−1.47	[2]		−0.04
2	5-butyl	0.77	[1]	0.69	0.08
3	5,5-diethyl	0.70	[1]	0.68	0.02
		0.65	[3]		−0.03
		0.65	[4]		−0.03
4	5-ethyl-5-_i_.propyl	0.97	[5]	1.09	−0.12
5	5-ethyl-5-butyl	1.89	[2]	1.74	0.15
6	5-ethyl-5-_i_.amyl	2.07	[5]	2.15	−0.08
7	5-ethyl-5-amyl	2.24	[2]	2.28	−0.04
8	5-ethyl-5-(1-methyl-butyl)	1.95	[3]	2.15	−0.20
		2.03	[5]		−0.12
		2.18	[6]		0.03
		2.10	[1]		−0.05
9	5-ethyl-5-phenyl	1.42	[4]	1.34	0.08
		1.47	[1]		0.13
10	5-ethyl-5-(3-hydroxy-1-methyl-butyl)	0.35	[2]	0.37	−0.02
11	5-ethyl-5-allyl	0.95	[5]	0.92	0.03
12	5,5-diallyl	1.19	[5]	1.15	0.04
13	5-ethyl-5-cyclohexenyl	1.20	[2]	2.15	−0.95
14	1-methyl-5-butyl	1.10	[3]	0.94	0.16
15	1,5-dimethyl-5-cyclohexenyl	1.49	[2]	2.39	−0.90
16	5-ethyl-5-_i_.amyl-2-thio	2.98	[2]	2.15[a]	0.83
17	5-allyl-5-(1-methyl-butyl)-2-thio	3.23	[2]	2.39[a]	0.84
18	5-ethyl-5-(methyl-allyl)-2-thio	2.19	[2]	1.44[a]	0.75
	Diethyl ether — Water				
1	none	−1.63	[7]	−1.74	0.11
2	5,5-diethyl	0.63	[8]	0.74	−0.11

TABLE X,1 (continued)

	Chloroform - Water				
1	none	-2.10	[9]	-2.87	0.77*
2	5,5-diethyl	-0.14	[10]	-0.11	-0.03
		-0.07	[11]		0.04
		-0.15	[9]		-0.04
		0.45	[12]		0.56*
		-0.10	[13]		0.01
3	5-ethyl-5-allyl	0.12	[10]	0.04	0.08
4	5-ethyl-5-i.propyl	0.20	[10]	0.40	-0.20
5	5,5-diallyl	0.33	[10]	0.18	0.15
		0.48	[13]		0.30*
6	5-ethyl-5-i.amyl	1.73	[10]	1.65	0.08
		1.30	[13]		-0.35*
		1.65	[14]		0.00
7	5-ethyl-5-(1-methylbutyl)	1.38	[10]	1.65	-0.27*
		1.41	[13]		-0.24*
		1.58	[15]		-0.07
8	5-ethyl-5-phenyl	0.65	[10]	0.65	0.00
		0.65	[13]		0.00
		0.73	[14]		0.08
9	5-allyl-5-(1-methylbutyl)	1.71	[13]	1.80	-0.09
10	5-ethyl-5-cyclohexenyl	0.13	[13]	1.50	-1.37*
11	1-methyl-5,5-diethyl	1.54	[10]	0.31	1.23*
		1.75	[13]		1.44*
12	1-methylpropyl-5-ethyl	1.00	[13]	0.68	0.32*
13	1-methyl-5,5-diallyl	2.15	[10]	0.60	1.55*
14	1-methyl-5-ethyl-5-phenyl	1.75	[13]	1.07	0.68*
		1.98	[10]		0.91*
15	1-methyl-5-ethyl-5-i.amyl	2.74	[10]	2.07	0.67*
16	1,5-dimethyl-5-cyclohexenyl	2.11	[10]	1.29	0.82*
		2.50	[14]		1.21*
		2.50	[15]		1.21*
17	1-methyl-5-ethyl-5-cyclohexenyl	2.48	[10]	1.92	0.56*
18	5-ethyl-5-allyl-2-thio	2.49	[10]	0.04[a]	2.45
19	5-ethyl-5-i.amyl-2-thio	2.51	[10]	1.65[a]	0.86
20	5-ethyl-5-(1-methylbutyl)-2-thio	2.22	[12]	1.65[a]	0.57
21	5-allyl-5-i.amyl-2-thio	2.84	[10]	1.80[a]	1.04

Refs. pg 347

TABLE X,1 (continued)

no	Substituents	obs.	ref.	calc.	res.
22	5-ethyl-5-cyclohexenyl-2-thio	2.27	[10]	1.50^a	0.77
	Benzene — Water				
1	5,5-diethyl	-0.77	[5]	-0.80	0.03
		-1.85	[10]		-1.05*
2	5,5-diallyl	-0.35	[5]	-0.58	0.23
3	5-ethyl-5-allyl	-0.51	[5]	-0.69	0.18
4	5-ethyl-5-_i_.propyl	-0.58	[5]	-0.26	-0.32*
5	5-ethyl-5-_i_.amyl	0.72	[5]	0.98	-0.26*
6	5-ethyl-5-(1-methylbutyl)	0.74	[13]	0.98	-0.24*
		0.51	[5]		-0.47*
7	5-ethyl-5-phenyl	-0.01	[5]	-0.13	0.12
		-0.16	[13]		-0.03
8	5-ethyl-5-(1-methylbutyl)-2-thio	0.51	[10]	0.98^a	-0.47
	Oil — Water				
1	5,5-dimethyl	-1.18	[15]	-1.85	0.67*
2	5,5-diethyl	-0.72	[16]	-0.68	-0.04
		-0.67	[15]		0.01
		-0.58	[17]		0.10
		-1.22	[18]		-0.54*
		-0.57	[19]		0.11
		-0.82	[20]		-0.14
		-0.96	[21]		-0.28*
3	5-ethyl-5-_i_.propyl	-0.14	[15]	-0.17	0.03
4	5,5-diallyl	-0.07	[15]	-0.19	0.12
		-0.12	[18]		0.07
		-0.07	[19]		0.12
		-0.24	[20]		-0.05
5	5-allyl-5-_i_.propyl	0.05	[15]	0.07	-0.02
		0.02	[20]		-0.05
6	5-ethyl-5-butyl	0.41	[15]	0.50	-0.09
		0.27	[20]		-0.23*
7	5-ethyl-5-sec. butyl	0.13	[15]	0.41	-0.28*
8	5-allyl-5-butyl	0.54	[20]	0.74	-0.20
9	5-ethyl-5-(1-methylbuten(1)yl)	-0.05	[20]	0.78	-0.83*
10	5-allyl-5-sec. butyl	0.39	[15]	0.66	-0.27*

TABLE X,1 (continued)

11	5-ethyl-5-phenyl	0.23	[16]	0.02	0.21*
		0.13	[15]		0.11
		0.00	[17]		-0.02
		0.01	[20]		-0.01
12	5-ethyl-5-amyl	0.46	[15]	1.09	-0.63*
13	5-ethyl-5-i.amyl	0.46	[15]	1.00	-0.54*
		0.69	[20]		-0.31*
14	5-ethyl-5-(1-methylbutyl)	0.64	[15]	1.00	-0.36*
		0.76	[17]		-0.24*
		0.78	[20]		-0.22*
15	5-allyl-5-(1-methylbutyl)	1.02	[20]	1.24	-0.22*
16	5-allyl-5-neopentyl	0.62	[20]	1.34	-0.72*
17	5-ethyl-5-cyclohexenyl	0.16	[20]	0.93	-0.77*
18	5-ethyl-5-cycloheptenyl	0.54	[20]	1.52	-0.98*
19	1,5-dimethyl-5-cyclohexenyl	0.88	[20]	0.60	0.28*
20	5,5-diethyl-2-thio	0.36	[17]	-0.68[a]	1.04
21	5-ethyl-5-(1-methylbutyl)-2-thio	1.95	[21]	1.00[a]	0.95
		1.80	[17]		0.80
	Carbon tetrachloride - Water				
1	5,5-diethyl	-1.46	[5]	-1.50	0.04
2	5-ethyl-5-allyl	-1.20	[5]	-1.25	0.05
3	5-ethyl-5-i.propyl	-1.21	[5]	-0.99	-0.22*
4	5,5-diallyl	-0.96	[5]	-1.00	0.04
5	5-ethyl-5-phenyl	-0.63	[5]	-0.74	0.11
6	5-ethyl-5-i.amyl	0.34	[5]	0.33	0.01
7	5-ethyl-5-(1-methylbutyl)	-0.03	[5]	0.33	-0.36*
8	1-methyl-5,5-diethyl	0.31	[5]	-1.22	1.53*
9	1-methyl-5,5-diallyl	0.84	[5]	-0.64	1.48*
10	1-methyl-5-ethyl-5-(1-methylbutyl)	1.95	[5]	0.61	1.34*
11	1-methyl-5-ethyl-5-phenyl	0.80	[5]	-0.38	1.18*
12	1,5-dimethyl-5-cyclohexenyl	0.88	[5]	-0.05	0.93*
13	1-methyl-5-ethyl-5-cyclohexenyl	1.49	[5]	0.61	0.88*
14	5-ethyl-5-allyl-2-thio	1.36	[5]	-1.24[a]	2.60
15	5-ethyl-5-i.amyl-2-thio	1.58	[5]	0.30[a]	1.28
16	5-ethyl-5-cyclohexenyl-2-thio	1.17	[5]	0.30[a]	0.87
17	5-allyl-5-i.amyl-2-thio	1.84	[5]	0.30[a]	1.54

Refs. pg 347

TABLE X,1 (continued)

no	Substituents	log P values			
		obs.	ref.	calc.	res.
	Oleyl alcohol — Water				
1	5,5-diethyl	0.14	[16]	-0.04	0.18
2	5,5-diallyl	0.38	[16]	0.56	-0.18
3	5-ethyl-5-phenyl	0.78	[16]	0.77	0.01
4	5-allyl-5-(1-methylbutyl)-2-thio	2.50	[22]	1.78[a]	0.72
	Heptane — Water				
1	5,5-diethyl	-2.15	[10]	-2.11	-0.04
2	5-ethyl-5-(1-methylbutyl)	-1.30	[23]	-0.42	-0.88*
3	5-ethyl-5-(1-methylbutyl)-2-thio	0.52	[9]	-0.42[a]	0.94
		0.19	[10]		0.61
		0.52	[23]		0.94
	n.Butanol — Water				
1	none	-1.16	[24]		

*) Outliers
a) Calculated as if the group C=O were present instead of C=S; the adhering residuals should therefore be considered as $\Delta(=S\ /\ =O)$ values.

greatest possible consistency for the series was established; these f_A values are given in Table X,2.

Octanol — water

Derivatives 1-12 were included in the calculation of f_A, which was found to be -1.78. There were no outliers for that value.

On application of $f = -1.78$ to 1-methyl-5-butylbarbituric acid (No. 14), excellent results were obtained. The values calculated for two 5-cyclohexenyl-substituted structures appeared, for uncertain reasons, to be far too high, however.

In the octanol — water series, three 2-thio structures were available, lending themselves to the calculation of a $\Delta(=S\ /\ =O)$ value. A mean of 0.81 was established.

Diethyl ether — water

Unfortunately, only two log P measurements had been reported. They could be applied in the calculation of $f_A = -2.16$.

Chloroform — water

Derivatives 1-9 appeared suitable in establishing an f_A value. We

TABLE X,2

INFLUENCE OF SOLVENT SYSTEM ON PARTITIONAL BEHAVIOUR OF BARBITURATES

Solvent system (·/water)	f_{obs} for A	key number	f_{calc} for A
octanol	−1.78		
diethyl ether	−2.16	0	−2.13
chloroform	−3.29	4	−3.37
benzene	−3.93	7	−4.04
oil	−3.47	4	−3.42
carbon tetrachloride	−4.58	7	−4.58
oleyl alcohol	−2.55	3	−2.59
heptane	−4.95	9	−5.02
n̲. butanol	−1.44[a]	0	−1.44

[a]) Only one structure available

found −3.29 and, as can be seen in Table X,1, most of the material is
consistent on this basis. A number of measurements must, however, be
regarded as outliers. In most instances, they refer to values determined
correctly by other workers. With the chloroform − water system, the
5-cyclohexenyl derivatives differ a good deal from the remaining structures, as they also did in the octanol − water system.

In this solvent system, 1-alkyl derivatives (which can be routinely
calculated in the octanol − water system) also deviate from the norm.
As regards the 2-thio derivatives, No. 18 would appear to be an arti-
fact, while for Nos. 19−22 a $\Delta(=S\ /\ =O)$ value of 0.81 with a rather
high standard deviation can be derived.

Benzene − water

In this solvent system, measurements were carried out with seven
barbiturates lending themselves to the recording of an f_A value, which
was found to be −3.93.

There appeared to be a fairly large number of outliers. The only
2-thio derivative whose f_A value was measured in this solvent system
(No. 8) gave a completely unacceptable $\Delta(=S\ /\ =O)$ value of −0.47.

Oil − water

This solvent system provided a large amount of data. For the calculation of f_A, 15 structures were available, the result being $f_A = -3.47$.
In this system also the f_A values of 5-cyclohexenyl derivatives proved
inconsistent with this value (Nos. 17−19), while that of the 5-neopen-
tyl derivative also deviated too much. Two 2-thio derivatives were

Refs. pg 347

measured, permitting the derivation of $\Delta(=S \; / \; =0) = 0.93$.

Carbon tetrachloride - water

For this solvent system an f_A value of -4.58 was found. The pattern outlined above repeated itself, both 1-methyl and 5-cyclohexenyl substitution leading to inconsistency with the skeleton value of -4.58. The measured partition values of the four 2-thio derivatives tended to be unreliable.

Oleyl alcohol - water

Measurements were performed with only three barbiturates, affording an f_A value of -2.55.

Heptane - water

In this solvent system, only two barbiturates have been measured, yielding two very different f_A values. As many as three workers investigated a 2-thio derivative of one of these barbiturates. Two of these measurements were found to be identical, which prompted us to regard their results as essential in commenting upon the partition values of the barbiturates No. 1 and No. 2, leading to the rejection of that of No. 2. Therefore, it seems very likely that f_A should be assigned a value of -4.95.

n.Butanol - water

Only the unsubstituted barbituric acid was used for a measurement.

At this point, the following generalizations can be made about the lipophilic behaviour of barbiturates in various solvent systems.

(1) For most barbiturates, the partition values can be calculated on the basis of an f value for the unsubstituted barbiturate skeleton (A). It must be stressed that each partitioning system has its own f_A value. These values are given in Table X,2 and, in general, they correspond fairly well with the f_A value for the octanol - water system:

$$f_A(a) \; = \; \rho'' \, f_A(oct) \; - \; kn.c_M(a) \qquad\qquad (X-1)$$

For the ρ'' and $c_M(a)$ values to be applied in eqn. X-1, see Table VII,7; for the $f_A(a)$ values so calculated and the pertinent key numbers, see columns 3 and 4 in Table X,2.

(2) The transformation of a barbiturate into a 2-thio derivative entails a greater lipophilicity. The material available suffers from inadequacy, because supplies are limited and also because several measurements were not correct. It is therefore not easy to draw sound

conclusions, but the following equation is tentatively put forward:

$$\Delta(=S \, / \, =0)_a \;=\; 3 \, kn.c_M(a) \qquad\qquad (X-2)$$

(3) Barbiturates substituted on the nitrogen atom (mainly $N-CH_3$ derivatives) do not follow the pattern that might have been expected from
previous experience. The transformation of an N-H into an $N-CH_3$ derivative would then have been associated with an increase in log P
by 0.24 ρ". The fact that not all measurements might have yielded
sufficiently reliable log P values made a judicious assessment difficult. It can be said, however, that the increase in log P is at
least five times as high as expected. This increase probably must
be coupled to the sharp decline in hydration, which may be associ-
able with a changed keto-enol pattern.

(4) The 5-cyclohexenyl derivatives are also inconsistent, with an un-
expected decrease in lipophilicity, which corresponds to about 3
key numbers. It remains uncertain to what this should be attributed.
Perhaps the bulkiness of the cyclohexenyl group imposes a deviating
enol form on the structure with changed hydration characteristics.
Further study of such deviating derivatives will, no doubt, provide
interesting details. In this context, it should be noted that the
cycloheptenyl and neopentyl substitutions produced similar effects.

(5) The effects referred under (3) and (4) are additive, that is, on
transformation of an N-H derivative into an $N-CH_3$ along with a simultaneous introduction of cyclohexenyl on C_5, the net effect con-
stitutes an increase in lipophilicity by a factor equal to the sum
of the values for the separate effects.

(6) The octanol - water system seems to be an exception in the pattern
outlined in (3); the experimental partition value of 1-methyl-5-bu-
tylbarbituric acid is as calculated and the value found for 1,5-di-
methyl-5-cyclohexenylbarbituric acid indicates that only cyclo-
hexenylanomaly plays a role. Further experiments will be necessary
to substantiate this exceptional behaviour of the octanol - water
system.

<u>Urea derivatives</u>

In Chapter IV, where the lipophilic behaviour of some complex functional groups was discussed, the $-NH-CO-NH_2$ group and a few derived acyl
structures were also dealt with. The material underlying the *f* values
listed there (see Table IV,8) can be extended by lipophilicity data on

TABLE X,3

LIPOPHILICITIES OF SOME UREA DERIVATIVES

no	compound	log P obs	ref.	log P calc	res.
	Derived from urea				
		octanol — water			
1	$NH_2-CO-NH_2$	-1.09	[25]	-1.35	0.26*
2	$CH_2=CH-CH_2-NH-CO-NH-CH_2-CH=CH_2$	0.64	[26]	0.63	0.01
3	$C_6H_5-CH_2-NH-CO-NH_2$	0.60	[27]	0.61	-0.01
4	$C_6H_5-NH-CO-NH_2$	0.83	[28]	0.95	-0.12
		0.82	[29]		-0.13
		0.97	[27]		0.02
5	$C_6H_5-NH-CO-NH-C_6H_5$	2.86	[27]	2.66	0.20
6	$C_6H_5-CH_2-NH-CO-NH-C_6H_5$	2.92	[27]	2.91	0.01
7	$CH_3-CH(CH_3)-CO-NH-CO-NH_2$	0.04	[30]	-0.03	0.07
8	$CH_3-CH(CH_3)-CH_2-CO-NH-CO-NH_2$	0.45	[30]	0.50	-0.05
9	$CH_3-CH_2-CH(C_2H_5)-CO-NH-CO-NH_2$	0.91	[30]	1.03	-0.12
10	$C_6H_5-CH_2-CO-NH-CO-NH_2$	0.87	[27]	0.75	0.12
11	$CH_3-CO-NH-CO-NH-CO-CH_3$	-0.68	[31]	-0.70	0.02
12	$CH_3-(CH_2)_2-CO-NH-CO-NH-CO-(CH_2)_2-CH_3$ (with CH_3 branch)	1.40	[4]	1.42	-0.02
		diethyl ether — water			
13	$NH_2-CO-NH_2$	-3.33	[33]	-3.30	-0.03
		-3.52	[34]		-0.22*
		-3.30	[33]		0.00
14	$CH_3-NH-CO-NH_2$	-2.92	[25]	-2.97	0.05
15	$CH_3-NH-CO-NH-CH_3$	-2.51	[25]	-2.51	0.00
16	$(CH_3)_2N-CO-NH_2$	-2.54	[25]	-2.51	-0.03
17	$CH_3-CH_2-NH-CO-NH_2$	-2.39	[25]	-2.44	0.05
18	$(CH_3-CH_2)_2N-CO-NH_2$	-1.72	[25]	-1.58	-0.14
19	$CH_3-CH_2-NH-CO-NH-CH_2-CH_3$	-1.72	[35]	-1.58	-0.14
20	$C_6H_5-NH-CO-NH_2$	0.04	[28]	-0.29	0.33*
		-0.26	[36]		0.03
		-0.32	[27]		-0.03
		chloroform — water			
21	$NH_2-CO-NH_2$	-3.85	[37]		
		oil — water			
22	$NH_2-CO-NH_2$	-3.82	[37]	-3.75	-0.07
23	$CH_3-NH-CO-NH_2$	-3.36	[37]	-3.46	0.10

TABLE X,3 (continued)

24	$CH_3-CH_2-NH-CO-NH_2$	-2.77	[37]	-2.87	0.10
25	$CH_3-CH_2-NH-CO-NH-CH_2-CH_3$	-2.12	[37]	-2.00	-0.12
	Derived from thiourea				
		octanol – water			
26	$NH_2-CS-NH_2$	-0.98	[38]	-1.06	0.08
		-1.14	[39]		-0.08
27	$C_6H_5-NH-CS-NH_2$	0.73	[29]	0.83	-0.10
28	$4-Cl-C_6H_4-NH-CS-NH_2$	1.64	[27]	1.54	0.10
		diethyl ether – water			
29	$NH_2-CS-NH_2$	-2.20	[25]	-2.12	-0.08
		-2.10	[33]		0.02
		-2.14	[34]		-0.02
30	$CH_3-NH-CS-NH_2$	-1.64	[34]	-1.79	0.15
31	$CH_3-CH_2-NH-CS-NH_2$	-1.35	[37]	-1.26	-0.09
32	$CH_3-CH_2-\ CH_2-NH-CS-NH_2$	-0.41	[34]	-0.73	0.32*
33	$C_6H_5-NH-CS-NH_2$	0.23	[40]		
		chloroform – water			
34	$NH_2-CS-NH_2$	-3.10	[33]		
35	$C_6H_5-NH-CS-NH_2$	0.54	[40]		
		oil – water			
36	$NH_2-CS-NH_2$	-2.92	[37]		

*) Outliers

a number of thio derivatives and is presented in Table X,3. Some further
partition data pertaining to a few other solvent systems are also included in this table.

Table X,4 shows the *f* values that can be derived from these series.
It is fairly clear that the lipophilic behaviour of an open urea structure differs a good deal from that of a barbiturate. The various link-
ing key numbers are also given in Table X,4; the key numbers pertaining
to the differences between aliphatic and aromatic involvement of the
groups are given on the left and the key numbers to be utilized for
conversion of octanol – water partition data into data in another solvent system, on the right of the column denoting the *f* values.

The over-all pattern is rather intricate but it is clear that the
behaviour of the urea fragment in the diethyl ether – water system is
peculear in that the aliphatic and the aromatic group need 7 and 3 key

TABLE X,4

f VALUES OF FRAGMENTS DERIVED FROM UREA AND THIOUREA DERIVATIVES

fragment (solvent system)		*f*
(al)NH–CO–NH$_2$ (octanol – water)	3 {	–1.81
(ar)NH–CO–NH$_2$ (octanol – water)		–0.94
(al)NH–CO–NH$_2$ (diethyl ether – water)	4 {	–3.89 — 7
(ar)NH–CO–NH$_2$ (diethyl ether – water)		–2.56 — 5
(al)NH–CO–NH$_2$ (chloroform – water)		–4.40* — 7
(al)NH–CO–NH$_2$ (oil – water)		–4.27 — 6
(al)NH–CS–NH$_2$ (octanol – water)	2 {	–1.52
(ar)NH–CS–NH$_2$ (octanol – water)		–1.07
(al)NH–CS–NH$_2$ (diethyl ether – water)	2 {	–2.71 — 3
(ar)NH–CS–NH$_2$ (diethyl ether – water)		–2.04* — 3
(al)NH–CS–NH$_2$ (chloroform – water)	6 {	–3.65* — 6
(ar)NH–CS–NH$_2$ (chloroform – water)		–1.81* — 2
(al)NH–CS–NH$_2$ (oil – water)		–3.44* — 5
(al)CO–NH–CO–NH$_2$ (octanol – water)		–1.67
(al)CO–NH–CO–NH–CO– (octanol – water)		–1.10
(al)CO–NH–CO–NH$_2$ (oil – water)		–2.97[a] — 3

*) Only one data point available for deriving the *f* value

numbers, respectively, for the conversion of octanol – water data into
diethyl ether – water data. It is worth noting that the *f* value for the
corresponding barbiturate conversion does not need any key number.

Another interesting feature of urea is that the transformation into
thiourea requires only 1 key number while with barbiturates such a procedure needs 3 key numbers.

For phenyl–substituted ureas and thioureas, the difference between
the two *f* values is no longer significant. On the other hand, the differences between ureas and thioureas observed in the diethyl ether –
– water and chloroform – water systems are much more in line with those
noted for the barbiturates.

Uracil derivatives

Uracil and its derivatives are to be considered as ring–closed
acylureas. Some of their log P values are presented in Table X,5.

The *f* value for the unsubstituted skeleton was established by as-

TABLE X,5

LIPOPHILICITIES OF URACIL AND SOME OF ITS 5-SUBSTITUTED DERIVATIVES
(octanol - water)

no	compound	log P obs	ref.	log P calc	res.
1	uracil	−1.07	29	−1.14	0.07
2	5-fluoro-uracil	−0.95	42	−0.92	−0.03
		−0.89	26		0.03
		−0.99	26		−0.07
		−0.85	29		0.07
3	5-chloro-uracil	−0.35	29	−0.40	0.05
4	5-bromo-uracil	−0.21	29	−0.19	−0.02
5	5-iodo-uracil	0.04	29	0.13	−0.09
6	5-trifluoromethyl-uracil	0.04	29	0.01	0.03
7	2-thiouracil	−0.28	29		

signing to the substituent the key number for aromatic substitution.
The greatest consistency was obtained for f = −1.32, which is 1 key
number higher than the value to be derived from the component parts of
the uracil skeleton: −CO−NH−CO−NH− + −CH=C−

    The difference in lipophilicity between uracil and its 2-thio deriv-
ative amounts to 3 key numbers, which indicates that the uracil structure does fit into the pattern of the barbiturates but not into that of
the ureas.

Benzodiazepines and ketimines

    The log P value of a benzodiazepine can be determined by simple f
summation. Diazepam is considered as an illustration:

$$\log P = f(C_6H_5) + f(C_6H_3) + f(Cl) + f(CH_3) + f(CH_2) +$$
$$f(NCO) + f(C=N) +$$
$$\left(kn_{Cl} + kn_{CON} + kn_{pe(1)} + kn_{cross-conj.} \right) c_M \qquad (X-3)$$

The only f value that is not yet available is that of the ketimine
group. Any attempt to trace literature data on the partitioning of sim-

TABLE X,6

f VALUE OF AROMATIC KETIMINE GROUP

compound	log P (obs)	f	
	3.71	−1.82	
	3.21	−1.82	(−2.11)
	3.31	−1.93	(−2.22)
	3.53	−1.65	(−1.94)

In parentheses: alternative f values

ple structures where the C=N group is present with an aromatic substit-uent attached to its carbon atom is bound to be unsuccessful. We were fortunately in the possession of four substituted benzophenoneimines[*], the log P values of which were measured in the octanol — 0.1N sodium

[*]We thank Dr. T. Bultsma and Dr. J.F.M. Meijer, Departm. of Med. Chem., Vrije Universiteit Amsterdam, for the donation of these ketimines.

hydroxide system. The results are shown in Table X,6 [27].

Even with these four log P values available, it remains rather difficult to derive a reliable $f(C=N)_{ar}$ value. Firstly, one has to reflect on the conformation that these structures will most probably possess. We were guided by useful information obtained previously about the conformations of methyl-substituted benzophenones from UV absorption measurements [32], and presume that the 4,4'-dimethyl derivative is comparable with 4,4'-dimethylbenzophenone so that its log P value will be construable as follows:

$$log\ P\ =\ 2\ f(CH_3)\ +\ 2\ f(C_6H_4)\ +\ f(H)\ +\ f(C=N)\ +\ 2\ c_M$$

Of the two key numbers introduced, one originates from the H bound to N(neg) and the other from the extra lipophilicity of the two carbon atoms indicated in Table X,6 by clarendon dots. The result is $f(C=N)_{ar}$ = -1.82.

The 2-methyl derivative is eligible for inclusion in Table X,6 with the conformation shown, the implication being that $f(C=N)_{ar}$ = -1.82 (an alternative value being -2.11).

The 2,2'- and 2,6-dimethyl derivatives should be considered to occur in a conformation where a marked de-coupling of the resonance of one of the phenyl rings with the C=N function has arisen. The f values will amount to -1.93 and -1.65 with alternative values of -2.22 and -1.94, respectively.

As regards the alternative values included above, there may be extra de-coupling of the resonance in an analogous manner to the suggestions given in Fig. V,11.

The most probable value would be -1.88, obtained by averaging the values -1.82, -1.82, -1.93 and -1.94, respectively. On the basis of this value, log P(diazepam) is calculated as follows:

$$log\ P\ =\ 1.886\ +\ 1.431\ +\ 0.061\ +\ 0.702\ +\ 0.530\ +\ (-2.894)\ +$$
$$(-1.88)\ +\ (3\ +\ 4\ +\ 3\ +\ 1)\ \times\ 0.287\ =\ 3.00$$

which agrees well with two of the three log P values reported (see Table X,7).

Correct log P values can likewise be calculated for bromazepam, medazepam, clonazepam and lorazepam. The values calculated for nitrazepam, oxazepam, prazepam and tetrazepam appear to differ from the experimental values (see Table X,7).

It is notable that the cyclohexenyl group shows the same deviating

Refs. pg 347

TABLE X,7

LIPOPHILICITIES OF BENZODIAZEPINES

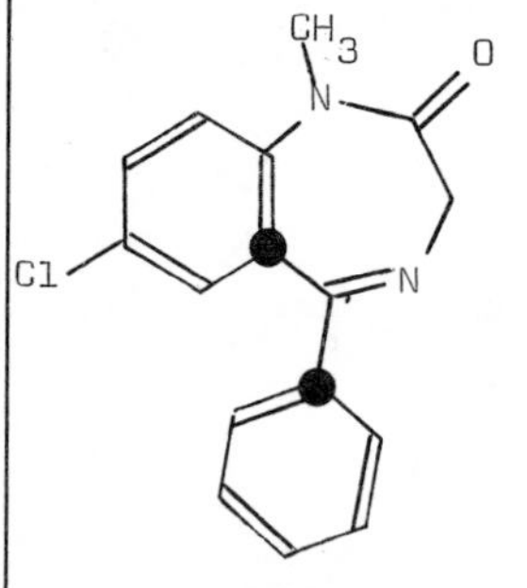

DIAZEPAM

obs: 2.66 [41];clc: 3.10
     2.82 [39]
     2.86 [27]

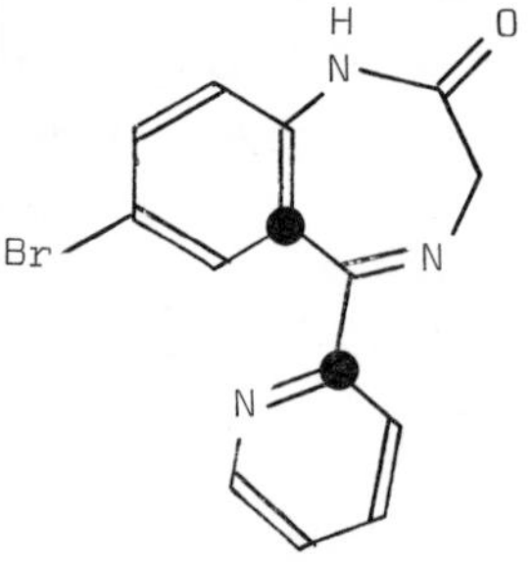

BROMAZEPAM

obs: 1.60 [41];clc: 1.61

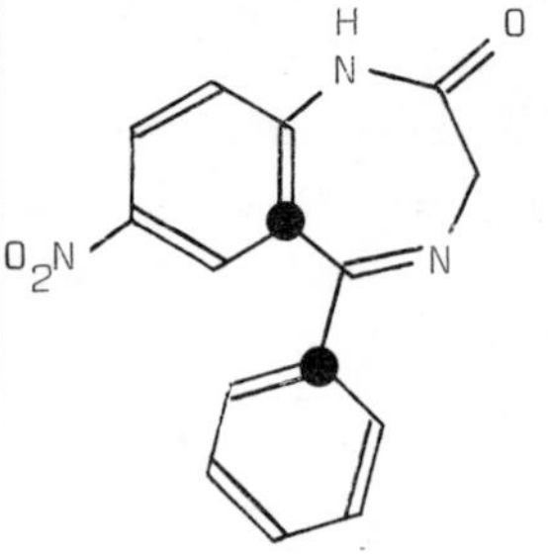

NITRAZEPAM

obs: 2.12 [41];clc: 1.76

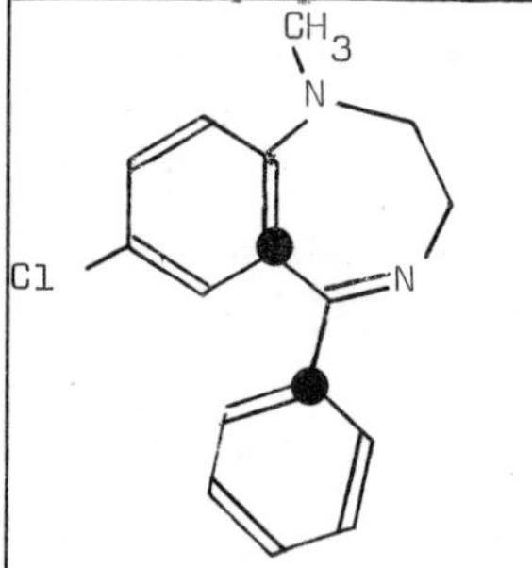

MEDAZEPAM

obs: 4.05 [41] ;clc: 3.97

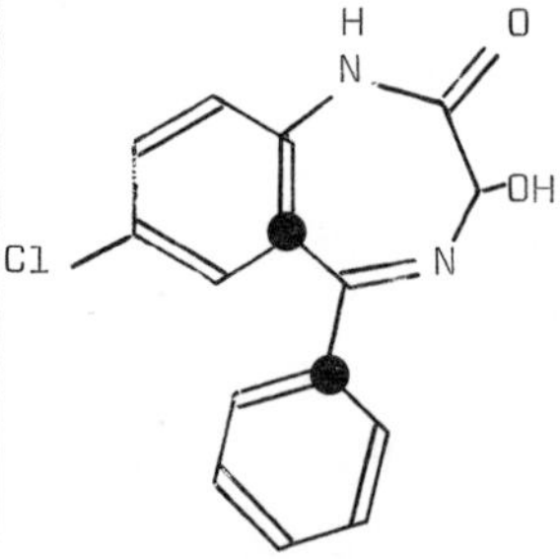

OXAZEPAM

obs: 2.17 [41] ;clc: 1.84

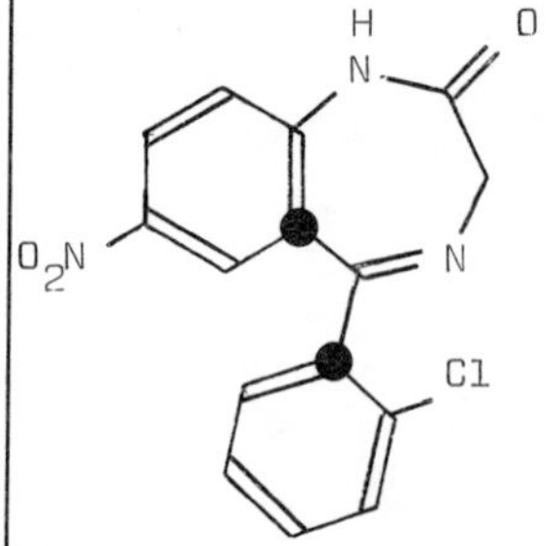

CLONAZEPAM

obs: 2.41 [41] ;clc: 2.48

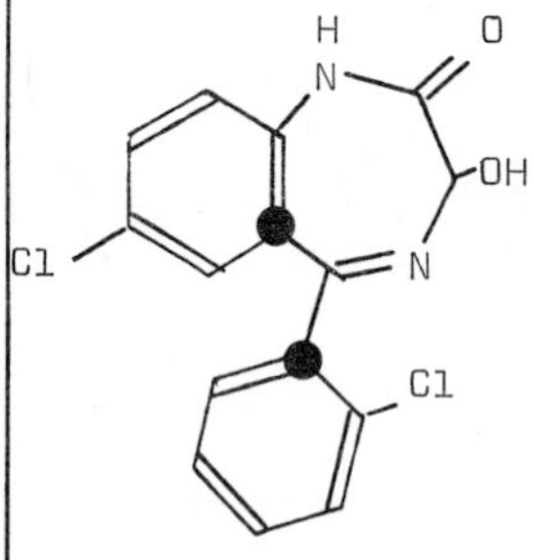

LORAZEPAM

obs: 2.38 [41];clc: 2.56

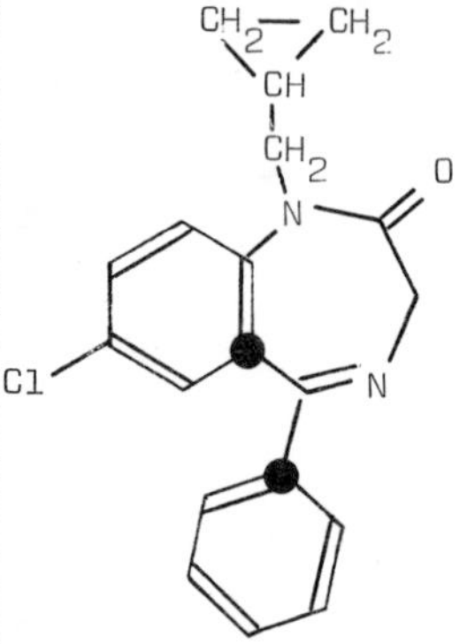

PRAZEPAM

obs: 3.73 [41];clc: 4.09

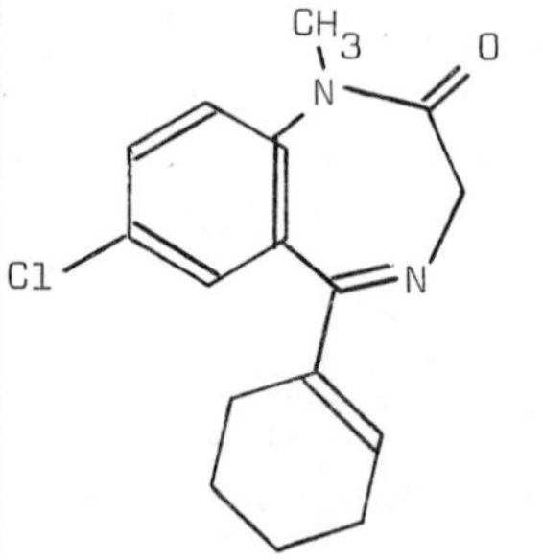

TETRAZEPAM

obs: 2.76 [41] ;clc: 3.52

behaviour as was observed in the barbiturates.

THE HANSCH APPROACH

In Chapter I, the HANSCH model was discussed as a means by which the
medicinal chemist attempts to correlate biological data with physico-
chemical parameters:

$$\log BR = f(p_n) \tag{X-4}$$

where BR = the biological response and p_n = a set of physicochemical
parameters.

The many parameters utilized in practice are divided into three
main groups:

(a) the linear free energy parameters (LFEPs), which include the parameters π (or f), σ and E_s, referring, respectively, to the hydrophobic, electronic and steric effects of the substituents introduced;

(b) the experimental physicochemical parameters, which need to be determined for each single structure involved in the investigation. This
implies that they are left out of consideration whenever the SARs
fall within the province of predictive evaluations because in that
event the structure involved is as yet unavailable. The parameters
in this group are of the electronic type, including dipole moments,
polarographic half-wave potentials, UV transition data, IR frequencies and NMR shifts. They may, however, also be kinetic in nature,
such as rates of hydrolysis and pK values. A third type is that
which is associated with the lipo-hydrophilic behaviour of a structure. Examples are log P values, retention times (determined by either GLC or HPLC techniques) and R_F values (determined by TLC).

(c) the quantum-chemical parameters, which may refer either to charge
(c1) or to energy (c2). The advantage of these parameters is that
they can be calculated independently of experimental data; in other
words, the structures need not be available.

Interrelations appear to be fairly numerous when, as is usual in
practice, parameters from the three indicated groups are applied simultaneously:

The partition coefficient (b group) is associated with π or f (a
group) as well as with the combination of superdelocalizability (c1

Refs. pg 347

group) and electronic charge (c1 group [14]; the dipole moment (b group) is related to the charge (c1 group) and the half-wave potential (b group to the LEMO-energy (c2 group)*.

One should also be prepared for these interrelations among parameters determined spectroscopically; thus, UV excitation data (b group) can be linked with the combination of LEMO- and HOMO-energy (c2 group)**. The NMR shift (b group) is clearly related to the charge (c1 group).

Furthermore, when it is borne in mind that σ (a group) can be coupled to IR frequencies (b group), to NMR shifts (b group) and to kinetic parameters (b group), the interweave of the whole pattern will be clear.

In practice, the chance of an interrelation of parameters will be smallest if the choice is confined to one of the main groups. Even with this restriction, it may be that interrelations do arise. We refer to Chapter V, where a possible interrelation in the group of LFEPs, that of f with E_s or f with σ, was discussed.

What can be said of the average quality of a biological regression equation? In practice, it appears to be highly dependent on the object that is chosen. Where an interaction of a compound with an isolated organ preparation is concerned, the value of the correlation coefficient, r, may be as high as 0.90 and sometimes r values as high as 0.95 are not unusual, which means that the quality of the regression equation hardly differs from that which is common for interactions between a compound and a cellular system. In that event, an r value of 0.98 will be regarded as highly satisfactory***.

Different situations are encountered when regression equations are elaborated on in vivo data. Surprisingly good results are then just possible (so that there is no reason to mistrust an r value as high as 0.90), but more frequently one will have r values as low as 0.65, which means that the regression equation leaves the greater part of the variancies unexplained.

The failure of a correct HANSCH approach in such instances can be attributed to several factors, such as shortage of or insufficient homo-

---

*)   LEMO-energy = energy of the lowest empty molecular orbital.
**)   HOMO-energy = energy of the highest occupied molecular orbital.
***) The quality of a regression equation is expressed in terms of r
    (correlation coefficient), s (standard deviation of the estimate),
    F (goodness of fit) and the Student t-values of the regressors.
    While for proper understanding of the quality of a regression equa-
    tion they are all essential, practice has demonstrated that r often
    comes before the others in initial testing.

geneity in the test objects; the intricate multicompartmental character
of transport and interactions in its entirety is likewise regarded as
an important reason that bad results occur so frequently. We made an attempt to visualize some aspects of the latter problem in Fig. X,1.

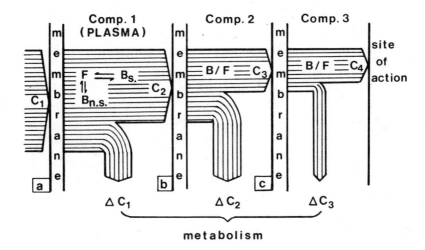

Fig. X,1: Drug transport visualized as a multicompartmental procedure; a, b and c: membrane systems serving
to separate the three compartments between the locus
of drug application and the site of drug action; F =
free drug and B = bound drug (for example drug-protein
binding, which may be specific or non-specific; see
the indications given)

   In this illustration, it is assumed that the site of action is three
membrane passages away from the locus of drug application. In each compartment, the concentration of the penetrating material will be reduced
by metabolic processes: Δc_1, Δc_2 and Δc_3 are the corrections to be
applied to the concentrations entering the three compartments. Each
following membrane system cannot be taken into consideration until the
relevant correction has been performed. The biological response, as it
will eventually become manifest at the site of action, is directly related to the concentration C_4, which is unknown, however, as it is tied
up inextricably with the originally applied concentration C_1. If in a
test series no significant metabolic differences do exist, the relationship

$$C_4 = c \times C_1 \qquad\qquad (X-5)$$

will be valid for all members of the investigated series with the appli-

Refs. pg 347

cation of the same correction factor (c) and a correlation study on the basis of eqn. X-4 will give no bad results. As soon as significant differences in the metabolic fates of the structures emerge, each structure will require its own correction factor and bad results with eqn. X-4 will become obvious.

Far from propagating the ignorance of even a severe influence of these metabolic differences, we propose that due consideration should be given to the following points.

As stated briefly in Chapter I, it is not certain that log P(oct) is indeed the value of choice to be introduced into the HANSCH equation, because it suggests without any proof that the octanol - water system not only models the complete drug transport with its various membrane passages, but is in addition exclusively ideal in bringing into the picture the hydrophobic aspects of drug-receptor interactions. The evidence presented in Chapter IX about the hyperdiscriminative behaviour of definite membrane systems makes it necessary to reflect again on the hydrophobic parameter in the HANSCH equation. A frequently employed form of this equation is

$$\log BR = -k_1 (\log P)^2 + k_2 \log P + \ldots + k_3 \qquad (X-6)$$

where all lipophilic aspects are described by log P(oct) values. In fact, this equation ought to be rewritten as follows:

$$\log BR = -k_1 (\log P_a)^2 + k_2' \log P_a + k_2'' \log P_b + \ldots + k_3 \qquad (X-7)$$

where subscript a represents a solvent system that is most in line with drug transport and subscript b is another solvent system, describing drug-receptor interactions most appropriately.

It has, however, appeared that any log P may be transformed into a log P(oct) value using the following equation:

$$\log P_s = \rho_s \log P(oct) - 0.268 \rho_s (\Sigma kn)_s \qquad (X-8)$$

where s represents any given solvent system and the other symbols have been explained in previous chapters.

By no means does the transformation of eqn. X-7 with two equations of type X-8 lead to forms that can be easily handled. For a few cases, we wish to indicate how an acceptable procedure can be arrived at.

I. Whenever hydrophobic substituents are used alone, no problem will arise and the HANSCH equation (eqn. X-7) can be written as

$$\log BR = -k_1 \, \rho_a^2 (\log P_{oct})^2 + k_2' \, \rho_a \log P_{oct} + k_2'' \, \rho_b \log P_{oct} + \cdots$$
$$+ \, k_3 \qquad (X-9)$$

or as

$$\log BR = -m_1 (\log P_{oct})^2 + m_2 \log P_{oct} + \cdots + k_3 \qquad (X-10)$$

which means that the log P(oct) values can be used as such and that any differences arising from the solvent a – octanol conversion are incorporated in the regressors m_1 and m_2.

II.   In cases when hydrophobic and hydrophilic substituents occur simultaneously, it will depend on the complexity of the hydrophobic aspects whether a practicable form can be found with minimal effort.

(a) If eqn. X-7 contains only one linear log P term, it can be simplified to

$$\log BR = k_1 \log P_a + \cdots + k_2$$
$$= k_1 \, \rho_a \left\{ \log P_{oct} - 0.268 \, (\Sigma kn)_a \right\} + \cdots + k_2 \qquad (X-11)$$

For all relevant substituents, the log P(oct) values can be decreased by 1, 2, 3, ... × 0.268, until the best fit is obtained.

(b) With two linear log P terms in eqn. X-7, we obtain the following form:

$$\log BR = k_1 \log P_a + k_2 \log P_b + \cdots + k_3 \qquad (X-12)$$

On application of the procedure under II(a), a regression equation is produced in which the two log P(oct) parameters are basically incompatible, because the $(\Sigma kn)_s$ terms do not run parallel. The trial and error procedure needed to trace the two sets of kn values will then be extremely timeconsuming but seems to be feasible in practice.

(c) With a squared log P term in addition to two linear log P terms, even more problems will arise and we fear that practice will result only in unresolvable problems.

     Unfortunately, we have not yet had a practical example at our disposal to illustrate any of the above considerations, and we therefore give the following hypothetical example.

     The example concerns a series of structures of type R-X, where R represents an aromatic unit and X a series of substituents as given in column 1 in Table X,8. In order to avoid undue complexity, the assump-

Refs. pg 347

tion has been made that only one log P type is operable and that other parameters can be excluded. It should be noted that the BR values have a much larger range than that usually encountered in actual practice. This is for intentional reasons to stress the hypothetical character of the example.

Application of f, f^2 and their combination resulted in the following regression equations:

$$\log BR = -4.29\, f(oct) + 6.30 \tag{X-13}$$
$$n = 12; \quad r = 0.766; \quad s = 3.29; \quad F = 14.2; \quad \underline{t} = -3.77$$

$$\log BR = -3.25\, \left[f(oct)\right]^2 + 7.29 \tag{X-14}$$
$$n = 12; \quad r = 0.952; \quad s = 1.56; \quad F = 97.8; \quad \underline{t} = -9.89$$

$$\log BR = -2.79\, \left[f(oct)\right]^2 - 1.07\, f(oct) + 7.36 \tag{X-15}$$
$$n = 12; \quad r = 0.962; \quad s = 1.47; \quad F = 55.9; \quad \underline{t}_1 = -6.40$$
$$\underline{t}_2 = -1.49$$

Eqn. X-14 is a significant improvement over eqn. X-13. The application of a linear f to eqn. X-14 as an extra parameter clearly demonstrates, however, the non-significance of the latter ($\underline{t}$ value -1.49) and indicates eqn. X-14 as an acceptable correlation.

Real outliers are absent in eqn. X-14, but a closer inspection of its residuals clearly points to a marked inhomogeneity of the data set. The next step in the improvement of the regression equation is preceded by considering the average key number of the applied substituents in solvent systems with hyperdiscriminative character (Chapter VIII will be of help) and a classification of the substituents in four groups:

$$kn = 0$$
$$kn = low$$
$$kn = medium$$
$$kn = high$$

Starting with kn = 3 for OH and $CONH_2$ groups, a trial and error procedure is performed by which one will gradually obtain more information and finally achieve the following equation:

$$\log BR = -1.36\, \left[f(oct) - kn\right]^2 - 3.78\, \left[f(oct) - kn\right] + 7.01 \tag{X-16}$$
$$n = 12; \quad r = 0.997; \quad s = 0.39; \quad F = 841; \quad \underline{t}_1 = -39.6$$
$$\underline{t}_2 = -33.3$$

For the sake of completeness, we give the two equations with f and f^2, respectively:

$$\log BR = -1.78 \left[f(\text{oct}) - kn \right] + 4.08 \tag{X-17}$$

$$n = 12; \quad r = 0.582; \quad s = 4.17; \quad F = 5.13; \quad \underline{t} = -2.27$$

$$\log BR = -0.34 \left[f(\text{oct}) - kn \right]^2 + 5.16 \tag{X-18}$$

$$n = 12; \quad r = 0.262; \quad s = 4.95; \quad F = 0.74; \quad \underline{t} = -0.86$$

TABLE X,8

ELABORATION OF A HYPOTHETICAL EXAMPLE

X	log BR obs	(oct)	key number classification	kn	log BR calc	res.
			$kn = 0$			
H	6.22	0.175		$\underline{0}$	6.31	−0.09
CH_3	3.62	0.702		$\underline{0}$	3.69	0.07
C_2H_5	0.06	1.232		$\underline{0}$	0.30	−0.24
$\underline{tert}.C_4H_9$	−8.30	2.256		$\underline{0}$	−8.42	0.12
			$kn = low$			
F	6.52	0.399		$0 \longrightarrow \underline{1}$	6.49	0.03
Cl	3.37	0.922		$0 \longrightarrow \underline{1}$	3.96	−0.59
Br	3.12	1.134		$0 \longrightarrow \underline{1}$	2.74	0.38
			$kn = medium$			
OCH_3	7.22	0.269		$\underline{1}$	7.01	0.21
COOH	9.64	−0.093		$1 \longrightarrow \underline{3}$	9.31	0.33
NO_2	9.61	−0.078		$1 \longrightarrow \underline{3}$	9.29	0.32
			$kn = high$			
OH	7.32	−0.343		$3 \longrightarrow \underline{8}$	8.01	−0.69
$CONH_2$	3.77	−1.110		$3 \longrightarrow \underline{9}$	3.49	0.28

The underlined key numbers appear in eqn. X-16.

The above results may indicate that in this hypothetical example
the biological data correlate best with a solvent system that does not
differ much from cyclohexane − water. It will be clear that the simultaneous operation of other parameters $(\sigma, E_s,$ etc.$)$ breaks the above proposed procedure down to an almost hopeless task.

REFERENCES

1 E. Kutter, unpublished analysis (see ref. [43]).
2 S. Anderson, unpublished analysis (see ref. [43]).
3 S. Roth and P. Seeman, Biochem. Biophys. Acta, 255(1972)207.

4   C. Hansch, A. Steward, S. Anderson and D. Bentley,
    J.Med. Chem., 11(1968)1.

5   K. Kakemi, Chem. Pharm. Bull. Japan, 15(1967)1708.

6   J. Schaeffer, private comm. to Hansch (see ref.[43]).

7   O.C. Dermer, W.G. Markham and H.M. Trimble, J. Amer. Chem. Soc.,
    63(1941)3524.

8   K.B. Sandell, Monatsh. Chem., 92(1961)1066.

9   C.A.M. Hogben, D.J. Tocco, B.B. Brodie and L.S. Schanker,
    J. Pharmacol. & Exptl. Ther., 125(1959)275.

10 S. Mayer, R.P. Maickel and B.B. Brodie,
    J. Pharmacol. & Exptl. Ther., 127(1959)205.

11 E.R. Garrett and P.B. Chemburkar, J. Pharm. Sci., 57(1968)1401.

12 K.O. Borg and G. Schill, Acta Pharm. Suecica, 10(1973)187.

13 A. Brodin and M.I. Nilsson, Acta Pharm. Suecica, 10(1973)187.

14 K.S. Rogers and A. Cammarata, J. Med. Chem., 12(1969)692.

15 D.L. Tabern and E.F. Shelberg, J. Amer. Chem. Soc., 55(1933)328.

16 K.H. Meyer and H. Hemmi, Biochem. Z., 277(1935)39.

17 L.C. Mark, J.J. Burns, L. Brand, C.J. Campomanes, N. Trousof,
    E.M. Papper and B.B. Brodie, J. Pharmacol. & Exptl. Ther.,
    123(1958)70.

18 R. Siebeck, Arch. Exp. Path. Pharmacol., 95(1922)93.

19 L. Velluz, Compt. Rend., 182(1926)1178.

20 I. Jansson, S. Orrenius, L. Ernster and J.B. Schenkman,
    Arch. Biochem. Biophys., 151(1972)391.

21 L. Brand, L.C. Mark, M.M. Snell, P.A. Vrindten and P.G. Dayton,
    Anesthes., 24(1963)33.

22 T. Irikura, Yakugaku Zasshi, 82(1962)356.

23 B.B. Brodie, H. Kurz and L.S. Schanker, J. Pharmacol. & Exptl.
    Ther., 130(1960)20.

24 J.F. Tinker and G.B. Brown, J. Biol. Chem., 173(1948)585.

25 R. Collander, Acta Chem. Scand., 3(1949)717.

26 Midwest Research Inst., communicated to Hansch (see ref.[43]).

27 H.M. de Kort and R.F. Rekker, unpublished results.

28 C. Church, unpublished analysis (see ref.[43]).

29 D. Nikaitani and C. Hansch, unpublished analysis (see ref.[43]).

30 R.I. Mrongovius, Eur. J. Med. Chem., 10(1975)474.

31 S. Anderson and C. Hansch, unpublished analysis (see ref.[43]).

32 R.F. Rekker and W.Th. Nauta, Rec. trav. Chim., Pays Bas,
    80(1961)747.

33 K.B. Sandell, Naturwissensch., 53(1966)330.

34 E.J. Ross, J. Physiol., 112(1051)229.

35 W. Ruhland and U. Heilmann, Planta, 39(1951)91.

36 K.B. Sandell, Monatsh. Chem., 92(1961)1066.

37 R. Collander, Phys. Plant., 7(1954)420.

38 J. Takeda, H. Morikawa, H. Kinoshita and M. Senda,
    Plant & Cell Physiol., 12(1971)949.

39 F. Helmer, unpublished analysis (see ref.[43]).

40 K.B. Sandell, Monatsh. Chem., 89(1958)36.

41 W.E. Müller and U. Wollert, N.S. Arch. Pharmacol., 278(1973)301.

42 National Cancer Inst., Drug Development Branch, communicated to
    Hansch (see ref.[43]).

43 A. Leo, C. Hansch and D. Elkins, Chem. Rev., 71(1971)525.

TABLE OF HYDROPHOBIC

No	fragment	hypodiscriminative systems			OCT.	OL.ALC.	ETHER
		s.BUT.	n.BUT.	n.PENT.			
1	C_6H_5		1.547	1.73	1.886		2.266
2	C_6H_4				1.688	2.009[1]	1.866
3	C_6H_3				1.431		1.677
4	CH_3	0.42	0.640	0.68	0.702	0.688	0.919
5	CH_2	0.32	0.404	0.51	0.530	0.535	0.531
6	CH		0.087	0.23	0.235	0.289	0.277
7	C(quart.)				0.15		
8	$CH_2=CH$				0.935		
9	$CH \equiv C$				0.73		
10	H			0.17	0.175		
11	H(neg)	0.36		0.44	0.462		0.59
12	(al)COOH	-0.33[2]	-0.527[1]	-0.71	-0.954	-1.289[1]	-1.238
13	(ar)COOH				-0.093		-0.347[1]
14	(al)COO				-1.292		-1.553
15	(ar)COO				-0.431	-0.901[2]	
16	(al)CO				-1.703		-1.937
17	(ar)CO				-0.842		-0.94
18	(al)O				-1.581		-1.867
19	(ar)O				-0.433	-0.512	-0.679[1]
20	(al)OH		-0.794[2]	-1.27	-1.491	-2.301[3]	-1.879
21	(ar)OH			-0.31	-0.343	-0.866[2]	-0.691[1]
22	(ar)COH				-0.38		
23	(al)OCH$_2$COOH				-1.155		
24	(ar)OCH$_2$COOH				-0.581		
25	(al)CF$_3$				0.757		
26	(ar)CF$_3$				1.331		

FRAGMENTAL CONSTANTS

		hyperdiscriminative systems					
OIL	CHLOR.	BENZENE	TOLUENE	XYLENE	TETRA	c.HEX.	No
2.091	2.348	2.230	2.00	2.06	2.30	2.312	1
1.906	1.995	1.768	1.78		2.15	2.123	2
						1.881	3
0.808	0.965	0.947	0.74	0.77	0.88	0.805	4
0.589	0.628	0.619	0.56	0.58	0.66	0.646	5
0.282	0.163	0.212	0.25	0.26	0.29	0.397	6
							7
							8
							9
							10
							11
−2.253[3]	−2.485[4]	−2.896[6]	−2.60[5]	−2.71[5]	−3.30[6]		12
−1.543[4]	−1.849[6]	−2.004[7]	−1.60[5]				13
−1.93[1]		−1.51			−1.44		14
							15
−2.326[1]	−1.136$_3$	−1.967			−2.11		16
−1.616[2]	−0.818$_1$						17
−2.124[1]							18
−0.704[1]	−0.25$_1$					−0.754[1]	19
−2.861[3]	−2.367[2]	−3.128[5]			−3.19[4]	−3.578[5]	20
−1.441[3]	−2.049[5]	−1.936[5]	−1.74[5]	−1.90[4]	−2.74[7]	−3.157[8]	21
						−1.058[2]	22
							23
							24
							25
							26

TABLE OF HYDROPHOBIC

| No | fragment | hypodiscriminative systems | | | OCT. | OL.ALC. | ETHER |
		s.BUT.	n.BUT.	n.PENT.			
27	(al)CCl$_3$				1.79		
28	(al)F				-0.462		
29	(ar)F				0.399		
30	(al)Cl				0.061		0.176
31	(ar)Cl				0.922		
32	(al)Br				0.270		0.43
33	(ar)Br				1.131		
34	(al)I				0.587		0.810
35	(ar)I				1.448		
36	C$_5$H$_4$N				0.526		
37	C$_3$H$_3$N$_2$				-0.119		
38	(al)NH$_2$		-0.940[1]	-1.22	-1.428		-2.563[3]
39	(ar)NH$_2$				-0.854	-0.685[1]	-1.375[1]
40	(al)NH		-1.318[1]		-1.825		-2.954[3]
41	(ar)NH				-0.964		
42	(al)N		-1.727		-2.160	-2.590[1]	-3.099[2]
43	(ar)N				-1.012		
44	(ar)SO$_2$NH$_2$				-1.530		-1.97[1]
45	(ar)SO$_2$NH				-1.992		-2.50
46	(ar)SO$_2$N				-2.454		-2.94
47	(al)CONH$_2$				-1.970		-3.403[4]
48	(ar)CONH$_2$				-1.109		
49	(al)CON				-2.894		
50	(al)NHCOO				-1.943		
51	(ar)NHCOO				-0.795		-3.309[9]
52	(al)OOCNH$_2$				-1.481		

FRAGMENTAL CONSTANTS

| hyperdiscriminative systems | | | | | | | |
OIL	CHLOR.	BENZENE	TOLUENE	XYLENE	TETRA	c.HEX.	No
							27
							28
						0.139[1]	29
-0.244[1]	-0.63[2]	-0.147[1]					30
		1.343_1			1.11	0.856[1]	31
-0.013[1]	-0.134[2]	0.094[1]	-0.08[1]	-0.22[1]			32
		1.584_1				0.909[1]	33
	0.16[2]	0.42[1]					34
							35
	0.899_1						36
							37
	-1.935[1]	-2.40[3]	-2.34[3]	-2.31[2]			38
	-0.981	-1.285[1]			-1.92[3]	-2.301[4]	39
	-2.386[1]	-3.242[4]	-2.70[3]	-2.69[2]		-2.766[1]	40
							41
			-3.02[3]	-3.07[2]			42
							43
	-2.567[2]						44
							45
							46
-3.945[5]	-2.98[2]						47
-2.525[4]	-2.21[3]						48
					-4.28[2]		49
							50
							51
-2.363[2]							52

TABLE OF HYDROPHOBIC

No	fragment	hypodiscriminative systems			OCT.	OL.ALC.	ETHER
		s.BUT.	n.BUT.	n.PENT.			
53	(al)NO_2				−0.939		
54	(ar)NO_2				−0.078		
55	(al)$C{\equiv}N$				−1.066		
56	(ar)$C{\equiv}N$				−0.205		
57	(ar)C=N				−1.88		
58	(ar)CH=CH−NO_2				0.395		
59	(ar)CH=C−NO_2				0.220		
60	(ar)CH=CH−COO				0.042		
61	(ar)CH=CH−CONH				−1.100		
62	(al)SH				0.00		
63	(ar)SH				0.62		
64	(al)S				−0.51		
65	(ar)S				0.11		
66	(al)S−S				0.37		
67	(al)SO				−2.75		
68	(ar)SO				−2.05		
69	(ar)SO_2				−1.87		
70	p.e. (1)		0.456	0.81	0.861	0.786	0.891
71	p.e. (2)		0.304	0.54	0.574	0.524	0.594
	$c_M(oct)$	0.29	0.275	0.27	0.268	0.211	0.219
	$c_M(a)$	0.19	0.221	0.24	0.268	0.227	0.263
	ρ'	1.52	1.226	1.135	1.000	0.932	0.834
	ρ''	0.66	0.813	0.879	1.000	1.072	1.195

Some of the f values are provided with an index; they will enable the user to establish at first sight how they are related with f (oct): 0.345^2 denotes that after multiplying the relevant f (oct) with ρ'' two key numbers have been added, whereas 0.345_2 denotes that two key numbers have been subtracted. The table is not complete; for a number of f values (for instance, those related to urea and its derivatives) reference is made to the relevant sections.

FRAGMENTAL CONSTANTS

hyperdiscriminative systems							
OIL	CHLOR.	BENZENE	TOLUENE	XYLENE	TETRA	c.HEX.	No
							53
	0.279[1]					−0.578[1]	54
	−1.13						55
							56
							57
						−0.569[3]	58
						−0.733[3]	59
						−0.555[2]	60
						−4.343[9]	61
							62
							63
							64
							65
							66
							67
							68
							69
1.065	0.954	0.894	0.90	0.93			70
0.710	0.636	0.596	0.60	0.62			71
0.301	0.253	0.246	0.297	0.321	0.277	0.261	
0.347	0.307	0.285	0.305	0.352	0.342	0.332	
0.873	0.826	0.866	0.979	0.916	0.812	0.792	
1.140	1.205	1.149	1.017	1.087	1.230	1.254	

f values not yet available in this table may be derived from their
f(oct) values by applying the correct ρ and $c_M(a)$ values denoted in the
lower part of the table; key numbers will have to be valued on sound
judgement.

AUTHOR INDEX

Many page numbers are followed (in parentheses) by extra numbers, which
    denote the number by which the compound is included in a table.

A

butyric acid, 2-hydroxy   208(41) — 214(18) — 257(12)
-     3-hydroxy   214(17)
-     2-(4-iodobenzenesulfonamide,
             3-hydroxy   219
-     4-methyl   222(24) — 228(18) — 233(30) — 242(18)
                     243(6) — 245(16) — 258(14) — 259(16)
-     2-phenyl   236(47)
-     3-phenyl   236(48)
-     4-phenyl   46(64) — 57(16) — 236(49)

C

carbazole   90(9)
carbon tetrachloride   102(2)
cardene   82
chloramphenicol   187(37) — 189(7)
chloroform   102(1)
chlorpromazine   35 — 319(18)
cinnamamide   95(14)
-     N-butyl   252(129)
-     -   4-chloro   252(127)
-     -   4-methoxy   252(137)
-     N-(4-methylbutyl)   252(136)
-         -   4-chloro   252(134)
-         -   4-methoxy   252(140)
-     N-ethyl   251(111)
-     -   4-methoxy   252(124)
-     N-heptyl   252(141)
-     N-methyl   251(100)
-     -   4-chloro   251(97) — 255(73)
-     -   4-methoxy   251(112)
-     N-pentyl   252(135)
-     -   4-chloro   252(133)
-     -   4-methoxy   252(139)
-     N-propyl   252(122)
-     -   4-chloro   251(117)
-     -   4-methoxy   252(131)
-     N-(3-methylpropyl)   252(130)
-         -   4-chloro   252(128)
-         -   4-methoxy   252(138)
-     N-i.propyl   252(123)
-         -   4-chloro   251(116)
-         -   4-methoxy   252(132)
cinnamic acid   95(13)
cinnamonitrile   95(12)
cinnamylalcohol   95(17)
clonazepam   340
cyclohexane   124(8)
cyclohexanol   44(32)
cyclohexanone   58(23)
cyclohexylamine   44(34)
cyclopentane   130
cyclopropane   124(4)
cyclopropylethylether   235(25)
cyclopropylmethylether   235(13)
cysteine   301(8)
cystine   301(10)

D

N(diphenylmethoxyethyl)-azetidine,   4-methyl   157(14)
              -          aziridine   153 - 157(9) - 161 - 172(5)
              -          -   4-methyl   157(10)
              -          hexahydroazepine 153 - 157(29) - 162 - 173(15)
              -          -   4-methyl   157(30)
              -          piperidine   153 - 157(25) - 162 - 172(13)
              -          -   4-methyl   157(26)
              -          pyrrolidine   153 - 157(19) - 162 - 172(10)
              -          -   4-methyl   157(20)
2-diphenylmethoxyheptylamine, N,N-dimethyl   152
2-diphenylmethoxyhexylamine, N,N-dimethyl   152
2-diphenylmethoxyoctylamine, N,N-dimethyl   152
2-diphenylmethoxypentylamine, N,N-dimethyl   152
2-diphenylmethoxypropylamine, N,N-dimethyl   152
diphenylsulfide   90(15) - 94(8)
dipropyleneglycol   235(37)
dipropylether   44(37)

E

ethane   124(3)
    -      1-bromo   125(31) - 232(4)
    -          -   2-phenyl   57(10)
    -      1-chloro   125(30) - 232(5)
    -          -   2-phenyl   57(7)
    -      1,2-dicarboxy   47(77) - 51(12) - 207(35) - 223(4)
                                      257(8) - 259(6)
    -          -   1-amino   301(14)
    -          -   1-bromo   220
    -          -   1,2-dibromo   220
    -          -   1,2-dihydroxy   213(16) - 214(16) - 257(9) - 260(5)
    -          -   1-hydroxy   102(3) - 207(36) - 213(15) - 239(3)
                                      259(7)
    -          -   1(4-iodobenzenesulfonamide)   219
    -      1,2-dicyano   226
    -      1,1-diethoxy   234(46) - 235(35)
    -      1,2-dihydroxy   166 - 205(11) - 234(2)
    -      hexafluoro   55
    -      1-iodo   57(12) - 125(32) - 188(41) - 205(8)
    -      1-nitro   54
    -          -   2-phenyl   54
ethanol   42(4) - 125(33) - 166 - 205(10) - 213(5)
                                   221(5) - 227(4) - 229(2) - 232(8)
                                   238(2) - 243(7) - 248(1) - 252(2)
                                   308(6)
    -      2-amino   47(75) - 51(14) - 205(14)
    -      2-amino (N,N-diethyl)   211(80) - 260(23)
    -      2-bromo   100(8)
    -      2-(2-butoxyethoxy)   212(100) - 234(56)
    -      2-chloro   100(7)
    -      1,1-dimethyl   27 - 53 - 166
    -      2-ethoxy   43(20) - 51(22) - 208(50) - 233(27) - 235(20)
    -      2-(2-ethoxyethoxy)   215(37) - 235(36)
    -      2-fluoro   100(6)
    -      2-methoxy   207(29) - 233(19)
    -      2-(2-methoxyethoxy)   209(61) - 233(36)
    -      2-methyl   166 - 213(11) - 221(12) - 233(18) - 235(9)
    -      2-phenoxy   100(13)

377

N

β-nitrostyrene, β-methyl, 2-nitro   250(74)
        -              -   3-nitro   250(76)
        -              -   4-nitro   250(75)
        -              -   4-*i*.propyl   255(88)
nonanoic acid   310
nonanol   319(5)

O

octane, 1,8-dicarboxy   212(106) — 223(57)
octanoic acid   310
octanol   45(51) — 236(44) — 319(4)
octylbenzoate, 4-amino   239(35)
octylcarbamate   170(12)
oxazepam   340
oxazolidine-2,4-dione, 3-methyl, 5-phenyl   60
        -                3,5,5-trimethyl   60 — 61

P

pentane, 1,5-diamino   209(63) — 259(17)
     -   1,5-dicarboxy   211(92) — 260(28)
     -   1-diethylamino, 4-amino   216(63) — 261(14)
     -   1,5-dihydroxy   209(60) — 235(28)
     -   1-fluoro   57(4)
2,4-pentanedione   222(22) — 223(7)
pentanoic acid   209(57) — 222(23) — 228(17) — 233(29) — 242(17)
                        245(15) — 258(13) — 259(15) — 261(5) — 310
        -   2-amino, 5-methyl   215(35) — 260(11) — 301(1)
        -   2-bromo   228(16) — 235(23) — 242(16)
        -   2-phenyl   236(50)
        -   4-phenyl   236(51)
        -   5-phenyl   236(52)
pentanol   44(26) — 228(20) — 233(32) — 243(10)
                        248(6) — 319(1)

2-pentanone, 5-phenyl   58(26)
pentobarbital   319(15)
pentylamine   44(29) — 245(18)
        -   N-butyl   251(90)
        -   N-methyl   249(22)
pentylbenzoate, 4-amino   238(24) — 238(26)
pentylcarbamate   170(8)
tert.pentylcarbamate   170(7)
phenanthrene   83 — 85(24) — 117
phenetole   254(38)
phenobarbital   319(14)
phenol   64(6) — 147 — 209(68) — 222(29) — 224(17)
                        228(27) — 233(37) — 242(19) — 244(15)
                        245(19) — 248(14) — 258(15) — 308(11)
                        319(8)

        -   2-amino   65(24) — 230(12) — 253(16)
        -   3-amino   72(35) — 80 — 215(33) — 248(16)
        -   4-amino   72(53) — 80 — 230(13) — 248(17)
        -   3-amino, N,N-dimethyl   74
        -   2-bromo   65(20)
        -   3-bromo   66(66)
        -   4-bromo   68(135) — 195 — 248(8) — 319(9)
        -   2-chloro   65(19)
        -   3-chloro   72(34) — 80

U

V

X